			IIIA	IVA	VA	VIA	VIIA	VIIIA 2 He 4.0026
			5 B 10.811	6 C 12.011	7 N 14.007	8 O 15.999	9 F 18.998	10 Ne 20.180
	IB	IIB	13 Al 26.982	14 Si 28.086	15 P 30.974	16 S 32.065	17 Cl 35.453	18 Ar 39.948
28 Ni 58.693	29 Cu 63.546	30 Zn 65.39	31 Ga 69.723	32 Ge 72.64	33 As 74.922	34 Se 78.96	35 Br 79.904	36 Kr 83.80
46 Pd 106.42	47 Ag 107.87	48 Cd 112.41	49 In 114.82	50 Sn 118.71	51 Sb 121.76	52 Te 127.60	53 I 126.90	54 Xe 131.29
78 Pt 195.08	79 Au 196.97	80 Hg 200.59	81 Tl 204.38	82 Pb 207.2	83 Bi 208.98	84 Po (209)	85 At (210)	86 Rn (222)
110 Uun (281)	111 Uuu (272)	112 Uub (285)		114 Uuq (289)				

63 Eu 151.96	64 Gd 157.25	65 Tb 158.93	66 Dy 162.50	67 Ho 164.93	68 Er 167.26	69 Tm 168.93	70 Yb 173.04	71 Lu 174.97
95 Am (243)	96 Cm (247)	97 Bk (247)	98 Cf (251)	99 Es (252)	100 Fm (257)	101 Md (258)	102 No (259)	103 Lr (262)

ESSENTIAL CONCEPTS of CHEMISTRY

THIRD EDITION

ALAN SHERMAN
Middlesex County College

SHARON SHERMAN
Rider University

Cover image and inside front cover periodic table © Shutterstock, Inc.

Kendall Hunt
publishing company

www.kendallhunt.com
Send all inquiries to:
4050 Westmark Drive
Dubuque, IA 52004-1840

Copyright © 1999, 2004 by Houghton Mifflin Company.
Copyright © 2015 by Alan Sherman and Sharon Sherman.

ISBN 978-1-4652-7363-5

Kendall Hunt Publishing Company has the exclusive rights to reproduce this work,
to prepare derivative works from this work, to publicly distribute this work,
to publicly perform this work and to publicly display this work.

All rights reserved. No part of this publication may be reproduced,
stored in a retrieval system, or transmitted, in any form or by any
means, electronic, mechanical, photocopying, recording, or otherwise,
without the prior written permission of the copyright owner.

Printed in the United States of America

BRIEF CONTENTS

CHAPTER 1:	THE ORIGINS OF CHEMISTRY: *Where It All Began*	1
CHAPTER 2:	SYSTEMS OF MEASUREMENT	9
CHAPTER 3:	MATTER AND ENERGY, ATOMS AND MOLECULES	39
CHAPTER 4:	ATOMIC THEORY, PART 1: *What's in an Atom?*	57
CHAPTER 5:	ATOMIC THEORY, PART 2: *Energy Levels and the Bohr Atom*	73
CHAPTER 6:	THE PERIODIC TABLE: *Keeping Track of the Elements*	95
CHAPTER 7:	CHEMICAL BONDING: *How Atoms Combine*	113
CHAPTER 8:	CHEMICAL NOMENCLATURE: *The Names and Formulas of Chemical Compounds*	135
CHAPTER 9:	CALCULATIONS INVOLVING CHEMICAL FORMULAS	161
CHAPTER 10:	THE CHEMICAL EQUATION: *Recipe for a Reaction*	181
CHAPTER 11:	STOICHIOMETRY: *The Quantities in Reactions*	205
CHAPTER 12:	THE CHEMISTRY OF SOLUTIONS	231
APPENDIX A:	*Basic Mathematics for Chemistry*	265
APPENDIX B:	*Important Chemical Tables*	283
APPENDIX C:	*Glossary*	287
Answers to Practice Exercises		295
Index		307

CONTENTS

Preface   xv

To the Student   xix

CHAPTER 1: THE ORIGINS OF CHEMISTRY: *Where It All Began*   1

LEARNING GOALS	2
INTRODUCTION	2
1.1 History	3
1.2 Chemistry Is a Diverse Field	6
1.3 How Chemistry Affects Our World	6
SUMMARY	7
SELF-TEST EXERCISES	7
EXTRA EXERCISES	8

CHAPTER 2: SYSTEMS OF MEASUREMENT   9

LEARNING GOALS	9
INTRODUCTION	10
2.1 Problem Solving	10
2.2 Polya's Method: The Four-Step General Model	10
2.3 Essential Estimation Skills	11
2.4 Significant Figures	11
RULE 1	13
RULE 2	13
RULE 1	15
RULE 2	15
RULE 3	15
RULE 4	16
2.5 Scientific Notation: Powers of 10	16
2.6 Area and Volume	18
2.7 The English System of Measurement	20
2.8 The Metric System of Measurement	21
2.9 Mass and Weight	27

2.10 Density	27
2.11 Temperature Scales and Heat	30
2.12 Converting Celsius and Fahrenheit Degrees	30
SUMMARY	31
KEY TERMS	32
SELF-TEST EXERCISES	32
EXTRA EXERCISES	37

CHAPTER 3: MATTER AND ENERGY, ATOMS AND MOLECULES — 39

LEARNING GOALS	40
INTRODUCTION	40
3.1 The Scientific Method	40
3.2 Matter and Energy	41
3.3 Law of Conservation of Mass and Energy	41
3.4 Potential Energy and Kinetic Energy	42
3.5 The States of Matter	42
3.6 Physical and Chemical Properties	43
3.7 Mixtures and Pure Substances	43
3.8 Solutions	44
3.9 Elements	45
3.10 Atoms	45
3.11 Compounds	45
3.12 Molecules	46
3.13 Molecular Versus Ionic Compounds	46
3.14 Symbols and Formulas of Elements and Compounds	47
3.15 Atomic Mass	48
3.16 Formula Mass and Molecular Mass	49
SUMMARY	50
KEY TERMS	50
SELF-TEST EXERCISES	51
EXTRA EXERCISES	53
CUMULATIVE REVIEW Chapters 1–3	54

CHAPTER 4: ATOMIC THEORY, PART 1: *What's in an Atom?* — 57

LEARNING GOALS	58
INTRODUCTION	58
4.1 What Is a Model?	58

4.2 Dalton's Atomic Theory	58
4.3 The Discovery of the Electron	59
4.4 The Proton	61
4.5 The Neutron	63
4.6 Atomic Number	64
4.7 Isotopes	64
4.8 What Next for the Atom?	68
SUMMARY	68
KEY TERMS	68
SELF-TEST EXERCISES	69
EXTRA EXERCISES	72

CHAPTER 5: ATOMIC THEORY, PART 2: *Energy Levels and the Bohr Atom* — 73

LEARNING GOALS	73
INTRODUCTION	74
5.1 Spectra	74
5.2 Light As Energy	76
5.3 The Bohr Atom	77
5.4 Quantum Mechanical Model	78
5.5 Energy Levels of Electrons	79
5.6 Electron Subshells	80
5.7 Electron Orbitals	81
5.8 Writing Electron Configurations	83
5.9 The Importance of Electron Configuration	86
SUMMARY	89
KEY TERMS	90
SELF-TEST EXERCISES	90
EXTRA EXERCISES	94

CHAPTER 6: THE PERIODIC TABLE: *Keeping Track of the Elements* — 95

LEARNING GOALS	96
INTRODUCTION	96
6.1 History	96
6.2 The Modern Periodic Table	99
6.3 Periodic Trends	100
6.4 Periodicity and Electron Configuration	100
6.5 Similarities Among Elements in a Group and Period	100

6.6 Atomic Radius	101
6.7 Ionization Potential	103
6.8 Electron Affinity	105
SUMMARY	105
KEY TERMS	106
SELF-TEST EXERCISES	106
EXTRA EXERCISES	108
CUMULATIVE REVIEW Chapters 4–6	109
CHAPTER 7: CHEMICAL BONDING: *How Atoms Combine*	**113**
LEARNING GOALS	113
INTRODUCTION	114
7.1 Lewis (Electron-Dot) Structures	114
7.2 The Covalent Bond: The Octet Rule	115
7.3 The Coordinate Covalent Bond	118
7.4 Ionic Bonding	119
7.5 Exceptions to the Octet Rule	120
7.6 Covalent or Ionic Bonds: The Concept of Electronegativity	120
7.7 Ionic Percentage and Covalent Percentage of a Bond	122
7.8 Shapes and Polarities of Molecules	124
SUMMARY	128
KEY TERMS	129
SELF-TEST EXERCISES	129
EXTRA EXERCISES	132
CHAPTER 8: CHEMICAL NOMENCLATURE	**135**
LEARNING GOALS	136
INTRODUCTION	136
8.1 Writing the Formulas of Compounds from Their Systematic Names	137
8.2 Writing the Formulas of Binary Compounds Containing Two Nonmetals	137
8.3 Writing the Formulas of Binary Compounds Containing a Metal and a Nonmetal	138
8.4 Polyatomic Ions	144
8.5 Writing the Formulas of Ternary and Higher Compounds	144
8.6 Writing the Names of Binary Compounds Containing Two Nonmetals	145
8.7 Writing the Names of Binary Compounds Containing a Metal and a Nonmetal	146
8.8 Writing the Names of Ternary and Higher Compounds	148
8.9 Writing the Names and Formulas of Inorganic Acids	148

8.10 Common Names of Compounds	151
SUMMARY	151
KEY TERMS	152
SELF–TEST EXERCISES	152
EXTRA EXERCISES	157
CUMULATIVE REVIEW Chapter 7–8	158

CHAPTER 9: CALCULATIONS INVOLVING CHEMICAL FORMULAS — 161

LEARNING GOALS	162
INTRODUCTION	162
9.1 Gram-Atomic Mass and the Mole	162
9.2 Empirical Formulas	167
9.3 Gram-Formula Mass and the Mole	169
9.4 Molecular Formulas	172
9.5 Percentage Composition by Mass	173
SUMMARY	175
KEY TERMS	175
SELF-TEST EXERCISES	175
EXTRA EXERCISES	179

CHAPTER 10: THE CHEMICAL EQUATION: *Recipe for a Reaction* — 181

LEARNING GOALS	181
INTRODUCTION	182
10.1 Word Equations	182
10.2 The Formula Equation	182
10.3 Balancing a Chemical Equation	183
10.4 Types of Chemical Reactions	186
10.5 The Activity Series	192
10.6 Oxidation–Reduction (Redox) Reactions	194
SUMMARY	196
KEY TERMS	197
SELF-TEST EXERCISE	197
EXTRA EXERCISES	202

CHAPTER 11: STOICHIOMETRY: *The Quantities in Reactions* — 205

LEARNING GOALS	206
INTRODUCTION	206

11.1 The Mole Method	206
11.2 Quantities of Reactants and Products	207
11.3 The Limiting-Reactant Problem	215
SUMMARY	220
KEY TERMS	220
SELF-TEST EXERCISES	220
EXTRA EXERCISES	226
CUMULATIVE REVIEW Chapters 9–11	226

CHAPTER 12: THE CHEMISTRY OF SOLUTIONS — 231

LEARNING GOALS	231
INTRODUCTION	232
12.1 Like Dissolves Like	232
12.2 Saturated, Unsaturated, and Supersaturated Solutions	234
12.3 Concentrations of Solutions by Percent	236
12.4 Molarity	240
12.5 Normality	243
12.6 Dilution of Solutions	246
12.7 Ionization in Solutions	247
12.8 Molality	249
12.9 Colligative (Collective) Properties of Solutions	251
12.10 Boiling-Point Elevation, Freezing-Point Depression, and Molality	251
12.11 The Processes of Diffusion and Osmosis	253
SUMMARY	256
KEY TERMS	257
SELF-TEST EXERCISES	257
EXTRA EXERCISES	264

APPENDIX A

Basic Mathematics for Chemistry	265
LEARNING GOALS	265
INTRODUCTION	265
A.1 Adding and Subtracting Algebraically	266
A.2 Fractions	266
A.3 Exponents	267
A.4 Working with Units	268
A.5 Decimals	270

A.6 Solving Algebraic Equations — 272
A.7 Ratios and Proportions — 274
A.8 Solving Word Equations — 275
A.9 Calculating and Using Percentages — 276
A.10 Using the Calculator — 277
SELF-TEST EXERCISES — 278

APPENDIX B
Important Chemical Tables — 283

APPENDIX C
Glossary — 287

ANSWERS TO SELECTED EXERCISES — 295

INDEX — 307

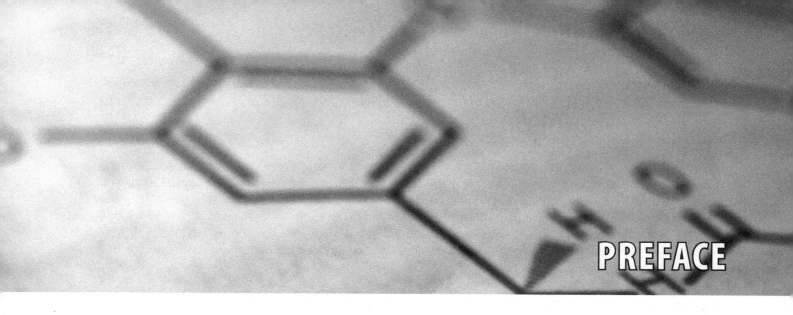

PREFACE

Essential Concepts of Chemistry, 3rd edition is a text intended for use in the one-quarter, two-quarter, or one-semester introductory or preparatory chemistry course. The text is a derivative of our well-known text *Basic Concepts of Chemistry*, which has been used by students for more than three decades. That text continues to receive favorable reviews from students and professors for the text's student-oriented approach, which has been its trademark throughout six editions. *Essential Concepts of Chemistry, 3rd edition* continues this tradition. However, because it contains twelve core chapters (instead of the twenty chapters of the longer text) and is published in paperback form, *Essential Concepts of Chemistry 3rd edition* is available to students at a much-reduced price.

Like its predecessor, *Essential Concepts of Chemistry, 3rd edition* is a text that students will enjoy reading. It is easy to understand and contains a sound problem-solving strategy, numerous examples worked out step-by-step, excellent figures, tables and photos, appendices, and references throughout that demonstrate how chemistry connects to everyday life. We have used a conversational writing style and many pedagogical aids to assist students in mastering the subject matter.

Features of the Book

Applications of Chemistry in Today's World

Essential Concepts of Chemistry, 3rd edition contains exciting and interesting examples of chemistry in everyday life to help students connect chemistry and the world around them. This is accomplished by the use of chapter opening vignettes that introduce the topic of the chapter. Many of these tell a story in which knowledge of basic chemistry plays a role in everyday life. One example is the case of a mother using the scientific method to determine the source of her child's allergy. Other vignettes relate the chapter's topic to a pivotal event in the history of chemistry, like Lavoisier's work to determine that air is a mixture of substances. Each opening vignette is an interesting story in its own right and leads the student into the chapter by demonstrating the relevance of chemistry to daily life.

Polya's Problem-Solving Framework

In addition to a strong program of problem-solving aids, we have also included the general problem-solving framework of the late George Polya of Stanford University. The four-step method is introduced in Chapter 2. We include estimation skills to stress the importance of examining a solution so the student can assess the reasonableness of an answer. In most chapters, at least one problem is solved using Polya's method. We encourage students to apply this method as they solve problems.

Cooperative Learning Opportunities

For several decades, research has shown that students working in small groups improve their academic performance. They can solve problems together and, as part of that process, they discuss the material being studied. Areas that need clarification are identified as students teach one another.

Opportunities for cooperative problem solving are found in the exercises at the end of each chapter. These exercises have been marked with a triangle (◄), indicating that they may be used as conventional problems or for cooperative problem solving. This may be done in class under the direction of the instructor or independently in small study groups.

Study Skills Section

A section on study skills appears in the frontmatter of this text. The section includes information about how to study chemistry, how to manage time effectively, and how to set realistic goals. An expanded section on study skills basics is followed by specifics on how to study chemistry. Other topics covered include how to take good notes, and how to take a chemistry test.

Readability

We maintain the readability of the text by using concise and direct sentences, a conversational tone, and paragraphs of manageable length. Students are provided with concrete examples of chemistry applications in everyday life in order to help them connect chemistry to society. In addition, line drawings, diagrams, and photographs are used to emphasize important points.

Problem Solving

In *Essential Concepts of Chemistry, 3rd edition*, we address the difficulties students encounter in problem solving. In this text we include a large number of worked-out examples, with solutions. Each example contains a practice exercise that offers an immediate opportunity to apply the skill just demonstrated. Answers to all practice exercises appear at the end of the book.

End-of-chapter problems are divided into two groups: self-test exercises and extra exercises. More difficult problems are marked with an asterisk. Self-test exercises are keyed to the specific learning goals of each chapter and appear in matched pairs. The odd-numbered problems, in most cases, are answered in the back of the book. These problems help reinforce the learning of individual skills. Extra exercises are not keyed to learning goals; these exercises help students review skills in a format resembling a quiz or an exam. The answers to many of these problems appear in the answers section at the end of the book.

Cumulative review problems appear after every two or three chapters. They are designed to test the student's knowledge of the chapters and to reinforce material from earlier in the course. These exercises are also useful in studying for tests, midterms, and final examinations.

Study Aids

Our text was planned with the backgrounds, needs, and learning styles of a wide variety of students in mind. Each chapter begins with a series of learning goals. Throughout each chapter the learning goals appear in the margin next to the discussion of the corresponding material. Key terms appear in boldface and their definitions can be found in the Glossary at the end of the book. A list of key terms appears at the end of each chapter, along with a summary that reviews the most important points discussed in the chapter.

A unit on mathematics (Appendix A) is included at the end of the book. It can be treated as a reference or used at the beginning of the course as a review of skills.

Complete Instructional Package

Essential Concepts of Chemistry, 3rd edition is part of a complete instructional package for introductory chemistry. Our *Laboratory Experiments for Basic Chemistry* (9th edition) contains twenty-five experiments and one laboratory exercise. The laboratory manual also contains important safety and chemical hazards guidelines, which discuss potential hazards and stress safety in the laboratory. Each experiment includes a chemical disposal guide with suggestions for disposing of various chemicals used in the experiment. There is also a section on keeping journals to accompany each experiment. Each journal page has a particular journal prompt, which helps clarify the thinking process.

Acknowledgments

We thank the following people who thoroughly reviewed and commented on portions of the manuscript over these many editions, so that we could continue to improve its quality. Without their time, effort, and helpful suggestions, we could not have forged ahead so easily. They are:

Alan Stevenson, Edgecombe Community College, Tarboro, North Carolina

Alfred Powell, East Arkansas Community College, Forest City, Arkansas

James M. Knight, Alabama Southern Community College, Monroeville, Alabama

Bob Boykin, Bossier Parish Community College

Scott Chaffee, West Virginia State College

Leslie DeVirdi, Colorado State University

Michael Dorneman, Mercer County Community College

John Flowers, City College of New York, Medgar Evers College

John Garland, Washington State University

Ed Groschwitz, Palomar Community College

Marie Hankins, University of Southern Indiana

LarRhee Henderson, Drake University

Kristin Kirschbaum, University of Toledo

John Konitzer, McHenry Community College

Miles Mackey, College of the Redwoods

Mary Taylor, Meridian Community College

Richard Wheet, Texas State Technical College

Once again, we thank those who have supported us, inspired our thoughts, and given us ideas: Professors Charles Oxman, Dominic Macchia, Linda Christopher, and Barbara Drescher of Middlesex County College; Russell Hulse and Christine Ritter of the Princeton University Plasma Physics Laboratory; Wayne Hoy of the Ohio State University; and Gregory Camilli of the Rutgers University Graduate School of Education.

Finally, we'd like to thank our grandchildren, Jake, Josee and Sienna Sherman for helping to illustrate some selected concepts in this textbook.

TO THE STUDENT

TIPS ON STUDYING CHEMISTRY

Chemistry is an exciting subject. It affects our lives each day in many ways. The foods we eat, the clothes we wear, the materials used to construct our homes, the medications that cure our illnesses, and the energy we depend on are all products of the science of chemistry coupled with technology.

As you learn chemistry, you will gain knowledge that will help you understand more about our technological society. You'll be able to apply your understanding of chemistry to help solve some of the problems you will face in the real world. You'll be able to approach familiar situations with new insight and be able to make decisions based on the knowledge acquired in this course.

Study Skills for Chemistry

How to Use This Book

Your textbook can be an important tool in helping you master the ideas and problem-solving skills you will learn in this course, especially if you know how to use it. Take a minute now to become familiar with the parts of the book. What follows are our suggestions for how you can best use the book to help you succeed in your course.

Begin each chapter by reading the learning goals and the introduction. The learning goals will help you organize the material you'll need to master; they identify topics that are likely to have special emphasis in class.

After you've read the introduction and learning goals, flip through the chapter to get a general sense of the topics it covers. Notice that the learning goals are repeated in the margins near the material to which they apply. The numbered headings in the text will alert you to the major topics of the chapter. Look also at the figures and tables in the chapter.

After your brief flip through the chapter, return to the beginning and start reading. Read each section thoroughly and take notes as you read. At the end of each section, pause to think about what you have just read; if you are not sure you understand everything in the section, go back and read again.

While some of the material in this course requires memorization, a large part of chemistry involves learning how to solve problems. *Essential Concepts of Chemistry, 3rd edition* has been written with a major focus on developing problem-solving skills. Throughout the text, you'll find worked-out examples. When you come to them, stop and read through them carefully. Each example presents a problem and then shows you the solution, step by step. Once you have read the worked-out example, try the practice exercise that follows. Check your answers against the answers given at the back of the book. Problem-solving skills are among the most important things you will learn in this course, and working the exercises in the chapter will give you practice and immediate feedback on how well you are mastering them.

At the end of the chapter, you'll find a summary and a list of key terms. Review these, testing yourself to see if you understand the terms. If you need help, the numbers following the terms tell you which section to review. (There is also a glossary at the back of the book giving definitions of the terms.) Finally, work the self-test exercises at the end of the chapter. You will see that they are grouped by learning goal, to help you organize the material. The answers to selected exercises are at the back of the book, to help you test your progress.

How can you be sure, as you study, that you have mastered the material in the chapter? If you are able to work through the examples, practice exercises, and self-test exercises at the end of the chapter, you will have mastered the necessary problem-solving skills. For sections that require you to memorize reactions or descriptive material, self-test exercises are designed to help you master this content.

You'll get the best results if you read the relevant chapter before going to the lecture. Survey the material to be covered and examine the types of problems to be solved. Know what the chapter is about. This preparation will help you get the most from your instructor's lecture.

If you find at the beginning of the course that you need some review of mathematics, turn to Appendix A at the end of this book. It reviews the mathematical skills you'll need in this course.

Taking Good Notes

Note taking is one of the most important skills for a student to develop. In class you will take notes. Review them the next time you study and reread each section of the textbook slowly for understanding. Combine what you learned in class with what you read in the textbook. Where problem solving takes place, go over the step-by-step solutions in the textbook. Then answer the practice exercises to be sure that you understand how to find the solutions yourself. If you're stuck, ask the instructor for help. Do not ignore material that you do not understand.

We have written *Essential Concepts of Chemistry, 3rd edition* in a simple, conversational, readable style. You should not have a difficult time understanding the book. Spend some time being sure that you can recall what you have learned. Sometimes it helps to write as you read, recalling what you have learned in your own words on paper. The last step is to reread the entire chapter and go over all of your notes.

Many students find it useful to keep a journal to write about the material being studied. A simple lined notebook can serve as a journal. After each lesson, summarize the point of the lesson. After reading a section in the textbook, write about what you thought was the point of the reading. Read your journal notes to a classmate or, if possible, to a teaching assistant or to the professor. Writing about a topic is an excellent way to find out if you understand the material being studied.

Your notes must be accurate to be useful. You will have to develop your own shorthand in order to take notes quickly. There are many symbols and abbreviations that you can use in devising your own shorthand. Mathematical symbols such as > (greater than) and < (less than), the plus sign (+), and the equals sign (=) are useful. Spelling a word phonetically by leaving out the vowels also speeds things up.

As you work through the exercises in each chapter of the book, clearly indicate each type of problem you are solving. Write a short description of the type of problem and explain why you are solving this particular problem. When there are several variables in a problem, be sure you can solve for each of the variables, depending on the information given.

Combine your lecture notes with your textbook notes. Be sure you understand your notes. To test your knowledge, turn to the end of the chapter and do the self-test exercises. If you can answer them, you're on your way.

The Importance of Attending Class

There is nothing more important than attending class to help you learn chemistry. Regular attendance allows you to have all of the lecture notes you'll need to organize your studies, and the experiments you perform in the laboratory give you hands-on experience in chemistry. Instructors are experts in teaching their subjects. They teach to a variety of learning styles, and you will benefit greatly from the classroom and laboratory experience and the personal interaction with the instructors and teaching assistants. If you need extra help, see your instructor after class and set up an appointment. Some instructors set aside time for weekly review sessions, and you will learn about these in class.

Making the Most of Your Study Group

Becoming part of a study group is one of the best ways to learn the material in the course. When people work together they can share their skills and resources. In this way they can get more done than by working alone. In addition, people often draw strength from groups. Setting a meeting with the study group is like making an appointment to study. It's an appointment that you are more likely to keep, as compared with a solo study session, which can be more easily skipped. With three or four classmates you can practice problem solving and reasoning. Study groups can be organized in many ways. Here are some suggestions for you to try.

The first step in making the most of your study group is to form a group that works for you. Find people who share some of your goals and who face similar challenges. Find people who are as serious about school as you are. This might include those who attend class regularly, take good notes, and keep up in class. If you are a single parent who supports a family, it might be helpful to form a group that contains at least one other member who has similar time constraints and obligations.

At the first meeting of the study group, set a method of operation. It is likely that one member will emerge as the leader. You might decide to have this person remain the leader, or you might share and rotate the leadership role. Decide what material the group will cover during each session. Set an assignment for the next session, so that everyone will be prepared and the time is used efficiently and effectively. Set time limits for discussion of each topic, so that all of the topics that must be covered are addressed. Practice teaching each other the material.

Solving problems as a group is an important task. At the end of each chapter in the self-test exercises, you will find certain problems marked with a triangle. These problems are good for discussion and are ideal for use in the study group. Once each member learns the material, the problems can be solved by the group. The next step is to have each member solve problems individually and then discuss them as a group.

Once the material is discussed and everyone understands the subject matter, you will be ready to test each other. Have each group member choose several questions, make a sample test of your own, and test each other. You might also try to predict what questions or types of questions will be on the test. Don't forget that even with a study group, it is necessary to do a great deal of studying on your own. Working individually is an important part of the learning process.

Taking a Test

Being well prepared and confident is the first step in taking a test. Cramming doesn't work in chemistry. You cannot open the book the night before the exam and learn a few weeks' worth of material in one sitting. Chemistry must be studied gradually. If you have trouble solving a particular type of problem, you may need to consult the instructor. That's difficult to do if you try to study all at once.

Try to study a little each day. You'll find this method more effective and less stressful.

Before an exam, it is important to speak with your instructor and find out what material will be covered on the test. Instructors emphasize important concepts in class. Some instructors will help guide your studies by offering information on material that deserves special attention.

Get a good night's sleep before an exam. Be as relaxed as possible so you can think easily. You'll know whether you're prepared or not. If you've studied properly, you should do well.

Once you begin the exam, glance through it to see how long it is and determine how to budget your time. If a problem gives you trouble, skip it and come back to it later. That way you'll have time to get to all of the questions.

Maintaining Good Study Habits

You can successfully complete basic chemistry if you develop good study habits. The sections that follow discuss some common-sense methods of achieving success in college and offer additional advice that can be applied to your day-to-day study of chemistry. As you become a well-prepared and successful student, you will approach examinations with confidence. The successful habits you form will stay with you as you complete your education and will benefit you throughout your life.

Time Management for College Students

You've got an exam on Friday and a paper due on Monday. You know you've got to get to work. Just then the phone rings. A group of friends from your dorm is going out for a snack. You know you shouldn't join them, but you've got to eat anyway, so you go along. When you return home you need a little time to unwind. You turn on the television. Two hours later you're relaxed, but you're also tired. You decide to call it a night. There's always tomorrow!

If this scenario sounds familiar, it should. It happens to all of us now and then. As a college student, however, it's very important to learn to manage your time effectively. The first step in managing your time effectively is to know where you're going. It helps to set goals for yourself. Although we may have vague notions of what we want from life, like being happy, or being a credit to society, or being financially secure, these generalized plans should be made concrete. Goals must be real. They must be examined closely. There are three different types that you should consider: long-range goals, medium-range goals, and short-range goals.

Long-range goals are usually personal wishes. They have to do with your career aims, your educational plans, and your social desires. Think about where you would like to be 5 or 10 years from now. The education you are now receiving in college should be a stepping-stone to help you achieve your long-range goals. Besides achieving the benefit of learning, a college education pays off in dollars. College graduates earn about $900,000 more during their lifetimes than their counterparts who have no degrees. Depending on your career plans; the grades you earn in your courses will help determine whether or not you will be able to fulfill your long-range goals. To achieve long-range goals, they need to be broken into smaller parts and examined closely.

Medium-range goals, sometimes called mid-term goals, can be accomplished in one to five years. They help you achieve your long-range goals. They can be set two or three times a year. For example, if you plan to enter medical school after graduation, you will need a considerable number of A's in your courses. A medium-range goal would be to get four or five A's in your courses for four years. Another medium-range goal might be to join a club or improve your skills in your favorite sport. Let's say that your grades last semester weren't the best. A medium-range goal for you might then be to improve your grades. If you're saving money to buy a car, then watching your budget more carefully might be a reasonable goal to set.

Short-range goals, also called short-term goals, can be accomplished in a year or less. These goals involve taking care of your daily tasks and keeping up with your assignments. Reading a chapter in a book, completing an assignment, or writing a paper are examples of short-term goals.

Achieving your goals often requires setting up a plan. Although we'd like to think that we have the willpower to do all that we have to, very often we become distracted. A well-planned, flexible schedule is a useful tool to help us get things done more effectively.

The first step to managing your time is to know yourself. When do you function the best? If you're a morning person, it's best to try and get done as much as you can in the morning, when you are at your best. If you function better late in the day, then you should try to schedule most of your work in the afternoon or evening.

Setting up a monthly plan is the next step in becoming organized. Obtain a calendar and write in, on the appropriate date, the important assignments and events you're responsible for each month of the semester (Figure 1). Term papers, exams, sports meets, and social activities can be handled better if you look at the overall picture. If you know you have two exams scheduled and a paper due close to each other, then you'll have to keep your schedule free to allow for your schoolwork.

After you've completed your monthly plan, devise a weekly plan. Map out a schedule for each day of the week. Think over how much time you need to complete each assignment and to study for your exams, in order to fit all of your responsibilities into your schedule (Figure 2).

It's usually best to find a quiet place to work. Accomplish your tasks in priority order, breaking the large tasks into smaller ones. Don't try to do too much in one day. It's better to do a little less than to do too much. Complete one task at a time and avoid working on two projects at once.

You may find that after a month or so, you can manage your time effectively without a written schedule. For some students having a written schedule is a necessity. When you are designing your plan, don't forget that you need time for enjoyment and for work. Giving yourself positive reinforcement is a good idea. Reward yourself when you've accomplished something worthwhile. It is important to enjoy what you're doing and to set aside enough time for rest and relaxation.

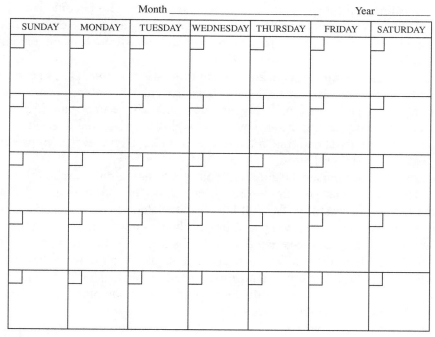

FIGURE 1 Monthly planning calendar.

Weekly plan
My main goal this week: _____

	Monday	Tuesday	Wednesday	Thursday	Friday	Saturday	Sunday
6 A.M.							
7 A.M.							
8 A.M.							
9 A.M.							
10 A.M.							
11 A.M.							
12 NOON							
1 P.M.							
2 P.M.							
3 P.M.							
4 P.M.							
5 P.M.							
6 P.M.							
7 P.M.							
8 P.M.							
9 P.M.							
10 P.M.							
11 P.M.							
12 MID							

FIGURE 2 Weekly plan.

While some people find it difficult to socialize, there are others who become overly involved in social activities. Attending college can be very exciting, and for some young adults the endless opportunities to work for a cause, socialize with friends, or attend frequent parties are irresistible. It is not uncommon for some college students to use study time for socialization.

Around the time of midterms panic sets in and lasts until final exam time. Don't get caught in this trap. Be sure that you are not becoming involved in too many social activities in order to avoid your studies. If you're serious about receiving a college education, then your study time will be important to you.

School-related stress usually results when you allow yourself to become overwhelmed and overloaded. Taking on too much responsibility at one time often results in a high stress level. This can cause a whole range of physical, emotional, and social problems.

When you register for your courses, be sure that you're not taking on too much. Most colleges schedule a full load for you. This means that you sign up for a set number of credits that school personnel believe a full-time student can handle successfully. The full load is calculated so that a full-time student can finish the requirements for an associate's degree in two years, or a bachelor's degree in four years.

Students who work diligently and have good study skills and few other demands on their time can normally handle a full load. Problems often arise when a student signs up for more credits than are recommended for a full load or when there are other important time-consuming responsibilities. When you have too much responsibility to handle at once, stress can readily develop. Instead of completing your course work with ease, you may find yourself playing "catch-up" all semester.

Arranging a course load to fit your needs is one of the most important items to think about as a college student. You can avoid overloading yourself by taking a night course during the summer at a local college or by attending summer school. Spending an extra semester or two in college is a reasonable price to pay in order to achieve success and avoid stress while you're in college.

Your college years are likely to be among the most demanding and enjoyable years of your life. During this time there are activities that will compete for your attention. Studying, developing relationships, and handling your financial affairs are among the most important challenges that will require your energy, creativity, and brainpower. Learning to deal successfully with the different facets of your life can be accomplished by developing good coping skills. These are skills that will remain with you throughout your life and can be applied to just about any situation you'll encounter.

Getting Yourself to Study

For many people it's difficult to sit and study. They find countless distractions to avoid getting to the task. If you have a problem getting started, convince yourself that you will be able to get through the material rather quickly if you break it up into small, manageable units. After learning each small unit, stop and take a short break, then continue.

Developing the ability to concentrate is another important study skill. Keep your mind on what you're doing. Don't allow yourself to be distracted. Turn off the cell phone and the television set and try to eliminate interruptions. Find a quiet place where you won't be disturbed. Sometimes low-level noise like instrumental music or a steady flow of traffic helps stimulate concentration. And don't try to study if you are hungry. That's a sure way to break your concentration.

Some people are procrastinators. They put things off and never get much done. If you tend to procrastinate, you might benefit from examining your behavior. Does procrastination keep you from reaching your goals? If so, being aware of the cost of procrastination might help you break the habit. Perhaps you are one who works best under pressure. If this produces a better work environment for you, perhaps you procrastinate to your benefit.

If you find that procrastination hurts your progress, here are some ways to break the habit. First, look at the benefits of doing the work at hand. Will you feel a sense of accomplishment when the job is done? Will you feel less stressed and more relieved? Next, break the job into smaller parts. Perhaps there is just too much to learn at once. Break the work into 15-minute segments. After 15 minutes, stop and take a rest. Work slowly to build up your capacity to work.

Perhaps you have tried the first two steps and you still can't get started. Contact a friend, roommate, parent, spouse, or child and set a mutual goal. Work for an hour, and then celebrate together. Another strategy is to reward yourself when you are done. If you simply can't break the habit, just sit down and force yourself to complete the task. You will certainly be relieved when it is completed. Finally, it might be necessary to accept the fact that procrastination is a barrier to your success. Be honest with yourself. Then move through the problem and solve it, instead of letting it drain you of your energy.

Final Thoughts

As you begin your study of chemistry, we hope that you do so with an open mind and a positive attitude. You are probably enrolled in a chemistry course because it is part of the path you have chosen in order to reach your career goals. We hope you will enjoy the course and find the textbook a useful tool to help you learn the subject matter. Our goal is to help you get the most out of this course. Chemistry may be new to you, and we hope that some of the tips for studying more effectively will be useful to you. If you have taken a chemistry course before, or if you know of a method of studying that works for you, we invite you to write to us and share your suggestions.

As you enter the fascinating world of chemistry, we hope your experience is a good one and that you come away from the course with an understanding of why we remain so enthusiastic about this exciting subject.

Sharon J. Sherman
Alan Sherman

CHAPTER 1

THE ORIGINS OF CHEMISTRY
Where It All Began

Imagine being present at the Princeton Plasma Physics Laboratory (PPPL) on the evening of one of the most important scientific experiments of the century—an experiment that if successful would take humans one step closer to harnessing fusion power as a commercial energy source. The scientists at PPPL had been working toward this night for over 40 years. Their goal was to harness the energy source of the stars.

Nuclear fusion is the process of combining nuclei of small atoms to form larger atoms. Our sun and other stars produce their energy by fusing hydrogen atoms into helium atoms. This process creates huge amounts of energy, on the order of tens-of-millions of degrees Celsius. Unlike nuclear fission (splitting atoms), fusion is a very clean process that produces no dangerous long-lasting radioactive isotopes, and the fuel used, hydrogen, is plentiful. If harnessed, fusion power could supply all of humankind's energy needs for thousands and thousands of years, long after fossil fuels have been depleted.

The technical difficulties of creating a device that could reproduce on earth the reactions that take place on the sun and stars appeared at first to be insurmountable. How could such fantastic heat be contained? What vessel or tools could handle it? Moreover, to duplicate the sun's fusion process on earth, the hydrogen atoms would have to form a *plasma* that must reach a temperature of about 100 million degrees Celsius for one second, with a density of about one one-hundred-thousandth that of air at sea level. Under these conditions, the hydrogen nuclei collide and produce helium nuclei and subatomic particles called neutrons. The energy of motion of these neutrons is then converted into electricity. This was the theory, but would it work?

Over the previous four decades, the scientists and engineers at PPPL built several prototype reactors. These reactors utilized magnets to hold the plasma in a "magnetic bottle." Scientists and engineers also developed techniques such as "neutral beam heating" and "radio frequency heating" to heat the plasma to very high temperatures. Each reactor was built to gain knowledge about the behavior of plasmas and to solve the problems associated with plasma heating and confinement. Each reactor helped scientists overcome various technical hurdles, bringing them ever closer to achieving controlled fusion.

In the mid-1970s the PPPL scientists and engineers constructed their largest reactor, the Tokamak Fusion Test Reactor (TFTR). It would take thousands of preliminary experiments over a period of two decades to ready this reactor for its greatest challenge—the production of large amounts of fusion power. The earlier reactors, and the preliminary experiments on TFTR, all used *protium,* a form of hydrogen, as the fuel. To achieve commercial fusion power, however, two different forms of hydrogen atoms, a 50–50 mixture of *deuterium* and *tritium,* would be required. Modifications to the reactor were made, and the researchers at PPPL were at last ready to attempt a fusion breakthrough.

It was a cool evening in Princeton, New Jersey, on December 9, 1993. We arrived at PPPL about 9:00 P.M. There were already about 200 people in the auditorium and main lobby area. There were television monitors everywhere, showing live pictures of TFTR and the project team in the control room. Staff members and their families were on hand to witness this historic event; scores of newspaper reporters and scientists from around the nation were also present. The excitement in the air was tempered with feelings of apprehension. The project team, located in a futuristic-looking control room, began to power up the systems of TFTR. Over the next two hours, these systems would be checked and run through several conditioning experiments.

Finally, at 11:07 P.M. all was ready. TFTR's systems were powered up. A 50–50 mixture of deuterium and tritium was injected into the reactor, and the neutral beam and radio frequency heaters were turned on. A whining sound from the reactor could be heard. A monitor beaming a picture from inside the reactor turned from intense white to intense black. All of the spectators, scientists, and engineers stood silent, as if frozen in time. In a few seconds, which seemed an eternity, it was over. The computer screens came to life, people began to move, while the spectators whispered as the scientists analyzed the information coming from the reactor. Within two minutes, the information was relayed to deputy laboratory director Dr. Dale Meade, who announced to the spectators that three million watts of fusion power had been obtained in a single burst. History was made! Spectators, scientists, and reporters cheered loudly. Dr. Ron Davidson, the director of PPPL said, "I think this is a historic milestone in fusion energy development, something analogous to the Wright Brothers' first flight."

The next day, newspaper headlines around the world announced the event. Humankind's dream of finding a clean, safe, and inexhaustible supply of energy took a quantum leap forward. We were at PPPL on the night of December 9 with our son Michael. It was a night that we will never forget.

Sharon and Alan Sherman

LEARNING GOALS

After you've studied this chapter, you should be able to:

1. Discuss the nature of science and scientific thought.
2. Name the substances that Greek philosophers thought were basic to the composition of the earth.
3. Discuss the role of the Egyptians in the development of chemistry.
4. Describe the major goals of the alchemists.
5. Explain how Robert Boyle's book laid the foundations of modern chemistry.
6. Name the two main branches of chemistry.
7. Discuss the uses of chemistry in today's world.

INTRODUCTION

Humans are curious. They ask questions and they try to find answers. The desire to understand their environment has made people search for explanations, principles, and laws, and modern chemistry is one result of that search.

Chemistry is the science that deals with matter and the changes it undergoes. Chemistry also focuses on energy and how it changes when matter is transformed. Chemists seek to understand how all matter behaves and to discover the principles governing this behavior.

Chemists ask fundamental questions and look for solutions in their attempt to unlock the secrets of nature. Many of their questions have been answered, and this information has built the body of knowledge that we call science.

Many of the basic concepts of chemistry have been incorporated into different fields of study. Physicists, biologists, engineers, medical and dental professionals, and nutritionists are just some of the professionals whose fields require chemical knowledge.

In this chapter we first look at the beginnings of chemistry and then briefly touch upon some areas of modern chemical research.

1.1 History

The earliest attempts to explain natural phenomena led to fanciful inventions—to myths and fantasies—but not to understanding. Around 600 B.C., a group of Greek philosophers became dissatisfied with these myths, which explained little. Stimulated by social and cultural changes as well as curiosity, they began to ask questions about the world around them. They answered these questions by constructing lists of logical possibilities. Thus Greek philosophy was an attempt to discover the basic truths of nature by thinking things through, rather than by running laboratory experiments. The Greek philosophers did this so thoroughly and so brilliantly that the years between 600 B.C. and 400 B.C. are called the "golden age of philosophy."

Some of the Greek philosophers believed they could find a single substance from which everything else was made. A philosopher named Thales believed that this substance was water, but another named Anaximenes thought it was air. Some earlier philosophers believed the universe was composed of four elements—earth, air, fire, and water. The Greek philosopher Empedocles took this belief a step further, maintaining that these four elements combine in different proportions to make up all the objects in the universe.

During this period, the Greek philosophers laid the foundation for one of our main ideas about the universe. Leucippus (about 440 B.C.) and Democritus (about 420 B.C.) were trying to determine whether there was such a thing as a smallest particle of matter. In doing so, they established the idea of the atom, a particle so tiny that it could not be seen. At that time there was no way to test whether atoms really existed, and more than 2,000 years passed before scientists proved that they do exist.

While the Greeks were studying philosophy and mathematics, the Egyptians were practicing the art of chemistry. They were mining and purifying the metals gold, silver, and copper. They were making embalming fluids and dyes. They called this art *khemia,* and it flourished until the seventh century A.D., when it was taken over by the Arabs. The Egyptian word *khemia* became the Arabic word *alkhemia* and then the English word *alchemy.* Today our version of the word is used to mean everything that happened in chemistry between A.D. 300 and A.D. 1600.

A major goal of the alchemists was to transmute (convert) "base metals" into gold. That is, they wanted to transform less desirable elements such as lead and iron into the element gold. The ancient Arabic emperors employed many alchemists for this purpose, which, of course, was never accomplished.

The alchemists also tried to find the "philosopher's stone" (a supposed cure for all diseases) and the "elixir of life" (which would prolong life indefinitely). Unfortunately they failed in both attempts, but they did have some lucky accidents. They discovered acetic acid, nitric acid, and ethyl alcohol, as well as many other substances used by chemists today.

The modern age of chemistry dawned in 1661 with the publication of the book *The Sceptical Chymist,* written by Robert Boyle, an English chemist, physicist, and theologian. Boyle was "skeptical" because he was not willing to take the word of the ancient Greeks and alchemists as truth, especially about the elements that make up the world. Instead Boyle believed that scientists must start from basic principles, and he realized that every theory had to be proven by experiment. His new and innovative scientific approach was to change the whole course of chemistry.

Table 1.1 chronologically outlines some of the major contributions in chemistry throughout history.

TABLE 1.1 Time Line of Chemistry

DATE	PERSON	EVENT
600 B.C.	Thales	Idea that water is the main form of matter
546 B.C.	Anaximenes	Idea that air is the main form of matter
450 B.C.	Empedocles	Idea that the four elements combine in different proportions
420 B.C.	Democritus	Idea of the atom
400 A.D.	Ko Hung	Attempts to find the elixir of life
1000	Avicenna	*Book of the Remedy*

(Continued)

4 ESSENTIAL CONCEPTS OF CHEMISTRY

TABLE 1.1 (Continued)

DATE	PERSON	EVENT
1330	Bonus	*Introduction to the Arts of Alchemy*
1500		*Little Book of Distillation*
1620	Van Helmont	Foundations of chemical physiology
1625	Glauber	Contributions to practical chemistry
1661	Boyle	*The Sceptical Chymist*
1766	Cavendish	Discovery of hydrogen
1775	Lavoisier	Discovery of the composition of air
1787	Lavoisier and Berthollet	A system of naming chemicals
1800	Proust	Law of Definite Composition
1800	Dalton	Proposes an atomic theory
1820	Berzelius	Modern symbols for elements
1829	Döbereiner	Law of Triads
1860	Bunsen and Kirchhoff	Spectroscopic analysis
1869	Mendeleev	Periodic Law
1874	Zeidler	Discovery of DDT
1874	Van't Hoff and Le Bel	Foundations of stereochemistry
1886	Goldstein	Naming of cathode rays
1897	Thomson	Proposes a structure of the atom
1905	Einstein	Matter–energy relationship: $E = mc^2$
1908	Gelmo	Discovers sulfanilamide
1911	Rutherford	Proposes the nuclear atom
1913	Bohr	Proposes energy levels in atoms
1922	Banting, Best, and Macleod	Discover insulin
1928	Fleming	Discovers penicillin
1932	Urey	Discovers deuterium
1942	Fermi	First atomic pile
1945		First atomic bomb; nuclear fission
1950	Pauling	Helical shape of polypeptides
1952		First hydrogen bomb; nuclear fusion
1953	Watson and Crick	Structure of DNA
1958	Townes and Schawlow	Develop laser beam
1969		Discovery of first complex organic interstellar molecule—formaldehyde
1970	Ghiorso	Synthesis of element number 105
1973	Cohen, Chang, Boyer, and Helling	Initiation of recombinant DNA studies
1974	Seaborg and Ghiorso	Synthesis of element 106
1979	A. Crewe	First color motion pictures of individual atoms
1982	Munzenberg and Armbruster	Synthesis of element 109

CHAPTER 1: THE ORIGINS OF CHEMISTRY 5

DATE	PERSON	EVENT
1984	Gallico	Development of test-tube skin
1985		Genetically engineered human insulin and human growth hormone become commercially available
1985	Diana	Development of compounds that kill cold viruses
1986		Nearly 800 genes mapped using genetic engineering techniques
1987		New class of high-temperature superconducting oxides with major implications for use in electronics discovered by researchers at the Uni. of Houston and at AT&T Bell Laboratories in Murray Hill, N.J.
1988	Zewail	First snapshots of chemical reactions
1989	Felix	Synthesis of human growth hormone
1990	Filisko	Development of chemical muscles
1991	R. Baughman	Chemists in the United States and several other countries develop techniques for synthesizing large amounts of soccer-ball-shaped C-60 molecules called buckminsterfullerenes ("Buckyballs")
1992	J. Rebek	Synthesis of molecules that mimic some of the essential features of living things
1993		Princeton University's Plasma Physics Laboratory produces 7 megawatts of fusion energy using deuterium and tritium as fuel in their Tokamak Fusion Test Reactor. These experiments represent a major breakthrough for the commercialization of fusion energy. (In 1994 the reactor produced 10.7 megawatts of fusion energy.)
1994		Fermi National Accelerator Laboratory announces that a team of 440 physicists has discovered the Top Quark. This elusive particle, sought after for 17 years, completes the gallery of six quarks. The quarks are some of the most fundamental particles of matter.
1995		Carl Wieman and Eric Cornell of the University of Colorado create a new state of matter, predicted by Albert Einstein over 70 years ago, called the Bose-Einstein Condensate. This type of matter was created at a temperature of 36 billionths of a degree above absolute zero.
1996		NASA scientists report that a meteorite, identified as originating from Mars, may contain fossils of ancient, primitive bacteria from that planet.
1997		The first mammal, a sheep, was cloned from an adult nucleus.
2001		The International Human Genome Sequencing Consortium published the first draft of the human genome in the journal *Nature* in February 2001 with the sequence of the entire genome's three billion base pairs some 90 percent complete.

(Continued)

TABLE 1.1 *(Continued)*

DATE	PERSON	EVENT
2005		Yves Chauvin of France and Americans Robert H Grubbs and Richard R Schrock on Wednesday won the Nobel Prize for a breakthrough in carbon chemistry that opens the way to smarter drugs and environmentally friendlier plastics.
2010		A new nanotech catalyst that offers industry an environmentally benign way to reduce toxic heavy metals from the chemical process through simple magnetic nanoparticles has earned McGill University researchers Chao-Jun Li, Audrey Moores and their colleagues a spot on Quebec Science's list of the Top 10 discoveries of 2010.
2014		A team of Harvard scientists and engineers has demonstrated a new type of battery that could fundamentally transform the way electricity is stored on the grid, making power from renewable energy sources such as wind and sun far more economical and reliable.

1.2 Chemistry Is a Diverse Field

During the 1700s and early 1800s, most chemists believed that there were two main branches of chemistry: organic and inorganic. Organic substances were thought to have been derived from living or once-living organisms; inorganic substances were said to have originated from nonliving materials. Sugars, fats, and oils were classified as organic chemicals. Salt and iron were classified as inorganic chemicals.

These two branches of chemistry still exist today, but the rules governing their classification have changed. Chemists now classify organic substances as those containing the element carbon. Inorganic substances are generally those that are composed of elements other than carbon. (Some inorganic substances are exceptions to this rule and do contain carbon.) Sugars, fats, and oils contain the element carbon and are indeed classified as organic substances. Salt and iron do not contain carbon, and both fall into the category of inorganic substances.

Other branches of chemistry include nuclear chemistry, biochemistry, pharmaceutical chemistry, analytical chemistry, environmental chemistry, and physical chemistry.

1.3 How Chemistry Affects Our World

Chemistry and chemical processes are keys to understanding the world in which we live today. Today's chemistry includes areas as diverse as lasers, superconductors, genetic engineering, environmental management, and the synthesis of new Pharmaceuticals.

Drawing upon such natural resources as the air, forests, oceans, mines, wells, and farms, the chemical industry produces more than 50,000 different chemicals. These chemicals are used to manufacture a variety of products and to provide raw materials for other industries to use in producing their goods. It is by this process that the items required to sustain our basic human needs are made available to us.

Chemistry has played an important role in the processing of foods. Foods are processed so that they remain fresh and free of harmful toxins for a longer period of time. Chemists are hard at work seeking and researching ways to alleviate a world food shortage at a time when human population is approaching 6 billion people. Thousands of drugs to help us treat disease have become available over the last several decades through the application of medical knowledge in the chemical and pharmaceutical industries. Just a few of the chemically based products we use today are plastics, cleansing agents, paper products, textiles, hardware, machinery, building materials, dyes and inks, fertilizers, and paints.

CHAPTER 1: THE ORIGINS OF CHEMISTRY 7

SUMMARY

Scientific thought had its beginnings in the contributions of the Greek philosophers of 600 B.C. to 400 B.C. The earliest chemistry was practiced at about that same time by the Egyptians. Later the art of chemistry—or alchemy—was taken over and extended by the Arabs. "Alchemy" now includes all chemistry from the fourth through the sixteenth centuries A.D. The alchemists made several important discoveries, although they were primarily interested in transmuting base metals into gold. Modern chemistry began with the work of Robert Boyle in the midseventeenth century. Boyle believed that theories must be founded on basic principles and must be supported by experiment.

Modern chemistry and its applications affect us in a variety of ways as we eat or sleep, work or play. Present chemical research is directed toward the solution of problems in many areas, such as physiology, medicine, and the environment.

SELF-TEST EXERCISES

All exercises whose numbers are in underscored, bold-faced italic are answered in the back of the book. Self-test exercises are arranged in matched pairs, one below the other. Difficult problems are marked with an asterisk. Problems marked with a triangle (◄) may be used for cooperative problem solving.

LEARNING GOAL 1

Nature of Science and Scientific Thought

◄ 1. Discuss the evolution of science from the golden age of philosophy to the birth of modern chemistry.
 2. What years are called the "golden age of philosophy"? Why are those years considered the beginning of scientific thought?

LEARNING GOAL 2

Influence of Greek Philosophy on Chemistry

 3. List the four elements that Empedocles thought made up the world.
 4. What did Thales think the world was composed of?
◄ 5. What contribution did Greek philosopher Democritus make to modern-day society?
 6. Who established the concept of the atom?
 7. If Empedocles had been right and the world were indeed made up of four elements, what two elements would compose paper?
◄ 8. If the world were made up of Empedocles' four elements, what elements would compose vegetation?

LEARNING GOAL 3

Egyptians' Role in the Development of Chemistry

 9. List some of the contributions of the Egyptians to the development of chemistry.
 10. The Egyptians mined and purified the metals gold, silver, and antimony. True or false?
 11. While the Greeks were studying philosophy and mathematics, the Egyptians were already practicing the art of chemistry, which they called _____ .
 12. Until what time in history did the art of *khemia* flourish?

LEARNING GOAL 4

Alchemists' Major Goals

◄ 13. What was the aim of chemistry from A.D. 300 to A.D. 1600?
 14. Name the metals that the alchemists tried to turn into gold.

8 ESSENTIAL CONCEPTS OF CHEMISTRY

15. The people who tried to turn lead into gold were known as _____.
16. What is meant by the term *elixir of life?*

LEARNING GOAL 5
Boyle's Influence on the Development of Modern Chemistry

17. What was the message behind Robert Boyle's book *The Sceptical Chymist?*
◄ 18. How do you think Boyle might have set about proving that water is not an element and that it can be broken down into simpler substances?

LEARNING GOAL 6
Branches of Chemistry

◄ 19. Name an organic chemical and an inorganic chemical.
20. Explain by what criterion chemicals are now classified as organic or inorganic.

LEARNING GOAL 7
Chemistry in Today's World

◄ 21. Choose one application of chemistry and discuss its impact on society today.
22. Weigh the benefits that chemical research and applications yield to society against the problems caused by the chemical industry. What legislative measures have been taken (or are pending) to solve some of these problems?

EXTRA EXERCISES

23. Make a list of the industries that are based on the science of chemistry.
24. From the "Time Line of Chemistry" (Table 1.1), choose those events in the history of chemistry that you believe have had the greatest social importance. Give reasons for your choices.
25. Explain the benefits of using Biodiesel over gasoline. Discuss the disadvantages of using this material.
◄ 26. If you could extend the "Time Line of Chemistry" 50 years into the future, what would you like to see happen? What advances in research do you believe will take place?
27. Can an educated public, better equipped with scientific knowledge, more easily assess the effects of science on society? Explain your answer.
28. With what is the science of chemistry concerned?
29. According to Empedocles, how would each of the following be classified?

    (a) copper   (b) rain   (c) lightning   (d) nitrogen gas

◄ 30. Explain in a few sentences how you believe chemistry has affected life in the context of one topic, such as food, clothing, or medicine.
◄ 31. Discuss the contributions of alchemy to the science of chemistry.
32. List four ways in which chemistry has had a positive, healthy effect on life in your home. Explain each.

CHAPTER 2

SYSTEMS OF MEASUREMENT

This is the story of Richard Novak, a jeweler in a large metropolitan city, who used something that he had learned in his chemistry class to help him with a business transaction.

One day a customer entered Mr. Novak's store carrying a small burlap bag. The customer approached Mr. Novak and explained that in the bag he had about 200 grams of pure gold nuggets that he wanted to sell. The customer asked Mr. Novak whether he would be interested in purchasing the nuggets. Mr. Novak responded by indicating that he would purchase the nuggets if he could have the material tested to verify that it was pure gold. The customer responded by saying that he didn't want the gold put into any kind of an acid solution. (The usual test for determining if a substance is pure gold is to immerse it in a solution of aqua regia, a mixture of hydrochloric and nitric acids.) Mr. Novak thought for a few minutes and said that he would test the nuggets by placing them in water, instead of acid. The customer consented. Mr. Novak's idea was to determine the density of the nuggets and compare that density to the published value for pure gold. First, Mr. Novak obtained the mass of the nuggets by placing them on a balance. Next, he obtained the volume of the irregularly shaped nuggets by water displacement. To do this he placed the nuggets into a graduated cylinder that was partially filled with water and measured the volume of water displaced. After determining the mass and volume of the nuggets, Mr. Novak calculated the density by dividing the mass of the nuggets by their volume. The nuggets had a density of 18.9 grams/cubic centimeter—exactly the value given for pure gold in the *Handbook of Chemistry and Physics*. Based on the physical appearance of the nuggets and the density information, Mr. Novak purchased the gold nuggets.

LEARNING GOALS

After you've studied this chapter, you should be able to:

1. Find the number of significant figures in a measurement.
2. Do calculations using the rules for significant figures.
3. Write numbers in scientific notation.
4. Find the area of any square, rectangle, circle, or triangle.
5. Find the volume of any cube, other rectangular solid, cylinder, or sphere.
6. Convert units of mass, length, and volume within the metric system using the factor-unit method.
7. Convert from a metric unit to the corresponding English unit using the factor-unit method.

8. Convert from an English unit to the corresponding metric unit using the factor-unit method.
9. Distinguish between the mass of an object and its weight.
10. Calculate the density, mass, or volume of an object when you are given the other two.
11. Convert temperatures from the Celsius to the Fahrenheit scale and vice versa.

INTRODUCTION

Chemists, as well as other scientists, use measurements when they do basic research. Measurements represent the dimensions, quantity, or capacity of things. Researchers ask questions and perform experiments. The ability to make accurate measurements of quantities such as mass, volume, temperature, or time enables them to gather the information that leads to the compilation of reliable scientific data. They analyze the data, often with the use of statistics, to look for patterns or regularities in nature. Sometimes the pattern or regularity is basic and can be stated simply to describe some natural phenomenon. The scientist may then devise a hypothesis, or speculative guess, that can explain the phenomenon. The problem-solving process continues as the scientist tests the hypothesis through further experimentation, which leads to the compilation and analysis of more measured data. Sometimes experimental results prove a hypothesis because it successfully explains a regularity or law of nature. The hypothesis then becomes a theory, or detailed explanation, which helps us describe and organize scientific knowledge.

In this chapter we look at systems of measurement so that you will have an understanding of how scientists record and manipulate quantitative data. We will begin with a look at problem solving so that you will be able to process the data you are given. Familiarity with problem-solving techniques will help you feel more comfortable as you work with units of measurement and tackle problems in all areas of chemistry.

2.1 Problem Solving

Humans have been given the gift of intelligence. We use this intelligence by solving problems. Each day we solve a variety of different problems as we deal with many types of obstacles. When a direct course of action is not clear, we devise ways to reach our goals, regardless of what it takes. Think of a typical day in your life and you will see that you deal with a whole range of problem types. You solve personal problems, social problems, scheduling problems, and financial problems, to name just a few. In your basic chemistry course, you will solve a variety of scientific problems. Like the problems we face in life, the problems you will encounter in basic chemistry require you to use judgment, originality, creativity, and independent thinking.

A great deal of research has gone into determining what makes a successful problem solver. George Polya, who is known as the "father of problem solving," tells us that when we solve problems we must search for some action to attain a goal. In other words, solving a problem does not mean just coming up with an answer. It means coming up with a strategy, or plan, to derive the answer. We will use Polya's four-step model as our general strategy. We will examine each step of Polya's model and show you how to proceed.

In preparation for taking an exam, you may have memorized many different formulas and definitions. Even though you know all the formulas and all the definitions, you may be unsure of where and when to use them. The general problem-solving framework will help you organize your thoughts and allow you to retrieve the relevant background information from your memory. It will help you sort relevant information from irrelevant information and will allow you to apply your knowledge.

One of the most important aspects of the problem-solving process is learning to ask yourself the correct questions. Questions such as "What is the unknown?" and "What information have I been given?" are a good starting point. Another important aspect is your own motivation to become adept at problem solving.

2.2 Polya's Method: The Four-Step General Model

Polya's four-step general model for problem solving will help you organize your thoughts and give you a general framework for solving any type of problem, whether it be a real life situation or a textbook problem.

STEP 1: Understand the Problem

This is a basic step in problem solving. It consists of asking yourself a variety of questions to diagnose the situation. Pinpoint what you are trying to determine. Is there unknown information? What is it? Is any information assumed? Is there any additional information needed? Can you restate the problem in your own words?

STEP 2: Devise a Plan

In this step you come up with a plan to solve the problem. You connect what you know with what is unknown to you. Do you know of a problem that is similar to this one? Is there a similar simpler problem that you can solve? Can you use a formula or write an equation? Can you make a table or draw a diagram to help you solve the problem?

STEP 3: Carry Out the Plan

In this step you actually solve the problem. If you will be using a formula or an equation, you will set it up, plug in the numbers, and solve.

STEP 4: Look Back

In this step you look back at your work and be sure that it is correct. Did you check to be sure that the conditions of the problem were satisfied by the solution? Were all relevant data used? Is your answer reasonable? Compare your solution to your estimate. How do they compare?

When we solve the textbook examples, we will work out at least one solution per chapter using Polya's four-step method. So that you can practice using the process on your own, examine the solutions that do not include Polya's method and try to follow the steps on your own.

2.3 Essential Estimation Skills

Estimation is a close cousin to problem solving. It is the process of producing an answer that is sufficiently close so that decisions can be made. Estimation is an essential skill because it helps make problem solving more meaningful. Instead of coming up with the exact answer to a problem, estimation enables us to determine a ballpark figure for the solution. It encourages us to look at the problem and decide what a sensible answer would be, which gives us a check on our final results. Sometimes when we solve equations we plug numbers into the calculator rather thoughtlessly. We are not always sure when an unreasonable answer has resulted because we don't routinely stop and think of what would be a sensible result and what would be a nonsensical result. Estimation can produce reasonableness about computation. It can encourage us to have a greater appreciation of number size and provide a complement to our routine use of the calculator. Overall, estimation can bring more meaning to the problem-solving process.

The language of estimation includes words such as *about, close to, just about, a little less than,* and *between*. There is no one correct estimate. When you solve the problem you will find the correct answer, which will be exact.

In chemistry, we look for exact answers to problems. To be sure that the answers we come up with are reasonable and sensible, we can use estimation. When the problem is solved and an exact solution is computed, the estimate can be compared with the exact solution. This enables the problem solver to have an additional system of checking the calculation and adds more meaning to the problem-solving process.

2.4 Significant Figures

Suppose that, in an experiment, you are measuring the temperature of a liquid. The thermometer you are using is calibrated, or marked off, only in whole degrees. Imagine that the mercury in the thermometer is halfway between 34 and 35 degrees (Figure 2.1a). You can then estimate that the temperature of the liquid is 34.5 degrees.

12 ESSENTIAL CONCEPTS OF CHEMISTRY

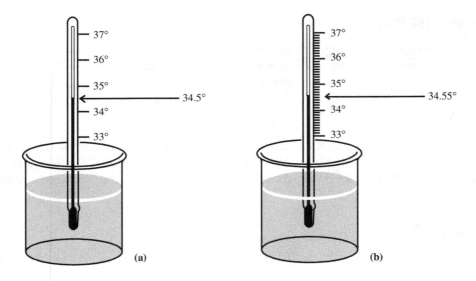

FIGURE 2.1

(a) We can measure the temperature of a liquid using a thermometer calibrated only in degrees. (b) We can also measure the temperature of a liquid using a thermometer calibrated in tenths of degrees.

Now you pick up another thermometer, which is calibrated in tenths of degrees (Figure 2.1b). When you read this thermometer, it looks as though the temperature of the liquid is 34.55 degrees. It seems that the two thermometers differ in sensitivity, but which measurement is correct?

To answer this question, we consider the number of significant figures in each measurement. **Significant figures** are *digits that express information that is reasonably reliable. The number of significant figures equals the number of digits written, including the last digit even though its value is uncertain.* For example, the temperature 34.5 degrees given by the first thermometer is known to be correct to three significant figures. These significant figures are the 3, the 4, and the 5. (Expressing the measurement as 34.50 degrees would mean that the temperature is known to four significant figures—the zero on the right is significant.)

The temperature 34.55 degrees given by the second thermometer shows a measurement with confidence to four significant figures. It may not be more accurate than the first measurement, but it is more *precise*. In general, the more significant figures there are, the more precise the measurement is.

Actually, neither of the two temperature measurements may be accurate. (They certainly wouldn't be accurate if the temperature of the liquid were actually, say, 47.51 degrees.) Thus **accuracy** involves *closeness to the actual dimension*. **Precision** is related to the *detail with which a measurement is known (expressed by the number of significant figures.)*

A measuring *instrument* is considered accurate if it consistently provides measurements that are close to the actual amount. The instrument is considered to have good precision if it consistently provides the same measurement when it is used to measure the same amount. To see the difference, consider several darts being thrown at a target. If the darts land very close together, the results exhibit precision. If all the darts land on the bull's eye, they exhibit accuracy as well.

Another experiment will show how significant figures play an important role in measurement. Suppose that we have to measure a certain piece of glass to find its perimeter (we do this by measuring the four edges of the glass and adding the lengths). We have two rulers, one calibrated in centimeters and the other calibrated in millimeters. We set out to measure each edge of the glass with the more precise ruler (the one calibrated in millimeters). However, when we get to the fourth side of the glass, we absent-mindedly pick up the less precise ruler (the one calibrated in centimeters). Looking at the glass, shown in Figure 2.2, we can see that three edges are measured to two decimal places and that the fourth is measured to one decimal place. What can we say about the perimeter of the glass? Is it 67.73 cm or 67.7 cm? The answer is 67.7 cm. We can report only what we know for sure—and in this case that is the answer to one decimal place. *Our least precise measurement determines the number of significant figures in our result.*

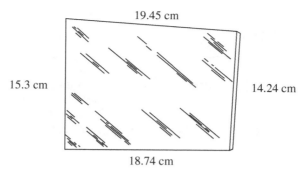

FIGURE 2.2 Measurement of a piece of glass.

Whenever we *add or subtract* measured quantities, we must report the results in terms of the least precise measurement. We do this by **rounding off** to the least number of decimal places. Our result must have no more decimal places than our least precise quantity.

To round off a number, we drop one or more digits at the right end of the number and, if necessary, adjust the rightmost digit that we keep. In some cases, dropped digits must be replaced with zeros. The rules we shall use in rounding numbers are as follows.

RULE 1

If the leftmost dropped digit is smaller than 5, simply drop the digits. Replace dropped digits with zeros as necessary to maintain the magnitude of the rounded number. Thus 7.431 rounded to two significant figures is 7.4. And 7,431 rounded to two significant figures is 7,400.

RULE 2

If the leftmost dropped digit is 5 or greater, increase the last retained digit by 1. Replace dropped digits with zeros as necessary to maintain the magnitude of the rounded number. Thus 93.56 rounded to three significant figures is 93.6. And 9,356 rounded to three significant figures is 9,360.

Here is a rounding rule that is useful in *multiplying or dividing* measured quantities.

STEP 1

Count the number of significant figures in each of the quantities to be multiplied or divided.

STEP 2

Report the result to the least number of significant figures determined in step 1. Round as required.

EXAMPLE 2.1

Add 18.7444 and 13 and report the result to the appropriate number of significant figures.

Solution

UNDERSTAND THE PROBLEM

We will add the numbers. Since 13 is the less precise measurement, we must round the sum to a whole number.

DEVISE A PLAN

Use Rule 2, which says if the leftmost dropped digit is 5 or greater, increase the last digit by 1. In this case the leftmost dropped digit is 7, so the 1 in 31 increases to 2, and we have 32.

CARRY OUT THE PLAN

$$18.7444$$
$$+13$$
$$\overline{31.7444} \quad \text{Round off to 32 (whole number).}$$

LOOK BACK

The answer makes sense since 18.7444 is close to 19, and 19 + 13 = 32.

Practice Exercise 2.1

Add 12.4432 and 15. Provide the appropriate number of significant figures in your answer.

EXAMPLE 2.2

Subtract 0.12 from 48.743 and use the appropriate number of significant figures in your answer.

Solution

UNDERSTAND THE PROBLEM

Because we are aware of only two decimal places, 0.12 is the less precise number. Therefore we subtract and round off to two decimal places.

DEVISE A PLAN

Use Rule 1, which says if the leftmost dropped digit is smaller than 5, simply drop the digits. In this case the leftmost dropped digit is the 3 in 48.623.

CARRY OUT THE PLAN

$$48.743$$
$$-\ 0.12$$
$$\overline{48.623} \quad \text{Round off to 48.62 (two decimal places).}$$

LOOK BACK

Use mental arithmetic, and you will see that the answer makes sense.

Practice Exercise 2.2

Subtract 1.23 from 54.667, giving the appropriate number of significant figures in your answer.

EXAMPLE 2.3

What is the area of a square whose side is measured as 1.5 cm?

Solution

The area of a square is equal to the length of its side squared, or

$$A = s \times s = 1.5 \text{ cm} \times 1.5 \text{ cm} = 2.25 \text{ cm}^2$$

Round off to 2.3 cm². (We report to only two significant figures, because the measurement has only two significant figures.)

Practice Exercise 2.3

Calculate the area of a square whose side is measured as 2.5 cm.

EXAMPLE 2.4

Divide 20.8 by 4 and give the result to the proper number of significant figures.

Solution

The result must be rounded to one significant figure because of the one-digit divisor, 4.

$$\frac{20.8}{4} = 5.2 \quad \text{Round off to 5.}$$

Practice Exercise 2.4

Divide 48.2 by 4, giving the result to the proper number of significant figures.

EXAMPLE 2.5

Multiply 20.8 by 4.1 and report the result to the proper number of significant figures.

Solution

The result must be rounded to two significant figures.

$$20.8 \times 4.1 = 85.28 \quad \text{Round off to 85.}$$

Practice Exercise 2.5

Multiply 25.5 by 3.2, giving the result to the proper number of significant figures.

Another problem arises when we use significant figures: What do we do about zeros? Are they significant figures or not? Here are some helpful rules. Read them carefully, work through Example 2.6, and then do Practice Exercise 2.6.

RULE 1

Zeros *between* nonzero digits are significant:

4.004 has four significant figures

RULE 2

Zeros to the *left* of nonzero digits are *not* significant, because these zeros show only the position of the decimal point:

0.00254 has three significant figures

0.0146 has three significant figures

0.06 has one significant figure

RULE 3

Zeros that fall at the *end* of a number are not significant unless they are marked as significant. If a zero does indicate the number's precision, we can mark it as significant by *placing a line over it*. Zeros to the right of the decimal place are always significant.

84,000 has two significant figures

84,$\overline{0}$00 has three significant figures

84,000.0 has six significant figures

16 ESSENTIAL CONCEPTS OF CHEMISTRY

RULE 4

Exactly defined quantities have an unlimited number of significant figures

$$4 \text{ qt} = 1 \text{ gal}$$
$$1 \text{ m} = 100 \text{ cm}$$

EXAMPLE 2.6

In this example, we tell you how many significant figures there are in different numbers. In Practice Exercise 2.6, *you* are asked to tell *us*.

Solution

(a) 0.00087 has two significant figures.

(b) 1.004 has four significant figures.

(c) 873.005 has six significant figures.

(d) 9.00000 has six significant figures.

(e) 320,000 has two significant figures.

(f) 18$\overline{0}$,000 has four significant figures.

(g) 180,000.0 has seven significant figures.

(h) 2,$\overline{000}$ has four significant figures.

Practice Exercise 2.6

Find the number of significant figures in each of the following numbers: (a) 0.0023 (b) 5.025 (c) 123.456 (d) 5,$\overline{000}$ (e) 5,$\overline{0}$00 (f) 12.000

2.5 Scientific Notation: Powers of 10

In science, we often deal with numbers that are *very* large or *very* small. Numbers like 100,000,000,000 (a hundred billion) or 0.0000008 (eight ten-millionths) arise frequently, and they are troublesome to work with in calculations. There is a shorthand method of writing such numbers, based on powers of 10. The number 100,000,000,000 can be written as 1×10^{11}, and 0.0000008 as 8×10^{-7}. In the first example, we moved the decimal point eleven places to the left:

$$1\,0\,0,0\,0\,0,0\,0\,0,0\,0\,0. = 1 \times 10^{11}$$

Moving the decimal point to the left is compensated for by multiplying by a positive power of 10. In the second example, we moved the decimal point seven places to the right.

$$0.\,0\,0\,0\,0\,0\,0\,8 = 8 \times 10^{-7}$$

Moving the decimal point to the right corresponds to multiplying by a negative power of 10. In both cases, the power that 10 is raised to is called the **exponent** of the **base number** 10. (If this causes you to hesitate, turn to Section A.3 in Appendix A and see that explanation, plus Examples A.1 through A.4.) We have

$$10^{11} \leftarrow \text{Exponent}$$
$$\phantom{10^{11}} \leftarrow \text{Base number}$$

Here's how to use the shorthand method: To express a number in **scientific notation**, write the number with only one significant figure to the left of the decimal point. Multiply it by 10 raised to the number of places you moved the decimal—positive if it was moved to the left, and negative if it was moved to the right.

$$3,800 = 3.8 \times 10^3$$

$$0.00625 = 6.25 \times 10^{-3}$$

$$100,000,000 = 1 \times 10^8$$

$$0.0000001 = 1 \times 10^{-7}$$

The number of significant figures is made clear by using scientific notation. For example, if we know the number 500 to three significant figures, we write 5.00×10^2. If we know the number 500 to only one significant figure, we write 5×10^2. Writing 5.00×10^2 is the same as writing $5\overline{00}$ (three significant figures). Writing 5×10^2 is the same as writing 500 (one significant figure).

With a little practice, you'll soon be using scientific notation as easily as you use the more standard notation, but do not deny yourself the practice you need!

EXAMPLE 2.7

Write the following numbers in scientific notation, showing the number of significant figures requested.

(a) 5,000 (two significant figures) 5.0×10^3
(b) 48,000 (three significant figures) 4.80×10^4
(c) 4,090,000 (three significant figures) 4.09×10^6
(d) 0.000087 (two significant figures) 8.7×10^{-5}

Solution

(a) $5,\overline{0}00. = 5.0 \times 10^3$ (two significant figures)

(b) $48,\overline{0}00. = 4.80 \times 10^4$ (three significant figures)

(c) $4,090,000. = 4.09 \times 10^6$ (three significant figures)

(d) $0.00008\,7 = 8.7 \times 10^{-5}$ (two significant figures)

Practice Exercise 2.7

Write the following numbers in scientific notation, showing the number of significant figures requested.

(a) 4,200 (two significant figures) 4.2×10^3
(b) 56,000 (three significant figures) 5.60×10^4
(c) 6,023,000 (four significant figures) 6.023×10^6
(d) 0.00123 (three significant figures) 3.21×10^{-4}

18 ESSENTIAL CONCEPTS OF CHEMISTRY

2.6 Area and Volume

Suppose you want to know the area of a tennis court. A flat rectangular surface like a tennis court has just *two* dimensions: length and width, abbreviated l and w. As Table 2.1 shows, the area of a rectangle is equal to its length l times its width w. So the area of the tennis court would be $l \times w$.

Now suppose you want to find the volume of a cereal carton. A geometric figure like a carton (a rectangular solid) has three dimensions: length l, width w, and height h. As Table 2.1 and Figure 2.3 show, the volume of your carton is $l \times w \times h$.

TABLE 2.1 Areas and Volumes of Various Geometric Figures

\multicolumn{3}{l}{*Areas (note that areas are always given in squared units: in^2, ft^2 and so on.)*}
1. Square	Area = side × side $A = s \times s$	
2. Rectangle	Area = length × width $A = l \times w$	
3. Circle	Area = $\pi \times$ (radius)^2 $A = \pi \times r^2$ $\pi = 3.14$	
4. Triangle	Area = $\frac{1}{2} \times$ base × height $A = \frac{1}{2} \times b \times h$	
\multicolumn{3}{l}{*Volumes (note that volumes are always given in cubed units: in^3, ft^3, and so on.)*}
5. Cube	$V = s \times s \times s$	

6. Carton (rectangular solid)	$V = l \times w \times h$	
7. Cylinder	$V = \pi \times r^2 \times h$	
8. Sphere	$V = \frac{4}{3}\pi \times r^3$	

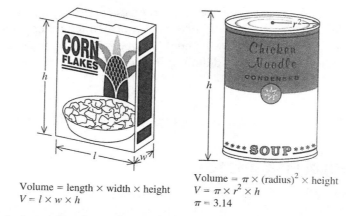

Volume = length × width × height
$V = l \times w \times h$

Volume = π × (radius)^2 × height
$V = \pi \times r^2 \times h$
$\pi = 3.14$

FIGURE 2.3 Formulas for calculating the volume of a cereal carton and a soup can.

Area is a *measure of the extent of a surface*. It is a two-dimensional measure that is always stated in squared units such as square feet (ft^2). **Volume** is a *measure of the capacity of an object*. It is a three-dimensional measure that is given in cubed units such as cubic feet (ft^3).

Table 2.1 lists formulas for the areas and volumes of several geometric figures. Example 2.8, which follows, shows you how to use them. Then, in Practice Exercise 2.8, you are asked to apply a volume formula on your own. You should take the time to work each practice exercise as you come to it. It will help you learn and apply what you have just read, and it will serve as a quick check of your understanding of each topic. Answers to practice exercises are provided in the back of the book.

20 ESSENTIAL CONCEPTS OF CHEMISTRY

EXAMPLE 2.8

The soup can shown in Figure 2.3 has a height of 4.00 inches and a radius of 1.00 inch. What is the volume of the can?

Solution

UNDERSTAND THE PROBLEM

Ask yourself what knowledge is required to arrive at a solution. A soup can has the shape of a cylinder, and therefore you need to know how to calculate the volume of a cylinder.

DEVISE A PLAN

Looking at Table 2.1, you find that the formula for the volume of a cylinder is $V = \pi r^2 h$.

$$\text{Given: } \pi = 3.14, r = 1.00 \text{ inch}, h = 4.00 \text{ inches}$$

CARRY OUT THE PLAN

$$V = \pi r^2 h$$
$$= (3.14)(1.00 \text{ in})^2 (4.00 \text{ in})$$
$$= 12.56 \text{ in}^3 \quad \text{or} \quad 12.6 \text{ in}^3 \text{ (Three significant figures)}$$

The can holds 12.6 cubic inches of soup.

LOOK BACK

See if the solution makes sense. Estimation can be used to give an approximate answer, telling us if our calculation is in the ballpark. We can say, "π is roughly 3, so 3 times 1 inch times 1 inch times 4 inches is about 12 cubic inches." Our estimate says that the answer should be about 12 cubic inches. Our exact calculation tells us that the answer is 12.6 cubic inches. We are definitely in the ballpark.

Practice Exercise 2.8

A cylinder has a height of 6.00 inches and a radius of 4.00 inches. What is its volume?

2.7 The English System of Measurement

The English system of units, still used in the United States today, has a built in problem. It is not an orderly system and is therefore difficult to use. We will look at three units in the English system and then discuss this inherent difficulty.

In the English system, the *foot* is the unit of length and is divided into 12 smaller units called *inches*. The inch can be used to measure short distances, and the foot can be used to measure longer distances. To measure still longer distances there are the *yard* (which equals 3 feet) and the *mile* (which equals 5,280 feet).

The unit of weight in the English system is the *pound*, and the pound is divided into 16 smaller units called *ounces*. To measure larger weights, the *ton* may be used. The ton is equal to 2,000 pounds.

In the English system as used in the United States today, the unit of liquid volume is the *quart*. To measure smaller quantities, the *fluid ounce* is used (32 fluid ounces equal 1 quart). Other units of liquid volume are the *pint* (2 pints equal 1 quart) and the gallon (4 quarts equal 1 gallon).

Table 2.2 lists these units. If you examine that table closely, the problem will be evident: there is no systematic relationship among units used to measure the same property. Consider the length unit. To convert inches to yards, it is necessary to know that 12 inches equal 1 foot and that 3 feet equal 1 yard. You cannot easily move from one unit to the other. The units for weight and volume are just as inconvenient to convert. Such problems led to the development of the metric system of measurement.

TABLE 2.2 The English System of Measurement (as Used in the United States)

LENGTH	WEIGHT	VOLUME
12 inches = 1 foot	16 ounces = 1 pound	16 fluid ounces = 1 pint
3 feet = 1 yard	2,000 pounds = 1 ton	2 pints = 1 quart
5,280 feet = 1 mile		4 quarts = 1 gallon

2.8 The Metric System of Measurement

In the late 1700s, the French decided to change their own system of measurement. To replace it they developed a logical and orderly system called the **metric system**. The advantages of this system led to its adoption in most countries of the world and in all branches of science. The British held out until 1965 and then began a changeover to the metric system.

In countries using the metric system, almost everything is measured in metric units— distances between cities, the weight of a loaf of bread, the size of a sheet of plywood. Nearly all countries have adopted the metric system. In 1975 the United States Congress passed a bill establishing a policy of voluntary conversion to the metric system and creating the U.S. Metric Board. A program of gradual conversion was the goal of this board, whose funding ended in 1982. The Office of Metric Programs, established by the U.S. Department of Commerce, now has the job of promoting the increased use of metric units of measurement by business and industry.

The metric system consists of (1) a set of standard units of measurement for distance, weight, volume, and so on, and (2) a set of prefixes that are used to express larger or smaller multiples of these units. The prefixes represent multiples of 10. This makes the metric system a decimal system of measurement. In 1960 an international group of scientists modified the metric system by adopting a system of units called the *Systèms International d'Unités* (International System of Units), or the SI system.

In the SI system there are seven basic units. Table 2.3 lists them. Other units, called *derived units,* are composed of combinations of basic units. (For example, the unit for speed, meters per second, is derived from the basic units *meter* and *second.*) The basic units were all carefully defined. Occasionally one of the definitions is modified to make it more precise or more useful. For example, the meter was first defined as one ten-millionth of the distance between the North Pole and the equator along a meridian of the earth. Later, to ensure that 1 meter meant the same thing everywhere, it was redefined as the distance between two scratches on a platinum bar that is kept at exactly the freezing point of water (zero degrees on the Celsius temperature scale) in a vault outside Paris. Still later, as more exacting measuring instruments were developed, the meter was defined as 1,650,763.73 wavelengths in vacuum of the orange-red line of the spectrum of krypton 86. You will learn more about this in Chapter 5.

Table 2.4 lists the more commonly used metric prefixes. To use one, simply "tack it on" to the front of a metric unit. That gives a related unit that is a multiple of the original unit. For example, the prefix *kilo* has a multiplier of 1,000, so a kilometer is equal to 1,000 meters. The prefix *centi* has a multplier of 0.01, so a centigram is equal to 0.01 gram. Conversely, there are 100 centigrams in a gram.

TABLE 2.3 Basic SI (Metric) Units

QUANTITY	UNIT	SYMBOL
Length	meter	m
Mass	kilogram	kg
Time	second	s
Electric current	ampere	A
Temperature	kelvin	K
Light intensity	candela	cd
Amount of substance	mole	mol

22 ESSENTIAL CONCEPTS OF CHEMISTRY

TABLE 2.4 Metric Prefixes

PREFIX	SYMBOL	MULTIPLIER
nano	n	0.000000001
micro	μ (Greek mu)	0.000001
milli	m	0.001
centi	c	0.01
deci	d	0.1
deka	da	10
hecto	h	100
kilo	k	1,000
mega	M	1,000,000

To make converting from one unit to another simpler, we will use a method called the factor-unit method, or dimensional analysis. With this approach, any problem that requires conversion from one unit to another can be set up and solved in a similar manner. We can say that:

$$\text{Quantity wanted} = \text{quantity given} \times \text{factor unit}$$

When you multiply the quantity given by the proper factor unit, some of the units cancel to give the desired quantity. Let's see how this works. We'll convert meters to centimeters as we begin to understand this method.

$$\text{Quantity wanted} = \text{quantity given} \times \text{factor unit}$$

$$\text{Centimeters} = \cancel{\text{meters}} \times \frac{\text{centimeters}}{\cancel{\text{meter}}}$$

Note that the factor unit expresses a relationship between the quantity wanted and the quantity given. The factor unit is written in such a way that the given units cancel when you multiply the quantity given times the factor unit. Then you're left with the quantity wanted:

$$\text{Centimeters} = \cancel{\text{meters}} \times \frac{\text{centimeters}}{\cancel{\text{meter}}}$$

$$\text{Centimeters} = \text{centimeters}$$

Of course any factor unit can be expressed in two ways. For example, the relationship 1 meter = 100 centimeters can be expressed as

$$\frac{100 \text{ centimeters}}{1 \text{ meter}} \quad \text{or} \quad \frac{1 \text{ meter}}{100 \text{ centimeters}}$$

You choose the factor unit that will make the proper terms cancel.

We'll now look at a variety of examples in which the factor-unit method is helpful. As you proceed through basic chemistry, you'll find this method useful in many cases. (Appendix A, especially Examples A.5 through A.9, provides many more examples that will help you understand this strategy.)

EXAMPLE 2.9

Change 40 meters to centimeters.

Solution

UNDERSTAND THE PROBLEM

This example requires conversion from one unit to another. We ask "How many centimeters are there in one meter?" Table 2.4 shows that *centi* means 0.01, so there are 100 centimeters in 1 meter.

DEVISE A PLAN

Use the factor-unit method to convert.

CARRY OUT THE PLAN

Remember that there are 100 centimeters in 1 meter. This can be written as

$$\frac{100 \text{ centimeters}}{1 \text{ meter}}$$

which reads "100 centimeters per meter"; the division line reads "per." Next we put in the numbers.

$$?\text{centimeters} = 40 \text{ meters} \times \frac{100 \text{ centimeters}}{\text{meter}} = 4{,}000 \text{ cm}$$

There are 4,000 centimeters in the 40 meters.

LOOK BACK

We want to see if this solution makes sense. We determined that one meter equals 100 centimeters. Does it make sense that 40 meters would be 40 times larger, or 4,000 centimeters? Yes, this does make sense.

Practice Exercise 2.9

Change 35 meters to centimeters.

EXAMPLE 2.10

Change $43\bar{0}$ milligrams to grams.

Solution

UNDERSTAND THE PROBLEM

You will see that it is similar to Example 2.9. Therefore we can use the same strategy to devise a plan.

DEVISE A PLAN

Again we use the factor-unit method. (This time we will use abbreviations.) We ask, "How many milligrams are there in 1 g?" Table 2.4 shows that *milli* means 0.001, so 1 g equals 1,000 mg. This means there is

$$\frac{1 \text{ g}}{1{,}000 \text{ mg}} \qquad \text{(which reads 1 g per 1,000 mg)}$$

CARRY OUT THE PLAN

We say that

$$g = mg \times \frac{g}{mg}$$

$$?g = 43\bar{0} \text{ mg} \times \frac{1 \text{ g}}{1{,}000 \text{ mg}} = 0.430 \text{ g}$$

There is 0.430 g in $43\bar{0}$ mg.

LOOK BACK

We ask if this solution makes sense. If 1,000 mg equals 1 g, how much of a gram would $43\bar{0}$ mg equal? Using estimation we can say that $43\bar{0}$ mg is a little less than half a gram, therefore our answer 0.430 g, which is a little less than half a gram is in the ballpark, and makes sense.

24 ESSENTIAL CONCEPTS OF CHEMISTRY

Practice Exercise 2.10

Change 35$\overline{0}$ mg to grams

EXAMPLE 2.11

How many centimeters are there in 8 meters?

Solution

$$cm = \cancel{m} \times \frac{cm}{\cancel{m}}$$

$$?\,cm = 8\,\cancel{m} \times \frac{100\,cm}{1\,\cancel{m}} = 800\,cm$$

There are 800 cm in 8 m.

Practice Exercise 2.11

How many centimeters are there in 18 meters?

EXAMPLE 2.12

Convert 580 millimeters to centimeters.

Solution

$$cm = \cancel{mm} \times \frac{cm}{\cancel{mm}}$$

$$?\,cm = 580\,\cancel{mm} \times \frac{1\,cm}{10\,\cancel{mm}} = 58\,cm$$

There are 58 cm in 580 mm.
  If you wonder where the term

$$\frac{1\,cm}{10\,mm}$$

came from, look at a metric ruler, and you'll see that there are 10 mm in 1 cm.

Practice Exercise 2.12

Convert 660 mm to centimeters.

EXAMPLE 2.13

Convert 75 millimeters to its corresponding length in (a) centimeters, (b) meters, (c) kilometers.

Solution

(a) $?\,cm = 75\,\cancel{mm} \times \dfrac{1\,cm}{10\,\cancel{mm}} = 7.5\,cm$

(b) $?\,m = 7.5\,\cancel{cm} \times \dfrac{1\,m}{100\,\cancel{cm}} = 0.075\,m$

(c) $?\,km = 0.075\,\cancel{m} \times \dfrac{1\,km}{1{,}000\,\cancel{m}} = 0.000075\,km$

Practice Exercise 2.13

Convert 55 millimeters to the corresponding length in (a) centimeters, (b) meters, (c) kilometers.

EXAMPLE 2.14

Convert 2.3 kilograms to the corresponding mass in (a) grams, (b) decigrams, (c) centigrams, (d) milligrams.

Solution

$$(a) \ ?g = 2.3 \ kg \times \frac{1{,}000 \ g}{1 \ kg} = 2{,}300 \ g$$

$$(b) \ ?dg = 2{,}300 \ g \times \frac{10 \ dg}{1 \ g} = 23{,}000 \ dg$$

$$(c) \ ?cg = 23{,}000 \ dg \times \frac{10 \ cg}{1 \ dg} = 230{,}000 \ cg$$

$$(d) \ ?mg = 230{,}000 \ cg \times \frac{10 \ mg}{1 \ cg} = 2{,}300{,}000 \ mg$$

Practice Exercise 2.14

Convert 4.4 kilograms to the corresponding mass in (a) grams, (b) decigrams, (c) centigrams, (d) milligrams.

You should now realize that conversion between units of the metric system is not difficult. If you want to convert metric units into English units, consult Table 2.5. (More precise conversion factors can be found in Appendix B, Table B.3.)

Although it will always be convenient to be able to convert between systems, it is even more important to be able to associate metric measurements with commonly used terms. Start now to "think metric," for only in this way will the metric system become meaningful to you. The following example offers a starting point.

EXAMPLE 2.15

A new car is described in an advertisement as having an overall length of 5.50 meters and an overall width of 1.50 meters. Find the dimensions of the car in feet.

Solution

Consulting Table 2.5, we see that to convert meters to feet it is necessary to multiply by 3.28. In other words, 1 meter = 3.28 feet. Therefore the length and width can be determined as follows:

$$\text{Length:} \quad ? \ ft = 5.50 \ m \times \frac{3.28 \ ft}{1 \ m} = 18.0 \ ft$$

$$\text{Widht:} \quad ? \ ft = 1.50 \ m \times \frac{3.28 \ ft}{1 \ m} = 4.92 \ ft$$

Practice Exercise 2.15

The dimensions of a new American car are as follows: overall length, 4.20 meters; overall width, 1.25 meters. Is this a compact car? Find its dimensions in feet.

Before we work the next example, we need to discuss the SI unit of volume. This unit, a derived unit, has the dimension (length)^3—for example, m^3 or cm^3. However, before the SI system was adopted, the metric unit of volume was the liter. One liter (1 L) is equal in volume to 1,000 cm^3, and 1 mL = 1 cm^3. For some measurements the liter is simpler to use than the SI units, and many people continue to use it in such cases. In this book, we shall use whichever unit best fits a particular situation. In the next example we work with liters, which are used in selling gasoline in all countries using the metric system.

26 ESSENTIAL CONCEPTS OF CHEMISTRY

TABLE 2.5 Conversion of Units

TO CONVERT	INTO	MULTIPLY BY
Length		
inches	centimeters	2.54 cm/in
centimeters	inches	0.39 in/cm
feet	meters	0.30 m/ft
meters	feet	3.28 ft/m
Weight (mass; see Section 2.9)		
ounces	grams	28.35 g/oz
grams	ounces	0.035 oz/g
pounds	grams	454 g/lb
grams	pounds	0.0022 lb/g
Volume		
liters	quarts	1.06 qt/L
quarts	liters	0.946 liter/qt

EXAMPLE 2.16

An American family visiting Mexico wants to fill its car with 15 gallons of gasoline. Mexico uses the metric system. How many liters of gasoline should the visitors ask the attendant to put in their tank?

Solution

We know that 15 gallons of gasoline are the same as $6\overline{0}$ quarts (4 quarts = 1 gallon). Consulting Table 2.5, we see that to convert quarts to liters we multiply by 0.946 liter/quart. Therefore we may solve the problem as follows:

$$? \text{ liters} = 6\overline{0} \text{ quarts} \times \frac{0.946 \text{ liter}}{1 \text{ quart}} = 57 \text{ liters}$$

Practice Exercise 2.16

To conserve water, the Simon family installed a reducing valve and a water-metering device in their shower. If Mr. Simon uses $2\overline{0}$ gallons of water in his shower, and Mrs. Simon uses 18 gallons in hers, how many liters of water do the Simons use?

Before we leave the topic of measurement and the factor-unit method, we should point out that sometimes we may want to use two (or more) factor units in solving a problem. Let's demonstrate how this works.

EXAMPLE 2.17

A laboratory bench measures exactly 3.00 yards in length. How many inches is this?

Solution

We know that there are exactly 3.00 feet in 1.00 yard. We also know that there are exactly 12.0 inches in 1.00 foot. Therefore, we may use these *two* factor units to solve the problem as follows:

$$? \text{ in} = (3.00 \text{ yd}) \frac{(3.00 \text{ ft})}{1.00 \text{ yd}} \frac{(12.0 \text{ in})}{1 \text{ ft}} = 108 \text{ in}$$

Notice how the various units canceled.

Practice Exercise 2.17

Given that a solar day has exactly 24.0 hours, use the following factor units to compute the number of seconds in a day:

$$1.00 \text{ hour} = 60.0 \text{ minutes}$$
$$1.00 \text{ minute} = 60.0 \text{ seconds}$$

It is valuable to know that a liter is just a little more than a quart. A meter is slightly longer than a yard (it is 1 yard and 3.375 inches). When you are dealing with weights, it is helpful to remember that a nickel weighs about 5 g. If you can keep these three approximations in mind, you'll find it easier to work in the metric system.

2.9 Mass and Weight

The mass of an object and its weight are often thought of as being the same, so the words *mass* and *weight* are frequently used interchangeably. This is incorrect because, by definition, the two terms have different meanings. **Mass** is *the quantity of matter in an object,* whereas **weight** is *the gravitational force that attracts an object.*

The definition of mass implies that your body's *mass* is constant no matter where you are. Your body's *weight,* on the other hand, varies from planet to planet and even varies slightly at different places on the earth. An astronaut who weighs 180 pounds on earth weighs only 30 pounds on the moon (which has only one-sixth of the gravitational pull of the earth) and has no weight at all in outer space. But *the astronaut's mass is the same in all places*. We can define weight mathematically as

$$\text{Weight} = \text{mass} \times \text{gravity} \quad \text{or} \quad W = m \times g$$

The force of gravity varies from planet to planet, so the weight of an object must vary too.

2.10 Density

Which is heavier: glass, iron, or wood from an oak tree? Naturally it depends on the size of each piece. However, what if all three were the same size? In other words, what if we had cubes of glass, iron, and oak wood, each with a volume of 1 cm³? (See Figure 2.4.) Suppose we weighed each cube to determine which was the heaviest. We would find that 1 cm³ of iron weighs 7.9 g, 1 cm³ of glass weighs 2.4 g, and 1 cm³ of oak wood weighs 0.6 g. We would conclude that, *for a particular volume,* iron has the greatest mass.

The concept of density enables us to express this relationship conveniently. **Density** can be defined as *the mass per unit volume of a substance or object* (Table 2.6). The density of a substance can be determined by using the formula

$$D = \frac{m}{V}$$

where D is the substance's density, m is its mass, and V is its volume. If the mass is measured in grams and the volume in cubic centimeters, then the unit of density is

$$\frac{\text{grams}}{\text{cubic centimeters}} \quad \text{or} \quad \frac{\text{g}}{\text{cm}^3}$$

(See Appendix A for a review of solving algebraic and word equations.)

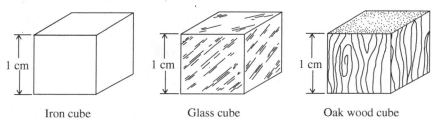

FIGURE 2.4 Three cubes of equal volume.

28 ESSENTIAL CONCEPTS OF CHEMISTRY

TABLE 2.6 Densities of Some Common Solids and Liquids

SUBSTANCE	DENSITY AT ROOM TEMPERATURE (g/cm³)
Ethyl alcohol	0.789
Water	1.00
Glycerol (glycerine)	1.261
Sucrose (table sugar)	1.58
Sodium chloride (table salt)	2.17
Aluminum metal	2.70
Iron metal	7.86
Copper metal	8.92
Mercury	13.59
Gold	18.88

EXAMPLE 2.18

A block of iron that is 5.0 cm long, 3.0 cm high, and 4.0 cm wide weighs 474 g. What is the density of iron?

Solution

We first calculate the volume of the block.

$$\text{Volume} = \text{length} \times \text{width} \times \text{height}$$

$$V = 5.0 \text{ cm} \times 4.0 \text{ cm} \times 3.0 \text{ cm} = 6\overline{0} \text{ cm}^3$$

We now know that $V = 6\overline{0}$ cm³ and $m = 474$ g, so all we have to do is solve the density formula for D.

$$D = \frac{m}{V} = \frac{474 \text{ g}}{6\overline{0} \text{ cm}^3} = 7.9 \frac{\text{g}}{\text{cm}^3}$$

Practice Exercise 2.18

A block of aluminum is 2.0 cm long, 3.0 cm high, and 5.0 cm wide, and it weighs 81.0 g. What is the density of the aluminum?

EXAMPLE 2.19

Suppose you are told that 400 g of alcohol occupy a volume of 500 mL. What is the density of alcohol?

Solution

In this example we are given the mass and the volume of the substance whose density we are asked to find. Note that the volume of the alcohol is given in milliliters (mL). But *1 milliliter is equal to 1 cubic centimeter.* That is, 1 mL = 0.001 L = 1 cm³. So it is acceptable to interchange the units mL and cm³. Now let's solve this problem.

$$D = \frac{m}{V} = \frac{400 \text{ g}}{500 \text{ cm}^3} = 0.8 \frac{\text{g}}{\text{cm}^3}$$

Practice Exercise 2.19

Suppose you are told that 880 g of a clear, colorless liquid occupy a volume of 110 mL. What is the density of this liquid?

EXAMPLE 2.20

Determine which is more dense, carbon tetrachloride or chloroform, from the following information:

(a) 16 g of carbon tetrachloride occupy a volume of $1\overline{0}$ mL.

(b) $3\overline{0}$ g of chloroform occupy a volume of $2\overline{0}$ mL.

Solution

We simply calculate the density of each liquid and see which is greater:

$$D \text{ (carbon tetrachloride)} = \frac{m}{V} = \frac{16\,\text{g}}{1\overline{0}\,\text{cm}^3} = 1.6 \frac{\text{g}}{\text{cm}^3}$$

$$D \text{ (chloroform)} = \frac{m}{V} = \frac{3\overline{0}\,\text{g}}{2\overline{0}\,\text{cm}^3} = 1.5 \frac{\text{g}}{\text{cm}^3}$$

The carbon tetrachloride is the more dense liquid.

Practice Exercise 2.20

Determine which is the more dense, liquid A or liquid B, given the following information:

(a) 55.0 g of liquid A occupy a volume of 10.0 mL.

(b) 25.0 g of liquid B occupy a volume of 12.5 mL.

The preceding calculations are straightforward uses of the density formula. But what if you were given the density of a material and its mass? Could you calculate its volume? Here's how to do it.

EXAMPLE 2.21

The density of alcohol is 0.8 g/cm³. Calculate the volume of 1.6 kg of alcohol.

Solution

First we have to solve the density formula,

$$D = \frac{m}{V}$$

for V. To do so, we multiply both sides of the equation by V. This gives us $D \times V = m$. Now we divide both sides of the equation by D. This gives us the formula we want:

$$V = \frac{m}{D}$$

We are given that $D = 0.8$ g/cm³ and $m = 1.6$ kg, or 1,600 g. We substitute these numbers into the formula and solve for V.

$$V = \frac{m}{D} = \frac{1{,}600\,\text{g}}{0.8\,\text{g/cm}^3} = 2{,}000\,\text{cm}^3 \text{ (or 2,000 mL)}$$

Practice Exercise 2.21

The density of chloroform is 1.5 g/cm³. Calculate the volume in cm³ of 2.0 kg of chloroform.

One important fact about density is that most substances *expand when heated*. Therefore, when a certain mass of a substance is hot, it occupies a larger volume than it does when it is cool. This means that the density of the substance decreases as it is warmed. Think about this for a moment, and make sure you understand why it is true. Most densities reported in chemical references, such as the *Handbook of Chemistry and Physics,* are the densities at 20°C, which is about room temperature. Water is an important exception.

2.11 Temperature Scales and Heat

When we heat a substance, we add a quantity of heat to that substance. We can then use a thermometer to measure the **temperature** of the substance. The thermometer measures the *intensity* of the heat; it tells us nothing about the quantity of heat that has entered the substance.

Three different temperature scales are commonly used in measuring heat intensity. Two of these—the Fahrenheit and Celsius scales—are in general use. The third, the Kelvin scale, is used mainly by scientists.

The **Fahrenheit temperature scale** was devised by Gabriel Daniel Fahrenheit, a German scientist, in 1724. On this scale (see Figure 2.5), the freezing point of pure water is at 32 degrees (32°F), and the boiling point of water is at 212 degrees (212°F). There are thus 180 Fahrenheit degrees between the freezing point and the boiling point of water.

The **Celsius temperature scale** was devised in 1742 by Anders Celsius, a Swedish astronomer. His objective was to develop an easier-to-use temperature scale; he did so by assigning a nice, round 100 Celsius degrees between the freezing and boiling points of pure water. On the Celsius scale (Figure 2.5), the freezing point of water is at zero degrees (0°C), and the boiling point of water is at 100 degrees (100°C). (The Celsius scale is also sometimes referred to as the centigrade scale.)

The **Kelvin temperature scale** is an *absolute* temperature scale. That is, its zero point (0 K) is at absolute zero, the lowest possible temperature theoretically attainable. The divisions of the Kelvin scale are the same size as Celsius degrees, but they are called kelvins (abbreviated K) rather than degrees.

2.12 Converting Celsius and Fahrenheit Degrees

On the Celsius scale, there are 100 divisions between the freezing point and the boiling point of water. On the Fahrenheit scale, there are 180 divisions between these two points. Therefore 100 Celsius degrees cover the same range as 180 Fahrenheit degrees, so that 1 Celsius degree = 1.8 Fahrenheit degrees. Moreover, 0°C is equivalent to 32°F. Formulas for converting from a temperature on one scale to a temperature on the other are based on these facts. The formulas are

$$°F = (1.8 \times °C) + 32 \quad \text{and} \quad °C = \frac{°F - 32}{1.8}$$

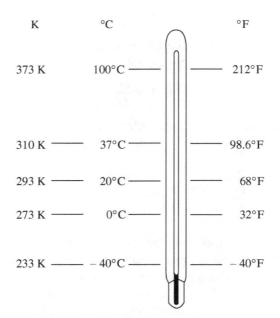

FIGURE 2.5 A comparison of Fahrenheit and Celsius scales.

Let us use these formulas to convert from Fahrenheit to Celsius degrees and from Celsius to Fahrenheit degrees. When using significant figures remember that exactly defined quantities have an unlimited number of significant figures.

EXAMPLE 2.22

Convert 122°F to degrees Celsius.

Solution

Substitute 122°F into the conversion formula for changing °F to °C.

$$°C = \frac{°F - 32.0}{1.8} = \frac{122 - 32.0}{1.8} = \frac{9\overline{0}}{1.8} = 5\overline{0}$$

$$122°F = 15\overline{0}°C$$

Practice Exercise 2.22

Convert 244°F to degrees Celsius.

EXAMPLE 2.23

Convert $1\overline{00}$°C to degrees Fahrenheit.

Solution

Substitute $1\overline{00}$°C into the conversion formula for changing °C to °F.

$$°F = (1.8 \times °C) + 32.0 = (1.8 \times 1\overline{00}) + 32.0 = 18\overline{0} + 32.0 = 212$$

$$1\overline{00}°C = 212°F$$

Practice Exercise 2.23

Convert $15\overline{0}$°C to degrees Fahrenheit.

SUMMARY

Measurements represent the dimensions, quantity, or capacity of things. Scientists record and analyze measured data as they do research to attain knowledge while engaging in the problem-solving process. Chemistry students, too, engage in problem solving. George Polya's four-step method facilitates this process. The process includes understanding the problem, devising a plan for solution, carrying out the plan, and looking back at the results. Estimation techniques can help students assess the reasonableness of a solution.

In working with measurements, significant figures are used to convey the precision of information. The results of measurements and computations should always be rounded to the appropriate number of significant figures. Scientific notation, in powers of 10, may be used to indicate significant figures; it also simplifies computations involving very large or very small numbers.

The metric (or SI) system is a system of measurement used in most parts of the world and in all scientific work. It consists of seven basic units and a number of derived units, along with a set of prefixes that indicate multiples of these units. The unit of length is the meter, the unit of mass is the kilogram, and the unit of volume is a derived unit (cubic centimeters or, sometimes, liters). The corresponding units in the English system, as used in the United States, are the foot (length), pound (weight), and quart (volume). The factor-unit method may be used to convert from unit to unit within either system or from system to system. Mass is a measure of the quantity of matter, and weight is a measure of the gravitational force on an object. The density of an object is its mass divided by its volume. Temperatures, which indicate heat intensity, are measured on the Fahrenheit, Celsius, or Kelvin temperature scale and may be converted from one scale to another.

32 ESSENTIAL CONCEPTS OF CHEMISTRY

KEY TERMS

We have listed the major terms that have been defined in this chapter. Be sure that you are familiar with each of these terms. If you need to refresh your memory, refer to the indicated section or use the glossary at the back of the book.

accuracy (**2.4**)
area (**2.6**)
base number (**2.5**)
Celsius temperature scale (**2.11**)
density (**2.10**)
exponent (**2.5**)
Fahrenheit temperature scale (**2.11**)
Kelvin temperature scale (**2.11**)
mass (**2.9**)

metric system (**2.8**)
precision (**2.4**)
rounding off (**2.4**)
scientific notation (**2.5**)
significant figures (**2.4**)
temperature (**2.11**)
volume (**2.6**)
weight (**2.9**)

SELF-TEST EXERCISES

LEARNING GOAL 1

Significant Figures

◀ 1. How many significant figures are there in each of the following measured quantities?

(a) 5,000.0 cm   (b) 300 m   (c) 0.204 dL   (d) 8.000 ft   (e) 340,000.0 mL   (f) 5.0×10^3 g
(g) 8×10^2 yd   (h) 6,$\overline{000}$ mm   (i) 6,$\overline{000}$ mm

◀ 2. How many significant figures are there in each of the following measured quantities?

(a) 412,$\overline{000}$,000 mm   (b) 4.00×10^3 km   (c) 0.8009 mi   (d) 0.00036 km   (e) 5.023×10^{-6} kg   (f) .703, $\overline{000}$ in   (g) 0.009 cL   (h) 5.0000×10^3 cm

3. Determine the number of significant figures in each of the following numbers:

(a) 3,$\overline{000,000}$ (b) 3,$\overline{000}$,000   (c) 3×10^6 (d) 3.00×10^6 (e) 3.000000×10^6 (f) 0.0000305
(g) 0.100054   (h) 6.720×10^{-8} (i) 305,075   (j) 35.00

4. Determine the number of significant figures in each of the following numbers:

(a) 4,$\overline{000,000}$ (b) 4,$\overline{000}$,000   (c) 2×10^5 (d) 0.00123   (e) 4.02×10^{-5} (f) 23.00
(g) 2×10^{-1} (h) 750

LEARNING GOAL 2

Calculations With Significant Figures

5. Find the perimeter of a four-sided object whose sides have the following dimensions: 4.234 cm, 3.8 cm, 5.67 cm, and 4.00 cm. How many significant figures are there in your answer?

6. Find the perimeter of a triangle whose sides are 3.23 cm, 5.006 cm, and 3 cm. How many significant figures are there in your answer?

7. Find the perimeter of a triangle whose sides are 4.00 cm, 5.000 cm, and 6.0 cm. How many significant figures are in your answer?

8. Find the area of a square whose sides are 4.2 cm. How many significant figures are there in your answer?

◀ 9. Find the perimeter of a four-sided object whose sides are 7.382 cm, 3.95 cm, 5.4342 cm, and 3.83 cm. (Use the proper number of significant figures.)

10. Find the area of a rectangle whose sides are 10.12 cm and 10.25 cm. (Use the proper number of significant figures.)
11. Find the area of a rectangle whose sides are 20.62 cm and 10.4 cm. How many significant figures are there in your answer?
◀ 12. Find the area of a square whose sides are 2.1 cm each. (Use the proper number of significant figures.)
13. State the number of significant figures in each number given in Exercise 19.
14. Find the perimeter of a square whose sides are 2.1 cm each. (Use the proper number of significant figures.)

LEARNING GOAL 3

Scientific Notation

15. Express the following numbers in scientific notation:

    (a) 600 (b) $6\overline{0}0$ (c) 600.0 (d) 320,000,000,000 (e) 0.002 (f) 0.0003007 (g) 0.00000015

16. Express the following numbers in scientific notation:

    (a) 900,000 (b) 45,000 (c) 2,970 (d) 2,546,000 (e) 0.00006 (f) 0.0122 (g) 0.056 (h) 20.0040

17. Express the following numbers in scientific notation:

    (a) 10,581 (b) 0.00205 (c) 1,000,000 (d) 802

◀ 18. Express the following numbers in scientific notation:

    (a) 45,000,000 (b) $4\overline{0}0$ (c) $4\overline{00}$ (d) 400.0 (e) $425,\overline{000}$

19. Express the following numbers in scientific notation:

    (a) 850,000,000 (b) 0.00000607 (c) 6,308,000 (d) 0.06005 (e) 500 (f) $5\overline{0}0$ (g) $5\overline{00}$ (h) 500.0 (i) $23,\overline{000},000$ (j) 0.0000000930

20. Express the following numbers in scientific notation:

    (a) 0.123 (b) 0.006 (c) 0.00601

LEARNING GOAL 4

Areas of Geometric Figures

◀ 21. Find the area of a rectangular room that measures 8.00 m by 3.00 m. Report your answer to the proper number of significant figures. If the height of the room is 4.00 m, what is the volume of the room?
22. Find the area of a rectangular room that measures 5.0 m by 4.0 m. Use the correct number of significant figures in reporting your answer.
23. Determine the area of this page in (a) square centimeters, (b) square inches.
24. Find the area of this page in (a) square meters, (b) square feet.
25. Find the area occupied by a rectangular house that measures $3\overline{0}$ m by $2\overline{0}$ m.
26. Find the area of a table that measures 1.5 meters by 2.0 meters.

LEARNING GOAL 5

Volumes of Geometric Figures

27. Find the volume of a milk carton that measures 50.0 mm long, 6.00 cm wide, and 1.50 dm high. Report your answer in cm³ to the proper number of significant figures.
28. Find the volume of a box that measures 5.0 m long, 4.0 m wide, and 2.0 m high.
29. Determine the volume of a cylinder-shaped can that measures 60.0 cm high and has a radius of 15.0 cm.

34 ESSENTIAL CONCEPTS OF CHEMISTRY

30. Determine the volume of a metal cylinder that is 25 cm high and has a radius of 3.0 cm.
◄ *31.* A residential swimming pool in the shape of a rectangle measures 16.0 feet by 32.0 feet. The average height of such a pool is 5.00 feet. What is the volume of this pool in ft^3? If 1.00 foot = 30.5 cm, what is the volume of the pool in cm^3? Report your answers to the proper number of significant figures.
32. There are 100 cm in 1 m. How many cubic centimeters (cm^3) are there in 2 cubic meters (2 m^3)?
33. If the pool in Exercise 31 is filled with water, how many liters of water can it hold? Knowing that 1.00 L = 1.06 quarts and that 4.00 quarts = 1.00 gallons, how many gallons of water can the pool hold? Report all answers to the proper number of significant figures.
◄ 34. Find the volume of a cylinder-shaped jar of apple-sauce that measures $1\overline{0}$ cm high and has a radius of 5.0 cm.
35. There are 12 inches in 1 foot. From this information, calculate the number of cubic inches in 1 cubic foot.
36. A cookie jar has the dimensions $4\overline{0}$ cm by 25 cm by 15 cm. Calculate its volume.

LEARNING GOAL 6

Conversion of Units Within the Metric System

37. Convert 0.25 kg to (a) grams, (b) decigrams, (c) milligrams, (d) micrograms.
38. Convert 4.17 kg to (a) grams, (b) decigrams, (c) milligrams, (d) micrograms.
39. Convert 3.1 m to (a) decimeters, (b) centimeters, (c) millimeters.
40. Convert 35,000 mm to (a) centimeters, (b) decimeters, (c) meters, (d) kilometers.
41. Convert 149 mm to (a) centimeters, (b) meters, (c) kilometers.
42. Convert 5.5 m to (a) decimeters, (b) centimeters, (c) millimeters.
◄ *43.* Convert 7.850 m to (a) millimeters, (b) centimeters, (c) decimeters, (d) kilometers.
44. Convert 125 mm to (a) centimeters, (b) meters, (c) kilometers.
45. Convert 0.34 km to (a) meters, (b) decimeters, (c) centimeters, (d) millimeters.
46. Convert 1.234 m to (a) millimeters, (b) centimeters, (c) decimeters, (d) kilometers.
47. Convert 2,185 mg to (a) centigrams, (b) decigrams (c) grams, (d) kilograms.
48. Convert 7.5 km to (a) meters, (b) decimeters, (c) centimeters.
49. Convert $3,5\overline{00}$ mL to (a) liters, (b) deciliters.
50. Convert $1,5\overline{00}$ mL to (a) liters, (b) deciliters.
51. There are $1\overline{00}$ cm in 1 meter. From this information, calculate the number of cubic centimeters (cm^3) in 1 cubic meter (m^3).
52. Convert $25,55\overline{0}$ mg to (a) centigrams, (b) decigrams, (c) grams, (d) kilograms.

LEARNING GOAL 7

Conversion of Metric to English Units

53. Express 25.0 m in (a) feet, (b) inches.
54. Express 100.0 m in (a) feet, (b) inches.
55. Express 1.2 L in (a) gallons, (b) quarts.
56. Express 10.0 L in (a) gallons, (b) quarts.
57. Express 5.00 g in (a) pounds, (b) ounces.
58. Express 100.0 kg in (a) pounds, (b) tons.
◄ *59.* Without looking at a conversion table, determine the number of cubic feet in a cubic meter.
◄ 60. Express $25\overline{0}$ m in (a) yards, (b) feet.
61. Express $1\overline{00}$ m in (a) yards, (b) feet.
62. Express 2.50 g in pounds.
63. There are 3.28 ft in one meter (1.00 m). From this information, calculate the number of square feet (ft^2) in one square meter (m^2).
64. Express 50.0 L in gallons.

CHAPTER 2: SYSTEMS OF MEASUREMENT   35

LEARNING GOAL 8

Conversion of English to Metric Units

65. Convert your height from feet and inches to meters.
66. Express $2\overline{0}$ gal in (a) liters, (b) milliliters.
67. Convert your weight from pounds to kilograms.
68. Express 100.0 lb in (a) kilograms, (b) grams.
◀ 69. Express $1\overline{0}$ ft in (a) meters, (b) centimeters, (c) millimeters.
70. Express 5.0 ft in (a) meters, (b) centimeters.
71. Express 6.00 ft in (a) meters, (b) centimeters.
72. Express 25 ft in (a) meters, (b) centimeters.
73. Express $1\overline{00}$ yd in (a) meters, (b) centimeters.
74. Express 15 gal in (a) liters, (b) milliliters.
75. Express 5.00 gal in (a) liters, (b) milliliters.
◀ 76. Express $22\overline{0}$ lb in (a) kilograms, (b) grams.

LEARNING GOALS 9 AND 10

Calculation of Density, Mass, and Volume

77. Determine the density of this textbook. You'll have to weigh it and determine its volume. Report your answer in grams per cubic centimeter.
78. A solid cube is 6.00 cm on each side and has a mass of 0.583 kg. What is its density in g/cm^3?
79. A cube is 5.00 cm on each side and has a mass of 600.0 g. What is its density?
80. A block of aluminum with a density of 2.70 g/cm^3 weighs 274.5 g. What is the volume of the block?
81. A block of aluminum with a density of 2.7 g/cm^3 and a mass of 549 g. What is the volume of the block?
◀ 82. The element barium has a density of 3.5 g/cm^3. What would be the mass of a rectangular block of barium with the dimensions 2.0 cm × 3.0 cm × 4.0 cm?
◀ 83. The element barium, which is a soft, silvery-white metal, has a density of 3.5 g/cm^3. What would be the mass of a rectangular block of barium with the dimensions 1.0 cm × 3.0 cm × 5.0 cm?
84. Calculate the mass of a cube that has a density of 2.50 g/cm^3 and is 1.5 cm on each side. Report your result to the proper number of significant figures.
*85. A spherical balloon is filled with helium, a gaseous element. Helium has a density of 0.177 g/L. The balloon has a radius of 3.0 cm. What is the mass of the helium in the balloon? (*Hint:* Volume of a sphere $=\frac{4}{3}\pi r^3$.)
86. A cube has a density of 0.250 g/cm^3 and a mass of 15.0 g. What is the volume of this cube?
87. A 1.00-L container holds a block of cesium that is 1.00 cm × 2.00 cm × 3.00 cm and whose density is 1.90 g/cm^3; 14.0 g of iron, whose density is 7.86 g/cm^3; and 0.500 L of mercury, whose density is 13.6 g/cm^3. The rest of the container is filled with air, whose density is 1.18×10^{-3} g/cm^3. Calculate the *average* density of the contents of the container. *Hint:* To calculate average density, use the formula

$$\text{Average density} = \frac{m_a + m_b + \ldots}{V_a + V_b + \ldots}$$

88. A small metal sphere has a mass of 75.0 g. The sphere is placed in a graduated cylinder containing water. The water level in the cylinder changes from 10.0 cm^3 to 20.0 cm^3 when the sphere is submerged. What is the density of the sphere?
*89. A fish tank whose dimensions are 1.50 ft × 1.00 ft × 0.500 ft contains six fish. Fish one and two weigh 2.00 g each, fish three weighs 2.50 g, and the combined weight of fish four, five, and six is 9.80 g. The bottom of the fish tank is covered with gravel, which occupies one-eighth of the volume of the tank. The density of this gravel is 3.00 g/cm^3. The rest of the fish tank is filled with water (density 1.00 g/cm^3). Assuming that the volume occupied by the fish is negligible, calculate the average density of the contents of the fish tank.

36 ESSENTIAL CONCEPTS OF CHEMISTRY

◀ 90. Determine the density of a rectangular solid that has the dimensions 8.0 cm × 2.0 cm × 3.0 cm and a mass of 192.0 g.

◀ *91*. Calculate the mass of a cube that has a density of 1.61 g/cm^3 and is 4.1 cm on each side. Express your answer to the proper number of significant figures.

92. A balloon contains a gaseous compound. The radius of the balloon is 1.06 cm, and the mass of the gaseous compound in the balloon is 9.80 g. What is the density of the gaseous compound?

*93. Suppose you are drinking a chocolate milk shake. Its average density is 2.00 g/cm^3. The volume of the container that holds it is 25$\overline{0}$ cm^3. The milk shake consists of 90.0% milk products and 10.0% chocolate powder by volume. Assuming that the density of the milk products is 1.50 g/cm^3, calculate the amount of chocolate powder that was used to make the drink.

94. Determine the length of a metallic cube that has a density of 10.5 g/cm^3 and a mass of 672 g.

95. A cube measures 3.50 cm on each side and has a mass of 4$\overline{00}$ g. What is its density?

96. The element copper has a density of 8.92 g/cm^3. What is the mass of a cylinder-shaped piece of copper that has a radius of 2.00 cm and a height of 10.00 cm?

97. Gold has a density of 18.9 g/cm^3. A certain block of gold is a rectangular solid that measures 30.0 cm by 5.00 cm by 10.0 cm. What is its mass?

98. An empty graduated cylinder has a mass of 80.00 g. The empty cylinder is filled with exactly 30.0 cm^3 of a liquid and weighed again. The cylinder with the liquid in it has a mass of 488 g. What is the density of the liquid? If you are told that this liquid is an element, can you determine what element it is?

99. The density of lead is 11.3 g/cm^3. What is the volume of a chunk of lead that has a mass of 25$\overline{0}$ g?

100. A small stone that appears to be an irregularly shaped diamond is weighed and found to have a mass of 21.06 g (about 105 carats if it is indeed a diamond). The stone is placed in a graduated cylinder that initially has 10.0 cm^3 of water in it. When the stone is immersed in the water, the water level rises to 16.0 cm^3. What is the density of the stone? Could the stone be a diamond? (*Hint:* Check the *Handbook of Chemistry and Physics*.)

101. A small rock has a mass of 55.0 g. The rock is placed in a graduated cylinder containing water. The water level in the cylinder changes from 25.0 cm^3 to 40.0 cm^3 when the rock is submerged. What is the density of the rock?

102. A piece of copper that has a mass of 120.0 g is melted and poured into some liquid silver that has a mass of 200.0 g. The two metals are mixed and allowed to solidify. What is the average density of this metal alloy? [*Hint:* D(copper) = 8.92 g/cm^3 and D (silver) = 10.5 g/cm^3.]

*103. An object in the shape of a sphere—for example, a hollow globe—is filled with ethyl alcohol. The sphere has a radius of 18.0 cm, and ethyl alcohol has a density of 0.800 g/cm^3. What is the mass of ethyl alcohol in the sphere? (*Hint:* Volume of a sphere = $\frac{4}{3}\pi r^3$.)

104. Determine the mass in grams of one gallon of gasoline if the density of gasoline at 25°C is 0.56 g/cm^3.

105. A cube has a density of 0.500 g/cm^3 and a mass of 25.0 g. What is the volume of this cube?

106. What mass of lead [D(lead) = 11.3 g/cm^3] occupies the same volume as 1$\overline{00}$ g of aluminum [D(aluminum) = 2.70 g/cm^3]?

LEARNING GOAL 11

Celsius and Fahrenheit Temperature Conversions

107. Convert 100.0°F to degrees Celsius and to Kelvins.

108. Convert 2,2$\overline{00}$°F to degrees Celsius and to Kelvins.

109. Convert –40°F to degrees Celsius. What is unusual about this temperature?

◀ 110. Convert –20.0°F to degrees Celsius.

111. An individual who is ill has a temperature of 40.0°C. Normal body temperature is 37.0°C. This represents a 3.0°C rise in temperature. What type of increase above normal body temperature does this represent in °F? What is the individual's body temperature in °F?

112. Convert 23.0°C to degrees Fahrenheit.

◀ **113.** Change each of the following temperatures from degrees Fahrenheit to degrees Celsius.

(a) 50°F   (b) −94°F   (c) 419°F   (d) −130°F

114. Convert −20°C to degrees Fahrenheit.

115. Change each of the following temperatures from degrees Celsius to degrees Fahrenheit.

(a) 95.0°C   (b) −80.0°C   (c) 80.0°C   (d) 210°C

116. Convert 15.0°C to degrees Fahrenheit.

EXTRA EXERCISES

117. Would a carefully weighed object weigh the same in Death Valley as it would on the top of Mount McKinley?

◀ **118.** In filling out an application form for a company that uses the metric system, a job applicant recorded his height as 4.00 m and his weight as 200.0 kg. Did he fill the application out correctly?

119. An art dealer needs to know the volume of a piece of irregularly shaped sculpture. A chemist friend tells the dealer to submerge the object in water and measure the amount of water displaced by it. The dealer finds that 48 g of water are displaced. If the density of the water is 1.0 g/cm³, what is the volume of the art object?

120. A chemistry student decides to cook a pizza. The instructions call for cooking the pizza for 10 minutes at 425°F. The oven dial, however, is set in degrees Celsius. At what temperature should the dial be set so that the pizza will be done in 10 minutes?

121. Choose a room where you live and measure its dimensions in English units; then convert each dimension to metric units.

122. Express the following relationships in scientific notation:

(a) _____ g = 1 kg

(b) _____ kg = 1 g

(c) _____ µg = 1 g

(d) _____ mg = 1 g

◀ **123.** Assuming that the following numbers are from experimental measurements, calculate the answer and state it to the proper number of significant figures.

$$\frac{(3.12 \times 10^6)(8.123 \times 10^{-4})}{3.1}$$

124. A temperature is measured as 16.0°F. What is this temperature in degrees Celsius?

125. Change the following temperatures to °F:

(a) 16°C   (b) $2\overline{00}$°C

126. Change the following temperatures to °C:

(a) $3\overline{00}$°F   (b) $-1\overline{50}$°F

127. The 1976 world's record for the 100-meter dash was 9.90 seconds. What would this speed be in miles per hour?

CHAPTER 3

MATTER AND ENERGY, ATOMS AND MOLECULES

We all learn the scientific method when we take a chemistry course, yet few of us think that we'll ever have to use it in our personal lives. This is the story of Joan Penn, who used the training and knowledge she received in a chemistry course to help her 11-year-old son Mike over a serious illness.

Our story begins on a Sunday afternoon in late September. Mike returned home from playing soccer complaining that he had a headache. By morning he felt better, so Mike went off to school. When he came home, he was coughing and running a low-grade fever. Two days later his mother took him to the doctor who diagnosed a viral infection. The doctor recommended that Mike rest until the virus passed.

After two days, his conditioned worsened. He was coughing and running a high fever. In addition, a rash appeared on his arms. A strep test proved positive and Mike was placed on an antibiotic. After two days on the medication, his condition continued to worsen, and he was now coughing severely and having bronchial spasms.

Joan was quite concerned. After all, the antibiotic Mike was on should have certainly wiped out any strep infection, yet Mike was still getting worse. She thought that perhaps her son was actually not suffering from a virus, but rather that he was having an allergic reaction to something environmental—for instance, something in the park, where he played every day. She decided to retrace Mike's schedule over the past two weeks to look for something—anything—that stood out from his regular schedule.

Joan checked Mike's school for any unusual incidents. Nothing. She checked the parks where he had played; she called the township to see if pesticides were being sprayed in the parks. All results proved negative. Joan decided to ask Mike if there was anything he had done over the past two weeks that he usually didn't do. One event stood out as slightly unusual.

On the day his initial symptoms appeared, Mike was at a friend's house. He and his friend had decided to build a tree house in the backyard, and they worked several hours cutting branches and putting the tree house together. Upon its completion, his friend's Mom bought them pizza to celebrate their fine work, which the boys gobbled down without having washed up.

Joan decided to find out more about the tree house and discovered that the boys built it using juniper branches. Joan called the county health department to find out if junipers could cause allergic reactions. The health officer stated that sticky sap from the pines *could* cause allergic reactions in humans, and he recited a list of symptoms that were identical to Mike's. The puzzle was solved! Mike was allergic to the juniper sap. When Joan later told her son about her findings, Mike recalled getting the sticky sap on his arms, where his rash had formed. He also said that he may have ingested some of the sap when he ate the pizza, which would explain his breathing problems.

Joan called Mike's doctor and told her about the information she had uncovered. Mike was immediately placed on Ventolin to relieve the bronchial spasms. He was also given a nebulizer to help him breathe more easily. Two weeks later Mike's fever was down, the rash had disappeared, and his breathing had returned to normal. Thanks to some good detective work by his mother, Mike was once again healthy.

LEARNING GOALS

After you've studied this chapter, you should be able to:

1. Explain what is meant by the scientific method.
2. Explain the Law of Conservation of Mass and Energy.
3. Explain the difference between physical and chemical properties.
4. Describe the difference between homogeneous and heterogeneous matter, between mixtures and compounds, and between compounds and elements.
5. Describe the difference between an atom and a molecule.
6. Explain the Law of Definite Composition (or Definite Proportions).
7. Explain the terms *atomic mass, formula mass*, and *molecular mass*.
8. Determine the formula or molecular mass of a compound when you are given the formula for the compound.

INTRODUCTION

This chapter is really the beginning of your study of chemistry. We start with discussions of the most elementary concepts, those of matter and energy. Then, in this chapter and later chapters, we build on and extend these concepts. As we do this, we shall be discussing the results of centuries of scientific research—the theories and laws of modern chemistry. These theories and laws are sometimes presented to students as though each resulted from a quick flash of insight on the part of some scientist. Actually they are the fruit of years—and sometimes decades or centuries—of hard work by many people.

3.1 The Scientific Method

Chemistry is an experimental science that is concerned with the behavior of matter. Much of the body of chemical knowledge consists of abstract concepts and ideas. Without application, these concepts and ideas would have little impact on society. Chemical principles are applied for the benefit of society through technology. Useful products are developed by the union of basic science and applied technology.

Over the past 200 years, science and technology have moved forward at a rapid pace. Ideas and applications of these ideas are developed through carefully planned experimentation, in which researchers adhere to what is called the **scientific method**. The scientific method is composed of a series of logical steps that allow researchers to approach a problem and try to come up with solutions in the most effective way possible. It is generally thought of as having four parts:

1. *Observation and classification.* Scientists begin their research by carefully observing natural phenomena. They carry out experiments, which are observations of natural events in a controlled setting. This allows results to be duplicated and rational conclusions to be reached. The data the scientists collect are analyzed, and the facts that emerge are classified.
2. *Generalization.* Once observations are made and experiments carried out, the researcher seeks regularities or patterns in the results that can lead to a generalization. If this generalization is basic and can be communicated in a concise statement or a mathematical equation, the statement or equation is called a *law*.
3. *Hypothesis.* Researchers try to find reasons and explanations for the generalizations, patterns, and regularities they discover. A hypothesis expresses a tentative explanation of a generalization that has been stated. Further experiments then test the validity of the hypothesis.

4. *Theory*. The new experiments are carried out to test the hypothesis. If they support it without exception, the hypothesis becomes a theory. A theory is a tested model that explains some basic phenomenon of nature. It cannot be proven to be absolutely correct. As further research is performed to test the theory, it may be modified or a better theory may be developed.

The scientific method represents a systematic means of doing research. There are times when discoveries are made by accident, but most knowledge has been gained via careful, planned experimentation. In your study of chemistry you will examine the knowledge and understanding that researchers using the scientific method have uncovered.

3.2 Matter and Energy

We begin with the two things that describe the entire universe: *matter* and *energy*. **Matter** is *anything that occupies space and has mass*. That includes trees, clothing, water, air, people, minerals, and many other things. Matter shows up in a wide variety of forms.

Energy is the *ability to perform work*. Like matter, energy is found in a number of forms. Heat is one form of energy, and light is another. There are also chemical, electrical, and mechanical forms of energy. And energy can change from one form to another. In fact, matter can also change form or change into energy, and energy can change into matter, but not easily.

3.3 Law of Conservation of Mass and Energy

The **Law of Conservation of Mass** tells us that when a chemical change takes place, no detectable difference in the mass of the substances is observed. In other words, mass is neither created nor destroyed in an ordinary chemical reaction. This law has been tested by extensive experimentation in the laboratory, and the work of the brilliant French chemist-physicist Antoine Lavoisier provides evidence for this conclusion. Lavoisier performed many experiments involving matter. In one instance he heated a measured amount of tin and found that part of it changed to a powder. He also found that the *product* (powder plus tin) weighed *more* than the original piece of tin. To find out more about the added weight, he heated metals in sealed jars, which, of course, contained air. He measured the mass of his starting materials (*reactants*), and when the reaction concluded and the metal no longer changed to powder, he measured the mass of the products. In every such reaction, the mass of the reactants (oxygen from the air in the jar plus the original metal) equaled the mass of the products (the remaining metal plus the powder). Today we know that the reaction actually stopped when all of the oxygen in the sealed jar combined with the metal to form the powder. Lavoisier concluded that when a chemical change occurs, *matter is neither created nor destroyed, it just changes from one form to another* (Figure 3.1), which is a statement of the Law of Conservation of Mass.

An experimenter puts a test tube containing a lead nitrate solution into a flask containing a potassium chromate solution. The experimenter weighs the flask and contents, then turns the flask upside down to mix the two solutions. A chemical reaction takes place, producing a yellow solid. The experimenter weighs the flask and contents again and finds no change in mass.

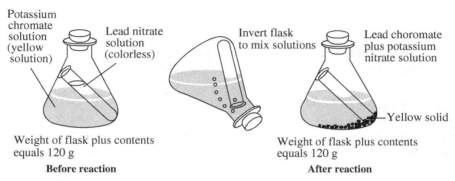

FIGURE 3.1 An experiment like Lavoisier's.

42 ESSENTIAL CONCEPTS OF CHEMISTRY

Whenever a chemical change occurs, it is accompanied by an energy transformation. In the 1840s, more than half a century after Lavoisier, three scientists—the Englishman James Joule and the Germans Julius von Mayer and Hermann von Helmholtz—performed a number of experiments in which energy transformations were studied. They provided experimental evidence that led to the discovery of the **Law of Conservation of Energy**. The law tells us that *in any chemical or physical change, energy is neither created nor destroyed, it is simply converted from one form to another.*

An auto engine provides a good example of how one form of energy is converted to a different form. *Electrical* energy from the battery generates a spark that contains *heat* energy. The heat ignites the gasoline-air mixture, which explodes, transforming chemical energy into heat and *mechanical* energy. The mechanical energy causes the pistons to rise and fall, rotating the engine crankshaft and moving the car.

At the same time, in the same engine, matter is changing from one form to another. When the gasoline explodes and burns, it combines with oxygen in the cylinders to form carbon dioxide and water vapor. (Unfortunately, carbon monoxide and other dangerous gases may also be formed. This is one of the major causes of air pollution.

To appreciate the significance of these facts, think of the universe as a giant chemical reactor or system. At any given time there are certain amounts of matter and energy present, and the matter has a certain mass. Matter is always changing from one form to another, and so is energy. Besides that, matter is changing to energy and energy to matter. But *the sum of all the matter (or mass) and energy in the universe always remains the same.* This repeated observation is called the **Law of Conservation of Mass and Energy**.

3.4 Potential Energy and Kinetic Energy

Which do you think has more energy, a metal cylinder held 1 foot above the ground or an identical cylinder held 5 feet above the ground? If you dropped them on your foot, you would know immediately that the cylinder with more energy was the one that was 5 feet above the ground. But where does this energy come from?

Work had to be done to raise the two cylinders to their respective heights—to draw them up against the pull of gravity. And energy was needed to do that work. The energy used to lift each cylinder was "stored" in each cylinder. The higher the cylinder was lifted, the more energy was stored in it—due to its position. *Energy that is stored in an object by virtue of its position* is called **potential energy**.

If we drop the cylinders, they fall toward the ground. As they do so, they lose potential energy because they lose height. But now they are moving; their potential energy is converted to "energy of motion." The more potential energy they lose, the more energy of motion they acquire. *The energy that an object possesses by virtue of its motion* is called **kinetic energy**. The conversion of potential energy to kinetic energy is a very common phenomenon. It is observed in a wide variety of processes, from downhill skiing to the generation of hydroelectric power.

3.5 The States of Matter

Matter may exist in any of the three physical states: solid, liquid, and gas.

A **solid** has a definite shape and volume that it tends to maintain under normal conditions. The particles composing a solid stick rigidly to one another. Solids most commonly occur in the **crystalline** form, which means they have a fixed, regularly repeating, symmetrical internal structure. Diamonds, salt, and quartz are examples of crystalline solids. A few solids, such as glass and paraffin, do not have a well-defined crystalline structure, although they do have a definite shape and volume. Such solids are called **amorphous solids**, which means they have no definite internal structure or form.

A **liquid** has a definite volume but does not have its own shape since it takes the shape of the container in which it is placed. Its particles cohere firmly, but not rigidly, so the particles of a liquid have a great deal of mobility while maintaining close contact with one another.

A **gas** has no fixed shape or volume and eventually spreads out to fill its container. As the gas particles move about they collide with the walls of their container causing *pressure*, which is a force exerted over an area. Gas particles move independently of one another. Compared with those of a liquid or solid, gas particles are quite far apart. Unlike solids and liquids, which cannot be compressed very much at all, gases can be both compressed and expanded.

Often referred to as the fourth state of matter, **plasma** is *a form of matter composed of electrically charged atomic particles.* Many objects found in the earth's outer atmosphere, as well as many celestial bodies found in space (such as the sun and stars), consist of plasma. A plasma can be created by heating a gas to extremely high temperatures or by passing a current through it. A plasma responds to a magnetic field and conducts electricity well.

3.6 Physical and Chemical Properties

Matter—whether it is solid, liquid, or gas—possesses two kinds of properties: physical and chemical. These unique properties separate one substance from another and ensure that no two substances are alike in every way. The **physical properties** are those that can be observed or measured without changing the chemical composition of the substance. These properties include state, color, odor, taste, hardness, boiling point, and melting point. A **physical change** is one that alters at least one of the physical properties of the substance without changing its chemical composition. Some examples of physical change are (1) altering the physical state of matter, such as what occurs when an ice cube is melted; (2) dissolving or mixing substances together, such as what happens when we make coffee or hot cocoa; and (3) altering the size or shape of matter, such as what happens when we grind or chop something.

Chemical properties stem from the ability of a substance to react or change to a new substance that has different properties. This often occurs in the presence of another substance. For example, iron reacts with oxygen to produce iron (III) oxide (rust). This is an example of a **chemical change**. The chemical properties can be observed or measured when a substance undergoes chemical change. The rusting of iron is an example of a chemical property of iron. When we pass an electric current through water, it decomposes to form hydrogen gas and oxygen gas. This reaction is an example of a chemical property of water.

Sometimes it is difficult to differentiate a chemical change from a physical change. In fact, physical changes almost always accompany chemical changes. Some of the signs of physical change that tell us that a chemical change has occurred include the presence of a large amount of heat or light, the presence of a flame, the formation of gas bubbles, a change in color or odor, or the formation of a solid material that settles out of a solution.

3.7 Mixtures and Pure Substances

Since matter consists of all the material things that compose the universe, many distinctly different types of matter are known. Matter that has a definite and *fixed composition* is called a **pure substance**, which is a substance that cannot be separated into any other form of matter by physical change. Some of the pure substances that you are familiar with are helium, oxygen, table salt, water, gold, and silver. Two or more pure substances can be combined to form a **mixture**, whose *composition can be varied*. The substances in a mixture can be separated by physical means; we can separate the substances without chemical change.

Matter can also be classified as heterogeneous or homogeneous (Figure 3.2). **Homogeneous matter** has the same parts with the same properties throughout, and **heterogeneous matter** is made up of different parts

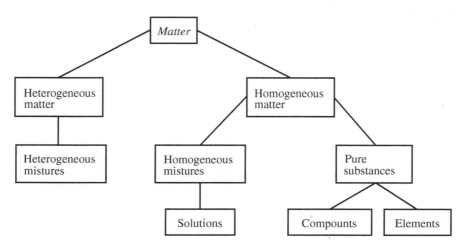

FIGURE 3.2 Classification of matter.

FIGURE 3.3 Separating sand from saltwater.

with different properties. A combination of salt and pepper is an example of heterogeneous matter, whereas a teaspoonful of sugar is an example of homogeneous matter. Another example of homogeneous matter is a teaspoonful of salt dissolved in a glass of water. We call this a **homogeneous mixture** because it is a uniform blend of two or more substances, and its proportion can be varied. In a homogeneous mixture every part is exactly like every other part. The salt can be separated from the water by physical means. Seawater and air are also examples of homogeneous mixtures.

We know that there are two types of homogeneous matter: pure substances and homogeneous mixtures. According to this classification scheme, matter can be broken down even further. Let's look at both homogeneous mixtures and heterogeneous mixtures. In a **heterogeneous mixture** different parts have different properties. A salt-sand mixture is heterogeneous because it is composed of two substances, *each of which retains its own unique properties*. It does not have the same composition or properties throughout, and its composition can be varied. A salt-sand mixture can be separated by physical means. If the mixture is placed in water, the salt will dissolve. The sand can be filtered out, and the salt can be recovered by heating the saltwater until the water evaporates (Figure 3.3).

We call any part of a system with uniform composition and properties a **phase**. A **system** is the body of matter being studied. A heterogeneous mixture is composed of two or more phases separated by physical boundaries. Additional examples of heterogeneous mixtures are oil and water (two liquids) and a tossed salad (several solids). It is important to note that although a pure substance is always homogeneous in composition, it may actually exist in more than one phase in a heterogeneous system. Think of a glass of ice water. This is a two-phase system composed of water in the solid phase and water in the liquid phase. In each phase the water is homogeneous in composition, but since two phases are present, the system is heterogeneous.

3.8 Solutions

Solutions are homogeneous mixtures. That is, they are uniform in composition. Every part of a solution is exactly like every other part. The salt-and-water mixture described in Section 3.7 is a solution. Even if we add more water to the solution, it will still be homogeneous, because the salt particles will continue to be distributed evenly throughout the solution. We would get the same result if we added more salt to the solution. However, we couldn't do this indefinitely. Homogeneity would end when the solution reached *saturation* (the point at which no more salt could dissolve in the limited amount of water). We will discuss this further in Chapter 12.

3.9 Elements

An **element** is *a pure substance that cannot be broken down into simpler substances, with different properties, by physical or chemical means.* The elements are the basic building blocks of all matter. There are now 118 known elements. Each has its own unique set of physical and chemical properties. (The elements are tabulated on the inside front cover of this book, along with their *chemical symbols*—a shorthand notation for their names. Some common elements are listed in Table 3.1.) The elements can be classified into three types: **metals**, **nonmetals**, and **metalloids**.

Examples of metallic elements are sodium (which has the symbol Na), calcium (Ca), iron (Fe), cobalt (Co), and silver (Ag). These elements are all classified as metals because they have certain properties in common. They have luster (in other words, they are shiny), they conduct electricity well, they conduct heat well, and they are malleable (can be pounded into sheets) and ductile (can be drawn into wires).

Some examples of nonmetals are chlorine, which has the symbol Cl (note that the second letter of this symbol is a lower-case "el" and not the numeral *one*), oxygen (O), carbon (C), and sulfur (S). These elements are classified as nonmetals because they have certain properties in common. They don't shine, they don't conduct electricity well, they don't conduct heat well, and they are neither malleable nor ductile.

The metalloids have some properties like those of metals and other properties like those of nonmetals. Some examples are arsenic (As), germanium (Ge), and silicon (Si). These particular metalloids are used in manufacturing transistors and other semiconductor devices (Table 3.1).

3.10 Atoms

Suppose we had a chunk of some element, say gold, and were able to divide it again and again, into smaller and smaller chunks. Eventually we could get a particle that could not be divided any further without losing its identity. This particle would be an atom of gold. An **atom** is *the smallest particle of an element that enters into chemical reactions.*

The atom is the ultimate particle that makes up the elements. Gold is composed of gold atoms, iron of iron atoms, and neon of neon atoms. These atoms are so small that billions of them are needed to make a speck large enough to be seen with a microscope. In 1970, Albert Crewe and his staff at the University of Chicago's Enrico Fermi Institute took the first black-and-white pictures of single atoms, using a special type of electron microscope. In late 1978, Crewe and his staff took the first time-lapse moving pictures of individual uranium atoms.

3.11 Compounds

A **compound** is *a pure substance that is made up of two or more elements chemically combined in a definite proportion by mass.* Unlike mixtures, compounds have a definite composition. Water, for instance, is made up of hydrogen and oxygen in the ratio of 11.1% hydrogen to 88.9% oxygen by mass. No matter what the source

TABLE 3.1 Names and Symbols of Some Common Elements

ELEMENT	SYMBOL	ELEMENT	SYMBOL
Aluminum	Al	Iodine	I
Bromine	Br	Magnesium	Mg
Calcium	Ca	Nickel	Ni
Carbon	C	Nitrogen	N
Chlorine	Cl	Oxygen	O
Chromium	Cr	Phosphorus	P
Fluorine	F	Silicon	Si
Helium	He	Sulfur	S
Hydrogen	H	Zinc	Zn

of the water, it is always composed of hydrogen and oxygen in this ratio. This idea, that *every compound is composed of elements in a certain fixed proportion*, is called the **Law of Definite Composition** (or the Law of Definite Proportions). It was first proposed by the French chemist Joseph Proust in about 1800.

The properties of a compound need not be similar to the properties of the elements that compose it. For example, water is a liquid, whereas hydrogen and oxygen are both gases. When two or more elements form a compound, they truly form a new substance.

Compounds can be broken apart into elements only by chemical means—unlike mixtures, which can be separated by physical means. More than thirty million compounds have been reported to date, and millions more may be discovered. Some compounds we are all familiar with are sodium chloride (table salt), which is composed of the elements sodium and chlorine, and sucrose (cane sugar), which is composed of the elements carbon, hydrogen, and oxygen.

3.12 Molecules

We have discussed what happens when a chunk of an element is continually divided: We eventually get down to a single atom. What happens when we keep dividing a chunk of a compound? Suppose we do so with the compound sugar. As we continue to divide a sugar grain, we eventually reach a small particle that can't be divided any further without losing the physical and chemical properties of sugar. This ultimate particle of a compound, *the smallest particle that retains the properties of the compound*, is called a **molecule**. Like atoms, molecules are extremely small, but with the aid of an electron microscope we can observe some of the very large and more complex molecules. Molecules are uncharged particles. That is, they carry neither a positive nor a negative electrical charge.

Molecules, as you might have guessed, are made up of two or more atoms. They may be composed of different kinds of atoms (for instance, water contains hydrogen and oxygen) or the same kind of atoms (for instance, a molecule of chlorine gas contains two atoms of chlorine). (See Figure 3.4.) *The Law of Definite Composition states that the atoms in a compound are combined in definite proportions by mass*. We can see in Figure 3.4 that they are also combined in definite proportions by number. For example, water molecules always contain two hydrogen atoms for every oxygen atom.

3.13 Molecular Versus Ionic Compounds

As we just learned, a molecule is the smallest *uncharged* part of a compound formed by the chemical combination of two or more atoms. Such compounds are known as *molecular compounds*, and we usually say that such compounds are composed of molecules. Water is a molecular compound composed of water molecules. Many molecular compounds are composed of atoms of nonmetallic elements that are chemically combined.

However, there are many compounds composed of oppositely charged *ions*. An ion is *a positively or negatively charged atom or group of atoms*. Compounds composed of ions are known as *ionic compounds*. These compounds are held together by attractive forces between the positive and negative ions that compose the compound. Ordinary table salt, sodium chloride, is such a compound. For these compounds, it is more proper to talk about a **formula unit** of the compound, rather than a molecule, as being *the smallest part of an ionic compound that retains the properties of the compound*. Ionic compounds are composed of metallic and nonmetallic elements. (We'll have much more to say about ions and ionic compounds when we discuss chemical bonding in Chapter 7.)

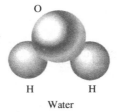
Water

Chlorine

FIGURE 3.4 Molecules of water and chlorine.

3.14 Symbols and Formulas of Elements and Compounds

We have already mentioned the chemical symbols—the shorthand for the names of the elements. In some cases, as in the symbol O (capital "oh") for oxygen, the chemical symbol is the first letter of the element's name, capitalized. Often, though, the symbol for an element contains two letters. In these cases only the first letter is capitalized; the second letter is never capitalized. For instance, the symbol for neon is Ne and the symbol for cobalt is Co. (Be careful of this: CO does not represent cobalt, but a combination of the elements carbon and oxygen, which is a compound.) Some symbols come from the Latin names of the elements: iron is Fe, from the Latin *ferrum*, and lead is Pb, from the Latin *plumbum*. (See Table 3.2.)

Because compounds are composed of elements, we can use the chemical symbols as a shorthand for compounds too. We use the symbols to write the **chemical formula**, which shows the elements that compose the compound. For example, sodium chloride (table salt), an ionic compound, contains one atom of sodium (Na) and one atom of chlorine (Cl). The formula unit for sodium chloride is NaCl. Water, a molecular compound, is another example. The water molecule contains two atoms of hydrogen (H) and one atom of oxygen, so we write it as H_2O. The number 2 in the formula indicates that there are two atoms of hydrogen in the molecule; note that it is written as a *subscript*. A molecule of ethyl alcohol contains two atoms of carbon, six of hydrogen, and one of oxygen. It is written as follows:

$$C_2H_6O \quad \leftarrow \text{One oxygen atom}$$

Two carbon atoms   Six hydrogen atoms

EXAMPLE 3.1

State the number of atoms of each element in a molecule or formula unit of the following compounds: (a) $C_6H_{12}O_6$ (b) $Ca(OH)_2$ (c) $C_3H_6O_2$ (d) $Al_2(SO_4)_3$

Solution

UNDERSTAND THE PROBLEM

We ask, "What does the subscript mean?" Subscripts tell us the number of atoms in a molecule or formula unit of the substance.

DEVISE A PLAN

Our plan will be to look at the subscripts noted in each case. We can then determine the number of atoms of each element in a molecule or formula unit of the compound.

TABLE 3.2 Elements with Symbols Derived from Latin Names

ELEMENT	SYMBOL	LATIN NAME
Copper	Cu	Cuprum
Gold	Au	Aurum
Iron	Fe	Ferrum
Lead	Pb	Plumbum
Mercury	Hg	Hydrargentum
Potassium	K	Kalium
Silver	Ag	Argentum
Sodium	Na	Natrium
Tin	Sn	Stannum

48 ESSENTIAL CONCEPTS OF CHEMISTRY

CARRY OUT THE PLAN

(a) There are 6 atoms of C, 12 atoms of H, and 6 atoms of O.

(b) There is 1 atom of Ca, 2 atoms of O, and 2 atoms of H. (In this case, the subscript 2 means multiply everything inside the parentheses by 2.)

(c) There are 3 atoms of C, 6 atoms of H, and 2 atoms of O.

(d) There are 2 atoms of Al, 3 atoms of S, and 12 atoms of O. (In this case, the subscript 3 means multiply everything inside the parentheses by 3.)

LOOK BACK

Recheck to be sure that you have followed the steps correctly.

Practice Exercise 3.1

State the number of atoms of each element in a molecule or formula unit of the following compounds: (a) C_2H_7N (b) $(NH_4)_2SO_4$

3.15 Atomic Mass

Suppose we want to find the relative masses of the atoms of the various elements. Suppose also that we have a double-pan balance that can weigh a single atom. We begin by assigning an arbitrary mass to one of the elements. Let's say we decide to assign a mass of 1 unit to hydrogen. (This is what chemists did originally, because hydrogen was known to be the lightest element even before the relative atomic masses were determined.)

Then, to find the relative mass of, say, carbon, we place an atom of carbon on one pan. We place hydrogen atoms on the other pan, one by one, until the pans are exactly balanced. We find that it takes 12 hydrogen atoms to balance 1 carbon atom, so we assign a relative atomic mass of 12 to carbon. In the same way for an oxygen atom, we find that it takes 16 hydrogen atoms to balance 1 oxygen atom. So we assign a relative mass of 16 to oxygen. By doing this experiment for all the other elements, we can determine all the masses relative to hydrogen (which is assigned mass 1).

Unfortunately this kind of balance has never existed, and more elaborate means had to be developed to find the relative atomic masses of the elements. But the logic we used here is the same as that used by the many scientists who determined the relative masses. (The atomic mass scale has been revised. The present scale is based on a particular type of carbon atom, called carbon-12. This carbon is assigned the value of 12 atomic mass units, amu.) The periodic table (inside the front cover) lists the relative atomic masses of the elements below their symbols. From now on, instead of referring to "relative atomic mass," we will use the simpler term **atomic mass**.

EXAMPLE 3.2

Using the periodic table (inside the front cover), look up the atomic masses of the following elements. (For this exercise, round all atomic masses to one decimal place.) (a) I (b) Ba (c) As (d) S

Solution

Look up the atomic mass of each element in the periodic table. Remember, the atomic mass is the number below the symbol of the element.

(a) I is 126.9

(b) Ba is 137.3

(c) As is 74.9

(d) S is 32.1

Practice Exercise 3.2

Using the periodic table (inside the front cover), look up the atomic masses of the following elements. (For this exercise, round all atomic masses to one decimal place.) (a) La (b) Fe (c) Ar (d) Sn

3.16 Formula Mass and Molecular Mass

The **molecular mass** of a compound is *the sum of the atomic masses of all the atoms that make up a molecule of the compound.* The term *molecular mass* is applied to compounds that exist as molecules. For example, the molecule P_2O_5 has two phosphorus atoms and five oxygen atoms. The atomic mass of each P (to one decimal place) is 31.0, and the atomic mass of each O (to one decimal place) is 16.0. Therefore the molecular mass of the compound is

$$(2 \times 31.0) + (5 \times 16.0) = 62.0 + 80.0 = 142.0 \text{ molecular mass}$$

(Check this calculation and see whether you get the same answer. If not, you may have forgotten that in an equation like this, multiplication and division are done *before* addition and subtraction.)

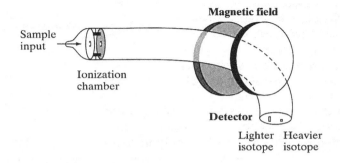

The masses of individual atoms are determined with a mass spectrometer. Electrons are removed from atoms (or molecules), and the resultant ions are accelerated through a magnetic field. The amount of bending in the path of the ions is related to the mass and the charge of the ions.

The **formula mass** of a compound is *the sum of the atomic masses of all the ions that make up a formula unit of the compound.* The term *formula mass* is applied to compounds that are written as formula units and exist mostly as ions (charged atoms or groups of atoms). For example, the formula unit of aluminum oxide is Al_2O_3. This means that a formula unit of aluminum oxide has two aluminum atoms (actually aluminum ions) and three oxygen atoms (actually oxide ions). The atomic mass of aluminum (to one decimal place) is 27.0, and the atomic mass of oxygen (to one decimal place) is 16.0. Therefore the formula mass of Al_2O_3 is

$$(2 \times 27.0) + (3 \times 16.0) = 54.0 + 48.0 = 102.0$$

Now let's try to determine the formula and molecular masses of some additional compounds.

EXAMPLE 3.3

Find the molecular or formula masses (to one decimal place) of the following compounds: (a) H_2O (b) NaCl (c) $Ca(OH)_2$ (d) $Zn_3(PO_4)_2$

Note: In a chemical formula, parentheses followed by a subscript mean that everything inside the parentheses is multiplied by the subscript. For example, in one formula unit of $Ca(OH)_2$, there is one Ca atom plus two O atoms and two H atoms.

Solution

We must find the atomic mass of each element in the periodic table and then add the masses of all the atoms in each compound.

(a) The atomic mass of H is 1.0, and the atomic mass of O is 16.0.

$$\text{Molecular mass of } H_2O = (2 \times 1.0) + (1 \times 16.0)$$
$$= 2.0 + 16.0 = 18.0$$

(b) The atomic mass of Na is 23.0, and the atomic mass of Cl is 35.5.

$$\text{Formula mass of NaCl} = (1 \times 23.0) + (1 \times 35.5)$$
$$= 23.0 + 35.5 = 58.5$$

(c) The atomic mass of Ca is 40.1, the atomic mass of O is 16.0, and the atomic mass of H is 1.0.

$$\text{Formula mass of } Ca(OH)_2 = (1 \times 40.1) + (2 \times 16.0) + (2 \times 1.0)$$
$$= 40.1 + 32.0 + 2.0 = 74.1$$

(d) The atomic mass of Zn is 65.4, the atomic mass of P is 31.0, and the atomic mass of O is 16.0.

$$\text{Formula mass of } Zn_3(PO_4)_2 = (3 \times 65.4) + (2 \times 31.0) + (8 \times 16.0)$$
$$= 196.2 + 62.0 + 128.0 = 386.2$$

Practice Exercise 3.3

Find the molecular or formula masses (to one decimal place) of the following compounds: (a) Na_2CO_3 (b) $CoCl_2$ (c) Cl_2O (d) N_2O_4

SUMMARY

The body of knowledge called science has been developed through the scientific method: observation and classification of data, generalization of observations, and testing of generalizations. Of importance in all the sciences is the idea that the universe is made up of only matter and energy. Matter and energy can be neither created nor destroyed, but they can be changed to other forms. And matter can be transformed into energy, and vice versa. Potential energy is energy that is stored in a body because of its position. Kinetic energy is energy that is due to motion.

Matter may exist in any of three states—solid, liquid, or gas—and may be either heterogeneous (nonuniform) or homogeneous (uniform). Mixtures are combinations of two or more kinds of matter, each retaining its own chemical and physical properties. Mixtures too may be either homogeneous or heterogeneous, and solutions are homogeneous mixtures. Elements are the basic building blocks of matter. The 109 known elements are, in turn, made up of atoms. A compound is a substance that is made up of two or more elements chemically combined in definite proportions by mass. Molecules are the smallest particles that retain the properties of a compound, and atoms are the smallest particles that enter into chemical reactions.

Each element has its own symbol, and every compound has its own chemical formula. Each element has a unique atomic mass. The atomic mass of an element is found in the periodic table. The formula mass or molecular mass of a compound is the sum of the atomic masses in a formula unit or molecule of the compound.

KEY TERMS

amorphous solid (**3.5**)
atom (**3.10**)
atomic mass (**3.15**)
chemical change (**3.6**)
chemical formula (**3.14**)
chemical property (**3.6**)
chemical symbols (**3.9**)
compound (**3.11**)
crystalline (**3.5**)
element (**3.9**)
energy (**3.2**)
formula mass (**3.16**)
formula unit (**3.13**)
gas (**3.5**)
heterogeneous matter (**3.7**)
heterogeneous mixture (**3.7**)
homogeneous matter (**3.7**)
homogeneous mixture (**3.7**)
kinetic energy (**3.4**)
Law of Conservation of Energy (**3.3**)
Law of Conservation of Mass (**3.3**)
Law of Conservation of Mass and Energy (**3.3**)

CHAPTER 3: MATTER AND ENERGY, ATOMS AND MOLECULES   51

Law of Definite Composition (**3.11**)
liquid (**3.5**)
matter (**3.2**)
metal (**3.9**)
metalloid (**3.9**)
mixture (**3.7**)
molecular mass (**3.16**)
molecule (**3.12**)
nonmetal (**3.9**)
phase (**3.7**)

physical change (**3.6**)
physical property (**3.6**)
plasma (**3.5**)
potential energy (**3.4**)
pure substance (**3.7**)
scientific method (**3.1**)
solid (**3.5**)
solution (**3.8**)
system (**3.7**)

SELF-TEST EXERCISES

LEARNING GOAL 1

The Scientific Method

1. (a) Explain the difference between a theory and a hypothesis.
   (b) Explain the difference between a theory and a scientific law.

2. Suppose that you are a researcher and you believe that you have found a vaccine that can prevent AIDS. How would you use the scientific method to determine whether the vaccine is effective?

LEARNING GOAL 2

Law of Conservation of Mass and Energy

◄ 3. According to the Law of Conservation of Energy, in any chemical or physical change, energy is neither created nor destroyed. Then why are we always worried about running out of energy in the future?

4. What is the significance of the Law of Conservation of Mass and Energy in terms of your study of chemistry?

LEARNING GOALS 3 & 4

Physical and Chemical Changes/Types of Matter

5. Match each word on the left with its definition on the right.

   (a) Homogeneous   1. The basic building block of matter
   (b) Heterogeneous   2. The word used to describe matter that is uniform throughout
   (c) Mixture   3. A type of matter in which each part retains its own properties
   (d) Compound   4. A chemical combination of two or more elements
   (e) Element   5. The word used to describe matter that is not uniform throughout

◄ 6. (a) Distinguish among an element, a compound, and a mixture.
   (b) What type of matter is uniform throughout?
   (c) What type of matter is not uniform throughout?

◄ 7. State whether each of the following processes involves physical or chemical changes:

   (a) Shredding paper   (b) Burning paper   (c) Cooking an egg   (d) Mixing egg whites with egg yolk   (e) Digesting food   (f) Toasting bread

8. Determine whether each of the following processes involves chemical or physical processes:

   (a) Ice melts.
   (b) Sugar dissolves in water.

52 ESSENTIAL CONCEPTS OF CHEMISTRY

(c) Milk sours.
(d) Eggs become rotten.
(e) Water boils.
(f) An egg is hard-cooked.

LEARNING GOAL 5

Difference Between Atom and Molecule

9. Describe the difference between an atom and a molecule.
◀ 10. (a) What is the smallest particle of matter that can enter into a chemical combination?
  (b) What is the smallest uncharged individual unit of a compound that is composed of two or more atoms?
<u>11</u>. Classify each of the following elements as metal, metalloid, or nonmetal:
  (a) Ba (b) Si (c) O (d) Hg (e) Ge (f) In (g) U
12. Classify each of the following elements as metal, metalloid, or nonmetal:
  (a) Mn (b) Nd (c) Al (d) At (e) Pt (f) Cl (g) Ra
<u>13</u>. Explain the difference between Co and CO.
14. Explain the difference between Si and SI.

LEARNING GOAL 6

Law of Definite Composition

◀ <u>15</u>. How does the following information obtained from several experiments confirm the Law of Definite Composition?

Experiment 1: $10\overline{0}$ g of water are decomposed by electrolysis into its elements: 88.9 g of oxygen gas and 11.1 g of hydrogen gas are obtained.
Experiment 2: 25.0 g of water are decomposed by electrolysis into its elements: 22.2 g of oxygen gas and 2.8 g of hydrogen are obtained.
Experiment 3: $50\overline{0}$ g of water are decomposed by electrolysis into its elements: 444.5 g of oxygen gas and 55.5 g of hydrogen are obtained.

16. Give an example of the Law of Definite Composition (Proportions).
<u>17</u>. State the number of atoms of each element in a molecule or formula unit of the following compounds:
  (a) $C_{12}H_{22}O_{11}$ (b) K_2CrO_4 (c) $H_8N_2O_3S_2$ (d) $Zn(NO_3)_2$
◀ 18. State the number of atoms of each element in a molecule or formula unit of the following compounds:
  (a) H_2SeO_4 (b) $C_{21}H_{27}FO_6$ (c) $(NH_4)_3PO_4$ (d) $Fe_3(AsO_4)_2$

LEARNING GOAL 7

Atomic Mass, Formula, Mass, and Molecular Mass

19. Using examples, explain the difference among atomic mass, formula mass, and molecular mass.
20. State which term—atomic mass, formula mass, or molecular mass—is best suited to describe each of the following substances:
  (a) $C_6H_{12}O_6$ (b) NaCl (c) Fe (d) CO_2 (e) H_2O (f) Al_2O_3 (g) Ca (h) $Ca(OH)_2$
◀ *<u>21</u>. If in the periodic table oxygen were assigned an atomic mass of 1, what would be the atomic mass of sulfur?

◀*22. If in the periodic table neon were assigned an atomic mass of 1, what would be the atomic mass of bromine?

23. Using the periodic table (inside front cover), look up the atomic masses of the following elements. (For this exercise, round all atomic masses to one decimal place.)

    (a) Rb (b) Cr (c) U (d) Se (e) As

24. Using the periodic table (inside front cover), look up the atomic masses of the following elements. (For this exercise, round all atomic masses to one decimal place.)

    (a) S (b) N (c) Li (d) Cs (e) Au

LEARNING GOAL 8

Formula or Molecular Mass of a Compound from the Formula

◀ <u>25</u>. Determine the molecular or formula mass of each of the following compounds. (For this exercise, round all atomic masses to one decimal place.)

   (a) FeO (b) Fe_2O_3 (c) CuI_2 (d) Na_3PO_4 (e) $Mg(OH)_2$ (f) $NiBr_2$ (g) $Hg_3(PO_4)2$
   (h) $(NH_4)_2CO_3$

26. Determine the molecular or formula mass of each of the following compounds. (For this exercise, round all atomic masses to one decimal place.)

   (a) H_2O (b) H_2SO_4 (c) NaCl (d) $Ca_3(PO_4)_2$ (e) P_2O_5 (f) $SrSO_4$ (g) C_2H_6O (h) SO_2

<u>27</u>. Determine the molecular or formula mass of each of the following compounds. (For this exercise, round all atomic masses to one decimal place.)

   (a) SiO_2 (b) H_2SO_3 (c) $Sr(OH)_2$ (d) RbF (e) $Cu(NO_3)_2$ (f) $CoBr_2$ (g) $(NH_4)_3PO_4$
   (h) $HC_2H_3O_2$

◀ 28. Determine the molecular or formula mass of each of the following compounds. (For this exercise, round all atomic masses to one decimal place.)

   (a) LiOH (b) Na_2CO_3 (c) $CoCl_2$ (d) NaBr (e) SO_3 (f) C_2H_6 (g) OF_2 (h) $(NH_4)_2SO_3$

EXTRA EXERCISES

29. State what each of the symbols and subscripts mean in the chemical formulas for (a) H_2O (b) $C_6H_{12}O_6$ (c) $Ca(OH)_2$ (d) H_2

◀ 30. Make a list of heterogeneous mixtures and homogeneous mixtures that you encounter in everyday life. Do the same for elements and compounds.

31. Write the names and symbols for the fourteen elements that have a one-letter symbol.

32. Write the names and symbols of all the metalloids.

◀ 33. How many metals are there in the periodic table? How many nonmetals? How many metalloids?

34. Write the names of the eleven elements whose symbols are not derived from their English names.

<u>35</u>. Name the elements present in each of the following compounds:

   (a) $MgCl_2$ (b) N_2O (c) $(NH_4)_2SO_4$ (d) H_3PO_4

36. Write the chemical formula of each of the following, given the number of atoms in a molecule or formula unit of the compound:

   (a) one nitrogen atom, two oxygen atoms (nitrogen dioxide) (b) two sodium atoms, one sulfur atom (sodium sulfide) (c) three potassium atoms, one arsenic atom, four oxygen atoms (potassium arsenate) (d) two phosphorus atoms, five oxygen atoms (diphosphorus pentoxide)

54 ESSENTIAL CONCEPTS OF CHEMISTRY

37. Classify each of the following as an element, a compound, or a mixture:

(a) gold
(b) air
(c) carbon dioxide
(d) wine
(e) table salt

38. State whether each of the following involves a physical or chemical change:

(a) toasting bread
(b) water freezing
(c) tearing paper
(d) burning wood

39. Determine the molecular or formula mass of each of the following compounds. (For this exercise, round all atomic masses to one decimal place.)

(a) OsO_4 (b) HNO_3 (c) $Fe(OH)_2$ (d) $Ba_3(PO_4)_2$

40. Explain the difference between Hf and HF.

CUMULATIVE REVIEW Chapters 1–3*

Indicate whether each of the following statements is true or false.

1. Chemistry is the science that deals with matter and the changes it undergoes.
2. Organic chemicals are those chemicals that have been derived from living or once-living organisms.
3. Sugar and salt are examples of inorganic chemicals.
4. The art of *khemia* flourished until A.D. 1600.
5. Alchemists were able to transmute lead and iron into gold.
6. The idea that every theory must be proved by experiment was advanced by Robert Boyle.
7. In 420 B.C., Greek philosophers were able to prove that atoms existed.
8. Some carbon-containing compounds are classified as inorganic chemicals.
9. The two main branches of chemistry that existed during the 1700s and 1800s still exist today.
10. Nuclear chemistry, biochemistry, and analytical chemistry are three subdivisions into which chemistry can be divided.

Answer the following questions to help sharpen your test-taking skills.

◀ *11.* Find the area of a rectangular room that measures 12.5 m by 10.2 m. Use the proper number of significant figures in reporting your answer.
12. Calculate the volume of a cardboard box that measures 15.25 cm long, 12.00 cm wide, and 24.85 cm high.
13. Determine the volume of a cylindrical solid whose radius is 14.50 cm and whose height is 25.05 cm.
14. Convert 0.28 kg to (a) grams, (b) decigrams, (c) milligrams.
15. Convert 6.8 m to (a) decimeters, (b) centimeters, (c) millimeters.
16. Convert 125 mm to (a) centimeters, (b) meters, (c) kilometers.
17. Convert 25,595 mL to (a) liters, (b) deciliters.
18. Express 50.0 m in (a) feet, (b) inches.
19. Express 25.5 g in (a) pounds, (b) ounces.
20. Determine the number of cubic centimeters in a cubic inch without using a conversion table.
21. Express 165 g in pounds.
22. Express 12.0 gallons in (a) liters, (b) ounces.
23. Determine the density of a cube that has a mass of 500.0 g and measures 12.0 cm on each side.
24. A spherical balloon with a volume of 113.0 mL is filled with an unknown gas weighing 20.0 g. Will the balloon float on air?

---

*Answers to cummulative-review questions are given in the back of the book.

CHAPTER 3: MATTER AND ENERGY, ATOMS AND MOLECULES   55

◀ 25. A small metal sphere weighs 90.0 g. The sphere is placed in a graduated cylinder containing 15.0 mL of water. Once the sphere is submerged, the water in the cylinder measures 30.0 mL. What is the density of the sphere?

26. A small metal sphere with a density of 3.50 g/cm³ has a mass of 937.83 g. Calculate the radius of the sphere.

27. Solve the following problems using the proper number of significant figures:

   (a) $28.64 + 3.2$ = 3.1
   (b) $125.4 \div 13.5$
   (c) 6.55×12.1
   (d) $98.4 - 0.12$

◀ 28. Express each of the following numbers in scientific notation:

   (a) 5,000   (b) 0.0005   (c) 602,300,000   (d) 35,000,000 3.5000×10^7
        5×10^3 5×10^{-3} 6.023×10^8

29. How many significant figures are there in each of the following numbers?

   (a) 5,500.0   (b) 0.5123   (c) 12.000   (d) 3,500

30. Convert each of the following temperatures from °F to °C:

   (a) 45.0°F   (b) −10.0°F   (c) 450°F   (d) −100.0°F

31. Convert each of the following temperatures from °C to °F:

   (a) 88.0°C   (b) −12.5°C   (c) 65.6°C   (d) 400°C

32. Distinguish between an amorphous solid and a crystalline solid.

33. Determine whether each of the following processes involves chemical or physical changes:

   (a) A match burns.
   (b) Glucose dissolves in water.
   (c) Bread becomes moldy.
   (d) A piece of wood is sawed.

34. Distinguish between heterogeneous and homogeneous matter.

35. State the number of atoms of each element in a molecule or formula unit of the following compounds:

   (a) $Zn(C_2H_3O_2)_2$ (b) $(NH_4)_2CrO_4$

36. Why do we use the term *molecular mass* for some compounds *and formula mass* for other compounds?

37. If in the periodic table calcium were assigned an atomic mass of 1, what would be the atomic mass of mercury?

38. Using the periodic table (inside front cover), look up the atomic masses of the following elements. (For this exercise, round all atomic masses to one decimal place.)

   (a) Yb   (b) At   (c) P   (d) Ag

39. Determine the molecular or formula mass of each of the following compounds. (For this exercise, round all atomic masses to one decimal place.

   (a) $Fe(C_2H_3O_2)_3$ (b) $(NH_4)_2SO_4$ (c) $C_9H_8O_4$ (d) $CO(SO_4)_2$

40. Explain in detail what the chemical formula $Al_2(SO_4)_3$ means.

41. Determine whether each of the following is an element, a compound, or a mixture:

   (a) air
   (b) arsenic
   (c) carbon dioxide
   (d) water
   (e) gold
   (f) root beer soda
   (g) gasoline

42. Determine whether each of the following is an example of a heterogeneous or homogeneous mixture:

    (a) beach sand
    (b) ethyl alcohol and water
    (c) tossed salad
    (d) soda water (club soda)

43. Using the periodic table (inside front cover), write the names and symbols for all elements whose symbol begins with the letter A. (*Hint:* There are eight.)

44. Write the names and symbols of all the nonmetallic elements.

45. Name the elements present in each of the following compounds:

    (a) K_2S (b) Ag_2CrO_4 (c) $KMnO_4$ (d) $Hg_3(PO_4)_2$

46. Write the chemical formula of each of the following, given the number of atoms in a molecule or formula unit of the compound:

    (a) two nitrogen atoms, one oxygen atom (dinitrogen monoxide)
    (b) two potassium atoms, one chromium atom, four oxygen atoms (potassium chromate)
    (c) one nitrogen atom, three hydrogen atoms (ammonia)
    (d) one strontium atom, one sulfur atom, four oxygen atoms (strontium sulfate)

47. State whether each of the following involves a physical or chemical change:

    (a) a candle burning
    (b) table sugar dissolving in water
    (c) tooth decaying
    (d) snow melting

48. Explain the difference between No and NO.

◄ 49. A piece of paper is weighed and is then burned. The resulting ash weighs less than the paper. Is this a violation of the Law of Conservation of Mass? Explain.

50. You are given a mixture of sodium chloride (table salt) and sand. Explain how you would separate this mixture.

CHAPTER 4

ATOMIC THEORY, PART 1
What's in an Atom?

Joe and Susan Townes, married for six years, had spent the initial four years of their marriage building their careers and saving for a home in which they could eventually raise their family. Joe was a successful engineer and Susan a well-established and respected real estate agent.

Then, two years ago, Susan, no longer content only with her career, told Joe she wanted to start a family. Suffering from extreme anxiety at work and unusually irritable when she lost a real estate deal, Susan was becoming depressed and tired and had already suffered a 7-kg (15.4-lb) weight loss. She told Joe that this "burn-out" could be cured by setting about to start the family they had always wanted.

Joe, sympathetic to Susan's problems and surprised Susan hadn't raised the issue before, nevertheless stated that he had no intention of starting a family at that time, much to Susan's dismay. Joe said that Susan's anxiety and irritability had affected their own previously solid marital relationship. Her erratic behavior, unpredictable mood swings, and the general breakdown in communication between them convinced Joe that starting a family should be postponed until Susan felt better and their relationship had returned to its former happy state.

Three days later, Susan, who had become even more depressed, lost a major real estate deal. She became extremely upset, with shaking hands, pounding in her chest, profuse sweating, and difficulty with breathing. Realizing that she was ill, Susan requested that a colleague take her to the nearest hospital emergency room.

At approximately 2:00 P.M. Dr. Philip Doherty met Susan in the Simon Memorial Hospital Emergency Room. He observed a fragile, pale young woman who was jittery and irritable. A complete history and physical revealed that Susan had a significant weight loss over the past three months accompanied by weakness, fatigue, emotional instability, and finally episodes of heart palpitations and breathlessness.

Based on the physical exam, Dr. Doherty immediately suspected that Susan was experiencing an endocrine disorder known as hyperthyroidism (Graves' Disease). His first course of action was to perform several blood tests to determine if Susan was, in fact, suffering from this condition. The test results came back positive; Susan had hyperthyroidism.

Treatment for Susan was directed toward reducing output of her thyroid hormone. The chosen method of treatment was radioactive iodine therapy. One of the most effective ways to administer this therapy is through the use of iodine-131 (I-131), a radioactive isotope of iodine. The isotope was given to Susan by mouth in the form of a single, one-time radioactive "cocktail." The I-131 atoms act chemically like the more common, nonradioactive iodine-126 (I-126) atoms. I-126 is an important nutritional component of our diets, because the thyroid gland requires iodine to produce the hormone thyroxine. The I-131 atoms, like the I-126 atoms, are quickly absorbed in the stomach and become concentrated in the thyroid. Unlike the I-126 atoms, however, radiation from the I-131 atoms destroys some of the cells of the thyroid, thereby decreasing the thyroid's activity.

58 ESSENTIAL CONCEPTS OF CHEMISTRY

Just three weeks after Susan was given the radioactive thyroid treatment her symptoms began to subside. Within two months, tests revealed that Susan's thyroid function was normal, and her behavioral and physical problems disappeared.

Last June, Susan and Joe Townes became the proud parents of a 4-kg (8.8-lb) baby boy.

LEARNING GOALS

After you've studied this chapter, you should be able to:

1. Describe Dalton's atomic theory, Thomson's model of the atom, and Rutherford's model of the atom.
2. Give the charge and mass of the electron, proton, and neutron.
3. Find the number of protons, electrons, and neutrons in an atom of an element.
4. Explain what isotopes are and give examples.
5. Use the standard isotopic notation for mass number and atomic number.
6. Calculate the average atomic mass of an element when you know the relative abundances of its isotopes.
7. Calculate the percentage abundance of two isotopes when you know the atomic mass of each isotope and the average atomic mass.
8. Use the table of relative abundances of isotopes to calculate the number of atoms or grams of a particular isotope present in a given sample.

INTRODUCTION

Around 420 B.C., Democritus developed the idea that the atom—indestructible and indivisible—was the smallest particle of matter. Twenty-two hundred years later, John Dalton reinforced this idea in his atomic theory. Yet less than a century after Dalton presented his theory, scientists had adopted a different model, in which the atom was made up of several even smaller particles. The development of that model is the subject of this chapter.

4.1 What Is a Model?

Before we try to discuss the atom, we should recall that until very recently no one had ever seen one. How, then, did we know what an atom was like? We really didn't, exactly. But scientists have put together observations, experimental results, and a good deal of reasoning and have come up with a *model*. This model of the atom is just a representation of what scientists believe the atom is like. Models can have different degrees of accuracy. Take a model of a car, for instance. It could be carved from a solid block of wood or plastic and represent just the shape of the car. Or a more detailed model could show the interior of the car as well. A still more detailed model might have a gasoline-powered motor and move just like a real car. This model would represent the car most accurately. The car model would be a physical model, whereas models of the atom are *intellectual* or *thought* models; however, they too have become more accurate as scientists have gathered more information about atoms.

4.2 Dalton's Atomic Theory

John Dalton (1766–1844) was a British schoolteacher who studied physics and chemistry. On the basis of facts and experimental evidence, he proposed an atomic theory. This theory was an attempt to explain all the different forms of matter. It has turned out to be one of the greatest contributions to chemistry since the time of the ancient Greeks. Dalton theorized that:

1. All elements are composed of tiny, indivisible particles called atoms. This idea was similar to Democritus' idea. Both men believed there was an ultimate particle—that all matter was composed of tiny indestructible and indivisible spheres.

2. All matter is composed of combinations of these atoms. In Chapter 3 we discussed how the atoms of different elements combine to produce formula units that make up compounds. For example, two atoms of hydrogen and one atom of oxygen combine to form a molecule of water.

3. Atoms of different elements are different. Dalton believed gold was different from silver because somehow the atoms of gold were different from the atoms of silver. For example, he thought they differed in mass.

4. Atoms of the same element have the same size, mass, and form. Dalton believed all gold atoms, for example, were the same as all other gold atoms in every respect.

Dalton's atomic theory laid the foundation for what is believed today. But his basic idea about the indivisibility of the atom was eventually shown to be incorrect, as was his idea that all atoms of the same element have the same size, mass, and form. Revisions of his theory have led to our current model of the atom. Let us review how our present-day model of the atom was developed.

4.3 The Discovery of the Electron

In the mid-1800s, scientists began to question whether the atom was really indivisible. They also wanted to know why atoms of different elements had different properties. Some of the answers came with the invention of the Crookes tube, or *cathode-ray tube* (Figure 4.1).

The scientists of the time knew that some substances conduct electric current (that is, they are conductors), whereas other substances do not. And, with enough electrical power, a current can be driven through any substance—solid, liquid, or gas. Electric current results from the accumulation of electric charge. Electric charge may be transferred from one object to another. We find that electric charge is either positive or negative. Experiments tell us that unlike charges attract, and like charges repel. In the cathode-ray tube, a high-voltage electric current is driven through a nearly empty tube, a near **vacuum**. The tube contains two pieces of metal called *electrodes*. Each electrode is attached by a wire to the source of an electric current. The source has two *terminals*, positive and negative. *The electrode attached to the positive electric terminal* is called the **anode**; *the electrode attached to the negative terminal* is called the **cathode**. Crookes showed that when the current was turned on, a beam moved from the cathode to the anode; in other words, the beam moved from the negative to the positive terminal. Because the beam originated at the *negative* terminal, and because it was attracted to the *positive* terminal, the beam had to be negative in nature.

What was this beam? Was it made of particles or waves? Did it come from electricity or from the metal electrodes? Physicists in Crookes's time were not sure about the answers to these questions, but they did make guesses. Whatever this beam was, it traveled in straight lines (they knew that because it cast sharp shadows). For lack of a better name, the German physicist Eugen Goldstein called the beams *cathode rays*, because they came from the cathode.

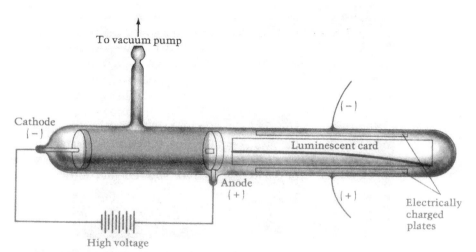

FIGURE 4.1 A Crookes or cathode-ray tube.

60 ESSENTIAL CONCEPTS OF CHEMISTRY

The German physicists in Crookes's time favored the *wave theory* of cathode rays because the beam traveled in straight lines, like ocean waves. But the English physicists favored the *particle theory*. They said that the beam was composed of tiny particles that moved very quickly—so quickly that they were hardly influenced by gravity. That was why the particles moved in a straight path. (Note how a single experimental observation led to two different theories.)

Crookes proposed a method to solve the dilemma. If the beam was composed of negative particles, a magnet would deflect them (Figure 4.2). But if the beam was a wave, a magnet would cause almost no deflection. Particles would also be more easily deflected by an electric field. In 1897 the English physicist J. J. Thomson used both these techniques—magnetic and electric—to show that the rays were composed of particles (Figure 4.3). Today we call these particles **electrons**. Moreover, we understand that an electric current is actually a flow of electrons.

Were the electrons in the beam coming from the metal electrodes or from the current source? Unless it was the metal that gave off these electrons, the atom could still be the smallest particle of matter. Proof that metals do give off electrons came from the laboratories of Phillipp Lenard, a German physicist. In 1902 he showed that ultraviolet light directed onto a metal makes the metal send out, or emit, electrons. This effect, known as the *photoelectric effect*, indicated that the metal atoms—and the atoms of other elements—contain electrons.

In 1911 a young American physicist named Robert Millikan calculated the mass of the electron: 9.11×10^{-28} grams. (To get an idea of how very small this is, consider the fact that it would take about 16,000,000,000,000,000,000,000,000,000, or 1.6×10^{28}, electrons to equal the weight of a standard half-ounce chocolate bar.)

FIGURE 4.2 A cathode-ray tube in operation. Compare these pictures with Figure 4.3. (Photos courtesy of the author.)

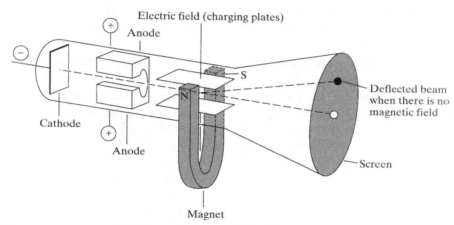

FIGURE 4.3 The Thomson experiment: (1) A beam of electrons (dashed line) moves from cathode toward anodes. (2) Some electrons pass between anodes. (3) Electric field causes beam of electrons to bend. This is visible on the screen. (4) But by adding a magnet to counteract deflection caused by electric field, one can make the beam follow a straight path. (5) One can then measure the strengths of the electric and magnetic fields and calculate the charge-to-mass ratio (e/m) of the electron.

4.4 The Proton

The discovery of the electron as part of the atom and as a *negative* particle of electricity was very important, but it raised many questions. Because electrons are part of all atoms, and because atoms in their normal state are electrically *neutral*, scientists reasoned, there must be something else in the atom that balances the negative electron. Otherwise, if matter had a negative charge, you would get a shock every time you touched anything.

This "something else," it seemed, would have to be some other sort of particle that has the same amount of charge as the electron but is positive in nature. Such reasoning led scientists to suppose that there were positive particles, which they called **protons**, in the atom. But then why could atoms give off only negative particles (electrons) and not positive ones (protons)? The British physicist J. J. Thomson suggested an answer. He thought of the atom as a sphere made up of positive electricity in which electrons were embedded (Figure 4.4). The English called this the *plum-pudding theory*, likening the electrons to the raisins in plum pudding. As Americans, we can find a more familiar analogy in a scoop of chocolate-chip ice cream. The chocolate chips represent the electrons. The ice cream represents a sea of positive electricity (composed of protons). Each element is different from the others because each has a different number of electrons and protons arranged in a different way—like different scoops of ice cream with different numbers of chocolate chips. For example, hydrogen atoms have one electron and one proton. Helium atoms have two electrons and two protons. Thomson seemed to have given scientists an idea they could agree on, but his theory still needed experimental proof.

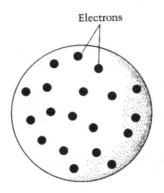

FIGURE 4.4 The Thomson model of the atom.

Lord Rutherford, a former student and colleague of Thomson, attempted to test Thomson's theory. He had been working with positive particles called **alpha particles** (α particles), which are actually helium atoms with their electrons removed. He reasoned that if he shot a stream of these positive particles through a gold foil, the neutral and symmetrical "plum-pudding" gold atoms would not greatly affect the path of the alpha particles. Most of the particles would move straight through the foil (Figure 4.5). Rutherford used gold foil in his experiment because it could be hammered into very thin sheets, making the thickness of the foil resemble a single layer of atoms. However, when Rutherford actually performed the experiment, he found that about one alpha particle in 20,000 ricocheted, or bounced back out of the foil! As Rutherford said, "It was almost as incredible as if you fired a fifteen-inch shell at a piece of tissue paper and it came back and hit you."

To explain these peculiar results, Rutherford devised a new model for the atom, called the nuclear model. We subscribe to this model today. Rutherford said that an atom must have a *center of positive charge*, where all the protons must be located. He called this center of positive charge the **nucleus** of the atom. Rutherford felt that the nucleus of the atom must be very small compared to the overall size of the atom. He also said that the nucleus must be where the mass of the atom is concentrated. Rutherford's reasoning was that some alpha particles were tremendously deflected (Figure 4.6) because they were repelled by a high concentration of positive charge with an immovable mass. He also decided that the electrons were located around the nucleus instead of in it and that the mass of the electrons was very small compared to the mass of the protons. His reasoning here was that for so many of the particles to go through the foil undeflected, the electrons had to be located in a relatively large space outside the nucleus but still within the atom.

Later experiments similar to Rutherford's showed that the diameter of the nucleus of an atom is approximately 10^{-13} cm, whereas the diameter of the whole atom is up to 100,000 times as great. Suppose we could expand an atom so that the nucleus was 1 millimeter in diameter (about the size of this dot: •). Then the nearest electron could be as far as 50 meters (half the length of a football field) away from the nucleus! And between the nucleus and the electrons is only vast empty space.

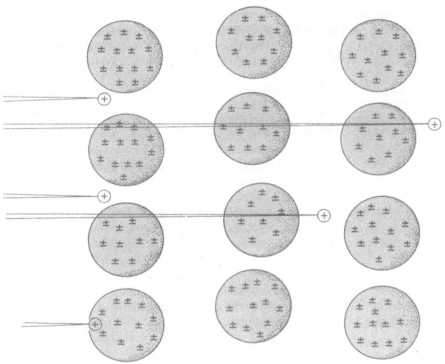

FIGURE 4.5 According to the Thomson model of the atom, positive particles ought to pass through a metal foil without being deflected.

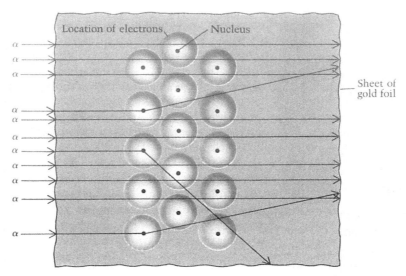

FIGURE 4.6 The actual way that positive particles pass through metal foil, as shown by Rutherford.

Rutherford's model of the atom, based as it was on new experimental evidence, made the Thomson model of the atom obsolete.

4.5 The Neutron

Thus, before 1920, scientists had experimentally confirmed the existence of two basic *subatomic* particles (particles that are fundamental constituents of an atom): electrons, which are negatively charged and have a very small mass, and protons, which are positively charged and contain most of the atom's mass. But a problem came up. Scientists had defined a unit of mass called the **atomic mass unit** (abbreviated amu). They knew that the proton had a mass of 1.0072766 amu and that the electron had a much smaller mass of 0.0005486 amu. These numbers all seemed fine at first. The mass of hydrogen—which is made up of one proton and one electron—was very close to the sum of the masses of a proton and an electron. (Remember that protons account for just about all of the atom's mass; when we are calculating the mass of an atom we can usually ignore the mass of the electrons.) The problem arose when scientists considered helium. The helium atom has two electrons and two protons, so the scientists thought it should have a mass of about 2 amu. But its mass is 4 amu. And when they looked at other elements, they were even more puzzled. Carbon has six electrons and six protons but a mass of 12 amu. Where was the extra mass?

To account for the extra mass in the atom, scientists proposed another particle, one that had no charge but had a mass equal to the proton's. In 1932, the British physicist James Chadwick (1891–1974) was able to prove the existence of a new subatomic particle called the **neutron**, which has a mass of 1.008665 amu. If this particle were also present in the atom, it would add to the mass but would not affect the charge (see Table 4.1).

TABLE 4.1 The Particles in the Atom

| | | | MASS | |
PARTICLE	SYMBOL	CHARGE	(amu)	(grams)
Proton	p	+1	1.0073	1.672×10^{-24}
Electron	e	−1	0.000549	9.107×10^{-28}
Neutron	n	0	1.0087	1.675×10^{-24}

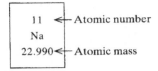

FIGURE 4.7 In the periodic table used in this text, an element's atomic number appears above its symbol.

Scientists could now picture helium as being composed of two electrons, two protons, and two neutrons, so they could understand helium's mass of 4 amu.

4.6 Atomic Number

The factor that makes one element different from another element is that atoms of different elements have different numbers of electrons and protons. *The number of protons in an atom of an element* is called the **atomic number** of that element and is symbolized by Z.

The atomic number of an element is also equal to the positive charge on the nucleus and to the number of electrons in the neutral atom of that element. On any periodic table there is a whole number identified with each element. This is the atomic number Z. In the periodic table inside the front cover, the atomic number is given above the symbol for each element. An atom of sodium (Na), which has the atomic number 11, has 11 electrons and 11 protons (Figure 4.7). An atom of magnesium (Mg), which has the atomic number 12, has 12 electrons and 12 protons; and an atom of uranium (U), which has the atomic number 92, has 92 electrons and 92 protons. Note that the elements are listed in the periodic table in the order of their atomic numbers.

4.7 Isotopes

If you look at the periodic table, you will see that the atomic masses of most elements are not whole numbers. For example, sodium (Na) has a mass of 22.990 amu, titanium (Ti) a mass of 47.90 amu, and rubidium (Rb) a mass of 85.468 amu. Why is this so? If each proton and each neutron gives a mass of almost exactly 1 amu to the atom, then the atomic masses of the elements certainly ought to be very close to whole numbers.

Investigation of this question led to the discovery that *each element can exist in more than one form*. When John Dalton said that the atoms of an element must be identical in size, mass, and shape, he did not know that *isotopes*, or different forms of an element, could exist. **Isotopes** are *forms of an element that have the same number of electrons and protons but different numbers of neutrons*.

Although the numbers of negative and positive charges of isotopes are identical, isotopes differ in mass. Hydrogen, for example, exists in three isotopic forms (Figure 4.8). The common form of hydrogen (called protium) contains one electron and one proton. Another form of hydrogen (called deuterium) contains one electron, one proton, and one neutron. A third form of hydrogen (called tritium) contains one electron, one proton, and two neutrons.

Note that all isotopes of an element have the same number of protons and thus have the same atomic number. However, it is possible to distinguish among isotopes by means of mass numbers. The **mass number** of an isotope is simply *the sum of the number of protons and the number of neutrons*.

A standard notation has been developed for giving the mass number and atomic number of an isotope. It consists of writing the symbol of the element with its mass number as a *superscript* (above the line) and its atomic number as a *subscript* (below the line). Both numbers are usually written to the left of the symbol for the element.

$$\text{Mass number} \rightarrow {}^{16}_{8}\text{O} \quad {}^{12}_{6}\text{C} \quad {}^{35}_{17}\text{Cl}$$
$$\text{Atomic number} \rightarrow$$

With this notation we can immediately tell the composition of the isotope in terms of protons, electrons, and neutrons. The number of protons (or, in the neutral atom, the number of electrons) is simply the atomic number of the element. The number of neutrons is the difference between the mass number and the atomic number. This notation allows us to distinguish among the isotopes of an element.

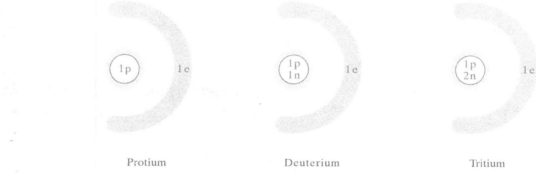

FIGURE 4.8 Isotopes of hydrogen.

EXAMPLE 4.1

Give the number of electrons, protons, and neutrons in each of the following neutral isotopes:
(a) $^{238}_{92}U$ (b) $^{235}_{92}U$ (c) $^{84}_{36}Kr$

Solution

The atomic number (subscript) gives the number of electrons in the neutral atom as well as the number of protons. Subtracting the atomic number from the mass number (superscript) gives the number of neutrons.

(a) Because the atomic number is 92, there are 92 electrons and 92 protons. And subtracting the atomic number from the mass number gives the number of neutrons: 238 − 92 = 146 neutrons. So there are 92 electrons, 92 protons, and 146 neutrons in an atom of $^{238}_{92}U$.

(b) Again there are 92 electrons and 92 protons. And 235 − 92 = 143 neutrons. So there are 92 electrons, 92 protons, and 143 neutrons in an atom of $^{235}_{92}U$. Note that $^{238}_{92}U$ and $^{235}_{92}U$ are an *isotope pair*.

(c) Here we have 36 electrons and 36 protons. And 84 − 36 = 48 neutrons. So there are 36 electrons, 36 protons, and 48 neutrons in an atom of $^{84}_{36}Kr$.

Practice Exercise 4.1

Give the number of electrons, protons, and neutrons in each of the following isotopes: (a) $^{28}_{14}Si$ (b) $^{31}_{15}P$ (c) $^{6}_{3}Li$
The following notations can be used to distinguish among the three isotopes of hydrogen:

$$^{1}_{1}H \qquad ^{2}_{1}H \qquad ^{3}_{1}H$$
Protium   Deuterium   Tritium

The deuterium and tritium forms of hydrogen are not as common as the protium form. If a random sample of all hydrogen atoms found on the earth were taken, it would be found to consist of 99.985% protium atoms, 0.015% deuterium atoms, and a vanishingly small amount of tritium atoms. The atomic mass given in the periodic table is the *weighted average of the atomic masses* of all three isotopes. *When calculating the average atomic mass for any element, we must take into account the relative abundance of each isotope of that element and its exact isotopic mass.* For example, the exact isotopic mass of $^{1}_{1}H$ is 1.0078 amu, and the exact isotopic mass of $^{2}_{1}H$ is 2.0141. It is for this reason that the atomic mass of hydrogen shown in the periodic table is 1.008 amu. This is nearly the same as the mass of protium, the most common isotope of hydrogen. Table 4.2 gives a breakdown of the isotopes of the first fifteen elements and their isotopic masses. This averaging process also explains why the atomic masses of the elements in the periodic table are not whole numbers.

If we know the percentage abundance of each isotope of an element, we can calculate the average atomic mass of the element. The following example shows how to do this. In the example we use the exact isotopic mass.

66 ESSENTIAL CONCEPTS OF CHEMISTRY

EXAMPLE 4.2

Using Table 4.2, calculate the atomic mass of boron.

TABLE 4.2 Naturally Occurring Isotopes of the First Fifteen Elements

NAME	SYMBOL	ATOMIC NUMBER	MASS NUMBER	ISOTOPIC MASS	PERCENTAGE NATURAL ABUNDANCE
Hydrogen-1	$^{1}_{1}H$	1	1	1.007825	99.985
Hydrogen-2	$^{2}_{1}H$	1	2	2.01410	0.015
Hydrogen-3	$^{3}_{1}H$	1	3	3.01605	Negligible
Helium-3	$^{3}_{2}He$	2	3	3.016	0.00013
Helium-4	$^{4}_{2}He$	2	4	4.003	99.99987
Lithium-6	$^{6}_{3}Li$	3	6	6.015	7.42
Lithium-7	$^{7}_{3}Li$	3	7	7.016	92.58
Beryllium-9	$^{9}_{4}Be$	4	9	9.012	100
Boron-10	$^{10}_{5}B$	5	10	10.013	19.6
Boron-11	$^{11}_{5}B$	5	11	11.009	80.4
Carbon-12	$^{12}_{6}C$	6	12	12.0000	98.89
Carbon-13	$^{13}_{6}C$	6	13	13.003	1.11
Nitrogen-14	$^{14}_{7}N$	7	14	14.003	99.63
Nitrogen-15	$^{15}_{7}N$	7	15	15.000	0.37
Oxygen-16	$^{16}_{8}O$	8	16	15.995	99.759
Oxygen-17	$^{17}_{8}O$	8	17	16.999	0.037
Oxygen-18	$^{18}_{8}O$	8	18	17.999	0.204
Fluorine-19	$^{19}_{9}F$	9	19	18.998	100
Neon-20	$^{20}_{10}Ne$	10	20	19.992	90.92
Neon-21	$^{21}_{10}Ne$	10	21	20.994	0.257
Neon-22	$^{22}_{10}Ne$	10	22	21.991	8.82
Sodium-23	$^{23}_{11}Na$	11	23	22.9898	100
Magnesium-24	$^{24}_{12}Mg$	12	24	23.9850	78.70
Magnesium-25	$^{25}_{12}Mg$	12	25	24.9858	10.13
Magnesium-26	$^{26}_{12}Mg$	12	26	25.9826	11.17
Aluminum-27	$^{27}_{13}Al$	13	27	26.9815	100
Silicon-28	$^{28}_{14}Si$	14	28	27.9769	92.21
Silicon-29	$^{29}_{14}Si$	14	29	28.9765	4.70
Silicon-30	$^{30}_{14}Si$	14	30	29.9738	3.09
Phosphorus-31	$^{31}_{15}P$	15	31	30.9738	100

Solution

From Table 4.2, the percentage abundance of ^{10}B is 19.6%, and its exact isotopic mass is 10.013 amu. That of ^{11}B is 80.4%, and its exact isotopic mass is 11.009 amu. Therefore the atomic mass of

$$B = (10.013 \text{ amu})(0.196) + (11.009 \text{ amu})(0.804)$$
$$= 1.96 \text{ amu} + 8.85 \text{ amu} = 10.81 \text{ amu}$$

For a review of changing percentages to decimal equivalents, see Example A.19 in Appendix A.

Practice Exercise 4.2

Using Table 4.2, calculate the atomic mass of lithium.

The concept of isotopes with different abundances allows us to solve two other kinds of quantitative problems. These are illustrated by the next two examples.

EXAMPLE 4.3

A hypothetical element Q exists in two isotopic forms, ^{270}Q and ^{280}Q. The atomic mass of Q is 276 amu. What is the percentage abundance of each isotope? (Assume that the exact isotopic mass is the same as the mass number for each isotope.)

Solution

UNDERSTAND THE PROBLEM

If we know the percentage abundance of each isotope of an element, we can calculate the average atomic mass of the element. This problem requires that we determine the percentage abundance of each element when given the atomic mass.

DEVISE A PLAN

We will use algebra and the ability to solve an algebraic equation. We will let x equal the fraction of ^{270}Q and $(1 - x)$ equal the fraction of ^{280}Q. Then we have

$$270x + 280(1 - x) = 276$$

CARRY OUT THE PLAN

$$x = 0.40$$
$$1 - x = 0.60$$

In other words, $4\bar{0}$% of the element is ^{270}Q and $6\bar{0}$% of the element is ^{280}Q.

LOOK BACK

We ask if the solution makes sense. Using estimation we can reason that if 50% of the element were ^{270}Q and 50% were ^{280}Q, then the atomic mass of the element should be 275. Since it is more than 275, there should be a slightly greater amount of ^{280}Q present, and that is the case with this solution.

Practice Exercise 4.3

A hypothetical element Y exists in two isotopic forms, ^{100}Y and ^{110}Y. The atomic mass of Y is 108. What is the percentage abundance of each isotope? (Assume that the exact isotopic mass is the same as the mass number for each isotope.)

EXAMPLE 4.4

Using Table 4.2, find how many ^{17}O atoms there are in a sample containing 1,000,000 oxygen atoms.

Solution

From Table 4.2, we find that the natural abundance of ^{17}O is 0.037%. Therefore the number of ^{17}O atoms in 1,000,000 O atoms equals

$$(0.00037)(1,000,000) = 370 \text{ atoms of } ^{17}O$$

Practice Exercise 4.4

Using Table 4.2, find how many ^{20}Ne atoms there are in a sample containing 1,000,000 neon atoms.

4.8 What Next for the Atom?

We have traced the development of a model for the atom from the time of Dalton, who believed that the atom was an indivisible unit of matter, to the time of Rutherford, who suggested that the atom has a nucleus composed of protons and neutrons and an outer part for the electrons. But many questions were still left unanswered. How were the electrons arranged in an atom? Where were the electrons? How were they moving? How fast were they going? Could they account for the formation of compounds? Why didn't these negative electrons "fall into" the positive nucleus of the atom? After all, opposite charges attract! We will answer these questions in Chapter 5.

SUMMARY

John Dalton's atomic theory was an attempt to explain all the different forms of matter. He theorized that (1) the atom is indivisible; (2) all atoms of the same element have the same mass, size, and form; (3) all matter is composed of combinations of atoms; and (4) atoms of different elements are different. Although the first two of these ideas were later shown to be incorrect, most of Dalton's theory is the basis for what is still believed today. Dalton's model of the atom was modified as various subatomic particles—protons, electrons, and neutrons—were discovered. Thomson's theory of the atom followed his "discovery" of electrons. This British physicist theorized that the atom is a sphere made up of positive electricity in which negatively charged electrons are embedded. Lord Rutherford, a colleague of Thomson, tested Thomson's theory with an experiment in which a stream of positive (alpha) particles was supposed to pass straight through the atoms in a gold foil with only a few particles being slightly deflected. Instead, some of the alpha particles bounced back. To explain this result, Rutherford theorized that an atom has a relatively very small center where all the positively charged protons and most of the atomic mass are located. Electrons, he postulated, are located outside the nucleus, perhaps relatively far from it.

One difficulty arises from the two-particle atomic model. For most atoms, the masses of all the protons and electrons did not account for the total mass of the atom. In 1932 James Chadwick was able to prove the existence of a new subatomic particle called the neutron. The neutron has about the same mass as a proton but carries no charge. With neutrons, scientists were able to account for the total mass of atoms.

The number of protons in the atom of an element is called the atomic number of that element. The sum of the number of protons and the number of neutrons is called the mass number of the element. All the atoms of any element have the same atomic number, but they may have different mass numbers. Two or more atoms with the same atomic number but different mass numbers are called isotopes.

KEY TERMS

alpha particle (**4.4**)
anode (**4.3**)
atomic mass unit (**4.5**)
atomic number (**4.6**)
cathode (**4.3**)
electron (**4.3**)

isotope (**4.7**)
mass number (**4.7**)
neutron (**4.5**)
nucleus (**4.4**)
proton (**4.4**)
vacuum (**4.3**)

CHAPTER 4: ATOMIC THEORY, PART 1   69

SELF-TEST EXERCISES

LEARNING GOAL 1

Atomic Theory and Models

1. Explain the idea of a model. Give an example of a physical model and an intellectual model.
◀ 2. The elements carbon and oxygen form two different compounds, carbon dioxide, which is a product of animal respiration, and carbon monoxide, which is a product of incomplete combustion when gasoline is burned. Discuss how Dalton's atomic theory explains these facts.
3. List the points in Dalton's atomic theory. Explain which ones had to be corrected after later experiments were performed.
◀ 4. Compare Dalton's atomic theory to the ideas set forth by Democritus in 420 B.C.
5. Contrast the Thomson model and the Rutherford model of the atom.
6. Give a general outline of the construction of a Crookes tube.
◀ 7. Explain how a television set is similar to a Crookes tube. (You may want to check an outside reference source for help in answering this question.)
8. Discuss the experiment that was conducted to test the wave and particle theories of cathode rays.

LEARNING GOAL 2

Mass and Charge of Electrons, Protons, and Neutrons

9. How many electrons would it take to equal the mass of one proton? Use Table 4.1 to obtain the masses of these subatomic particles.
10. Which of the three subatomic particles has the greatest mass? Which has the least mass?
11. Name the three major subatomic particles. Give their charges and approximate atomic masses (in amu).
◀ 12. Compare the charges of the three subatomic particles and briefly explain how each charge was determined.

LEARNING GOAL 3

Number of Protons, Electrons, and Neutrons

13. Find the number of protons (p), electrons (e), and neutrons (n) in neutral atoms of the following:
   (a) $^{244}_{94}Pu$ (b) $^{48}_{22}Ti$ (c) $^{262}_{103}Lr$

14. Find the number of protons (p), electrons (e), and neutrons (n) in neutral atoms of the following:
   (a) $^{75}_{33}As$ (b) $^{266}_{109}Mt$ (c) $^{222}_{86}Rn$

15. Find the number of protons, electrons, and neutrons in neutral atoms of the following:
   (a) $^{16}_{8}O$ (b) $^{17}_{8}O$ (c) $^{18}_{8}O$ (d) $^{20}_{10}Ne$ (e) $^{21}_{10}Ne$ (f) $^{22}_{10}Ne$

16. Find the number of protons, electrons, and neutrons in neutral atoms of the following:
   (a) $^{31}_{15}P$ (b) $^{22}_{10}Ne$ (c) $^{24}_{12}Mg$

17. From the information that follows, write the isotopic notation for each atom, then decide which one is not an isotope of the other two atoms:
   (a) 8p, 8e, 8n   (b) 7p, 7e, 8n   (c) 8p, 8e, 9n

◀ 18. Which of the following describes an atom that is not an isotope of the other two?
   (a) 12p, 12e, 14n   (b) 12p, 12e, 12n   (c) 13p, 13e, 12n

70 ESSENTIAL CONCEPTS OF CHEMISTRY

19. Find the number of protons, electrons, and neutrons in neutral atoms of the following:

    (a) $^{57}_{25}Mn$ (b) $^{60}_{27}Co$ (c) $^{80}_{36}Kr$ (d) $^{128}_{52}Te$

20. Determine the number of protons, electrons, and neutrons in neutral atoms of the following hypothetical isotopes:

    (a) $^{35}_{16}X$ (b) $^{300}_{90}Y$

◀ 21. The alchemists wanted to turn lead into gold. Suppose that you had the isotope of lead $^{208}_{82}Pb$, and you wanted to turn it into the isotope of gold $^{197}_{79}Au$. How many protons, electrons, and neutrons would you have to remove from the lead atom to turn it into this gold atom?

22. Which of the following describes an atom that is not an isotope of the other two?

    (a) 14p, 14e, 15n (b) 15p, 15e, 14n (c) 15p, 15e, 15n

◀ 23. Fill in the blanks in the following table. (Use the periodic table when necessary.)

SYMBOL	PROTONS	ELECTRONS	NEUTRONS	MASS NUMBER	ATOMIC NUMBER
$^{174}_{70}Yb$	70	70	104	174	70
$^{141}_{59}R$	59	59	82	141	59
$^{104}_{44}U$	44	44	60	104	44
$^{45}_{21}C$	21	21	24	45	21
$^{50}_{22}I$	22	22	28	50	22
$^{25}_{12}Mg$	12	12	13	25	12

◀ 24. State the number of protons, electrons, and neutrons in $^{42}_{20}Ca$.

LEARNING GOAL 4

Definition of an Isotope

◀ 25. (a) Can two different isotopes of a particular element have the same mass number? Explain.
    (b) Can atoms of two different elements have the same mass number? Explain.

26. Atoms of an element that have the same number of electrons and protons but different numbers of neutrons are called ___Isotopos___.

27. Isotopes are atoms with the same number of ___Protons___ and ___Electrons___ but different numbers of ___neutron___. (Place the words *protons, electrons*, and *neutrons* in the proper places.)

28. (a) The number of protons in an atom is called the ___atomic number___ of that element.
    (b) The sum of the numbers of protons and neutrons in an isotope is called the ___mass number___ of that isotope.

LEARNING GOAL 5

Standard Notation for Mass Number and Atomic Number

29. Write the standard isotopic notation for the following atoms:

    (a) 80p, 80e, 120n (b) 7p, 7e, 8n (c) 13p, 13e, 14n (d) 107p, 107e, 155n

◀ 30. Write the standard isotopic notation for the following atoms:

    (a) 3p, 3e, 3n (b) 9p, 9e, 10n (c) 14p, 14e, 16n

CHAPTER 4: ATOMIC THEORY, PART 1

LEARNING GOAL 6

Average Atomic Mass of an Element

◀ *31*. The element gallium (Ga) exists in two isotopic forms with the following abundances: 60.16% ^{69}Ga and 39.84% ^{71}Ga. Calculate the atomic mass of gallium. (The exact isotopic mass of ^{69}Ga is 68.9257. The exact isotopic mass of ^{71}Ga is 70.9249.)

32. The element carbon (C) exists in two isotopic forms with the following abundances: 98.89% ^{12}C and 1.11% ^{13}C. Calculate the atomic mass of carbon.

33. The element chlorine (Cl) exists naturally in two isotopic forms with the following abundances: 75.53% ^{35}Cl and 24.47% ^{37}Cl. Calculate the atomic mass of chlorine. (The exact isotopic mass of ^{35}Cl is 34.9689. The exact isotopic mass of ^{37}Cl is 36.9659.)

34. The element magnesium (Mg) exists in three isotopic forms with the following abundances: 78.70% ^{24}Mg, 10.13% ^{25}Mg, and 11.17% ^{26}Mg. Calculate the atomic mass of magnesium. Use Table 4.2.

35. The element chromium (Cr) exists naturally in four isotopic forms with the following abundances: 4.31% ^{50}Cr, 83.76% ^{52}Cr, 9.55% ^{53}Cr, and 2.38% ^{54}Cr. Calculate the atomic mass of chromium. (The exact isotopic mass of ^{50}Cr is 49.9461. The exact isotopic mass of ^{52}Cr is 51.9405. The exact isotopic mass of ^{53}Cr is 52.9407. The exact isotopic mass of ^{54}Cr is 53.9389.)

36. The element copper exists in two isotopic forms with the following abundances: 69.09% ^{63}Cu and 30.91% ^{65}Cu. Calculate the atomic mass of copper. (The exact isotopic mass of ^{63}Cu is 62.9296. The exact isotopic mass of ^{65}Cu is 64.9278.)

LEARNING GOAL 7

Percentage Abundance of Isotopes

37. Bromine exists in two isotopic forms, ^{79}Br and ^{81}Br. The atomic mass of bromine is 79.90 amu. Calculate the percentage abundance of each isotope. (The exact isotopic mass of ^{79}Br is 78.9183. The exact isotopic mass of ^{81}Br is 80.9163.)

◀ 38. Boron (B) exists in two isotopic forms, ^{11}B and ^{10}B. The atomic mass of boron is 10.81 amu. Calculate the percentage abundance of each isotope. Use Table 4.2.

39. The element antimony occurs naturally in two isotopic forms, ^{121}Sb and ^{123}Sb. The atomic mass of antimony is 121.75 amu. Calculate the percentage abundance of each isotope. (The exact isotopic mass of ^{121}Sb is 120.9038. The exact isotopic mass of ^{123}Sb is 122.9041.)

◀ 40. Chlorine exists in two isotopic forms, ^{35}Cl and ^{37}Cl. The atomic mass of chlorine is 35.453 amu. Calculate the percentage abundance of each isotope. (The exact isotopic mass of ^{35}Cl is 34.9689 and that of ^{37}Cl is 36.9659.)

LEARNING GOAL 8

Table of Relative Abundances of Isotopes

41. Using Table 4.2, determine how many ^{22}Ne atoms you will find in 1.00×10^9 neon atoms.

42. Using Table 4.2, determine how many ^{29}Si atoms you will find in 5.00×10^6 silicon atoms.

43. Using Table 4.2, determine how many atoms of oxygen you would need to get 7,400 atoms of ^{17}O.

44. Using Table 4.2, determine how many atoms of oxygen you would need to get 102 atoms of the isotope ^{18}O.

45. Using Table 4.2, determine how many ^{6}Li atoms you would find in 100,000,000 lithium atoms.

46. Using Table 4.2, determine how many ^{7}Li atoms you would find in 1,000 lithium atoms.

◀ *47*. Using Table 4.2, determine how many ^{26}Mg atoms you would find in 5.00×10^6 Mg atoms.

48. Using Table 4.2, determine how many ^{14}N atoms you would find in 1.00×10^{14} nitrogen atoms.

49. Using Table 4.2, determine how many ^{13}C atoms you would find in 2.00×10^4 C atoms.

◀ 50. Using Table 4.2, determine how many atoms of ^{22}Ne you would find in a sample of neon that contains 1×10^6 neon atoms.

72 ESSENTIAL CONCEPTS OF CHEMISTRY

51. Using Table 4.2, determine how many ^{15}N atoms you would find in 1.0×10^{15} N atoms.
52. Using Table 4.2, determine how many ^{6}Li atoms you would find in 1,000 atoms of Li.
53. Using Table 4.2, determines how many atoms of hydrogen you would need to get 150,000 atoms of ^{2}H.
54. Using Table 4.2, determine how many ^{14}N atoms you would find in exactly 100,000,000 atoms of N.

EXTRA EXERCISES

55. The existence of isotopes became known to scientists around the year 1900, about 100 years after Dalton first proposed his atomic theory. The proof that isotopes existed were associated more with physics experiments than with chemistry experiments. What does this tell you about the chemical properties of isotopes? Explain.
56. Consider the following unknown atoms.

$$^{200}_{84}A \quad ^{208}_{82}B \quad ^{222}_{86}C \quad ^{184}_{74}D$$

   (a) Which atom has the most electrons in the neutral state?
   (b) Which atom has the fewest neutrons?
   (c) What is the mass number of element C?
   (d) Which atom has the fewest protons?
   (e) Identify element A, using a periodic table.

57. What is the relationship between the atomic number and the number of electrons in a neutral atom?
58. A student produces a beam of rays and can't decide whether these rays are cathode rays or alpha particles. Explain a simple experiment that the student might perform to determine what these rays are.
59. What experiment gave the proof that atoms of all elements contain electrons?
60. Indicate the number of protons, electrons, and neutrons that each of the following atoms contains:

   (a) $^{13}_{6}C$ (b) $^{10}_{5}B$ (c) $^{14}_{7}N$

61. What was the problem with Dalton's statement that all the atoms of a particular element are identical in every respect?
62. How many neon-21 atoms are there in 1 million neon atoms?
63. How many electrons are needed to equal the mass of one proton?
64. Can the atomic number of an element exceed its mass number? Explain.
65. Make a list of some of our modern conveniences that would not function if atoms were indeed indivisible.
66. Referring to Table 4.2, determine all the possible molecular masses for H_2O, using the various isotopes of hydrogen and oxygen.

CHAPTER 5

ATOMIC THEORY, PART 2
Energy Levels and the Bohr Atom

Let's journey back to the beginning of the nineteenth century. It is the year 1814 and the German optician Joseph von Fraunhofer is testing a prism of excellent quality he has just manufactured. Fraunhofer allows light to pass through a slit and then through his triangular prism. When the light is projected onto a white screen, it forms a series of colored lines that are separated by dark lines. Fraunhofer counts almost 600 lines and notes the position of each relative to the others.

It is 30 years later: Robert Kirchhoff and Robert Bunsen use Fraunhofer's technique to show that when specific elements are heated in a flame they produce a unique series of lines—*a line spectrum*. The device they've developed is called a *spectroscope*, used by Bunsen and Kirchhoff for "fingerprinting" elements by the light patterns they produce. Of even more importance, Bunsen and Kirchhoff are able to work backward and deduce the elements in an unknown substance from its line spectrum.

The spectroscope is also used to identify new elements. In 1860 Bunsen and Kirchhoff test a mineral that produces spectral lines different from all known elements. Additional chemical testing shows that this new element is similar to sodium and potassium. They name the element *cesium*, from the Latin *sky blue*, which is the most prominent line in its spectrum.

Other chemists also use this technique and discover new elements. In 1875, the French chemist Emile Lecoq de Boisbaudran discovers a new element contained in zinc ore. He names the element gallium (see Chapter 6). Chemists at the time are not sure why each element has its own unique line spectrum, but many chemists feel that it has something to do with the internal structure of the atom. The nineteenth century, which began with Dalton's atomic theory of an indivisible atom with no internal structure, is about to undergo a major refinement.

LEARNING GOALS

After you've studied this chapter, you should be able to:

1. Discuss the difference between a line spectrum and a continuous spectrum.
2. Describe the Bohr model of the atom.
3. Explain the significance of shells and energy levels.
4. Differentiate between the ground state and an excited state of an atom.
5. Describe the quantum mechanical model of the atom.
6. Explain the significance of energy sublevels.

74 ESSENTIAL CONCEPTS OF CHEMISTRY

7. Define and discuss electron orbitals.

8. Write the electron configurations for various elements using the filling-order diagram.

9. Explain the importance of electron configuration.

INTRODUCTION

By the early 1900s, scientists had developed a model of the atom with protons and neutrons in the center, or nucleus, and electrons outside the nucleus. But where, exactly, were the electrons? And why weren't the negative electrons attracted into the nucleus by the positive protons? In 1913 the Danish physicist Niels Bohr proposed a model to answer these and other questions. His model has since been modified in the light of later observations and research, but it still provides an excellent basis for understanding atomic structure. In this chapter, we discuss the observations that led Bohr to propose his model, as well as the model itself.

5.1 Spectra

In the mid-1600s the English scientist, astronomer, and mathematician Sir Isaac Newton allowed a beam of sunlight to pass through a prism. He found that the prism separated the light into a series of different colors, which we call the *visible spectrum* (Figure 5.1). His results were published in *Opticks* in 1704. Later experiments showed that a specific amount of energy is associated with each color of light, because light is a form of energy.

In the 1850s the German physicist Gustav Robert Kirchhoff and the German chemist Robert Bunsen developed an instrument called the spectroscope. Its main parts were a heat source (the Bunsen burner), a prism, and a telescope (Figure 5.2). Kirchhoff and Bunsen heated samples of various elements. In each case, as an element became hotter and hotter and began to glow, it produced light of its own characteristic color. When the light from the heated element was allowed to pass through the prism, it separated into a series of bright, distinct lines of various colors.

Some light from the heated element passes through a collimator, which is simply a brass tube with a narrow slit at both ends. The collimator directs a fine beam of light onto the prism. When light passes through the prism, it is broken into its different parts—a spectrum—and observers can see the spectrum through a telescope.

The spectrum produced by sunlight is called a **continuous spectrum**, because one color merges into the next without any gaps or missing colors. The spectrum that Bunsen and Kirchhoff found is called a **line spectrum** (Figure 5.3), because it is a series of bright lines separated by dark bands. Each of the elements produces its own unique line spectrum.

Scientists find these line spectra very useful for identifying elements in unknown samples. Analysis of line spectra has enabled scientists to identify the elements that compose the sun, the stars, and other extraterrestrial bodies. This technique is also used in chemical laboratories to identify the elements present in unknown samples.

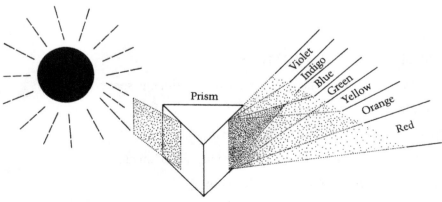

FIGURE 5.1 Sunlight passing through a prism.

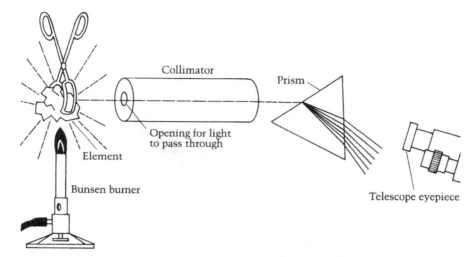

FIGURE 5.2 A Spectroscope works as follows: An element is heated in the flame of a Bunsen burner.

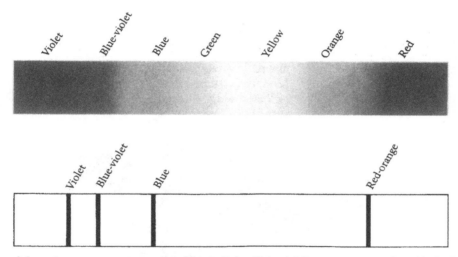

FIGURE 5.3 (Above) A continuous spectrum produced by sunlight. (Below) A line spectrum produced by hydrogen.

The visible spectrum produced by sunlight is only a small part of what is called the **electromagnetic spectrum** (Figure 5.4). The existence of electromagnetic waves was predicted by the British physicist James Clerk Maxwell in 1864. He assumed that electric and magnetic fields acting together produced radiant energy. He believed that radiant energy took the form of electromagnetic waves and that visible light was only one form of electromagnetic wave (Figure 5.5). We distinguish one form of electromagnetic radiation from another in terms of *wavelength*. The wavelength is the distance from one point on a wave to the corresponding point on the next wave. For electromagnetic waves, the shorter the wavelength, the higher the energy. Included in the electromagnetic spectrum were both visible waves and invisible waves, each associated with a different amount of energy. For example, x rays are invisible waves with enough energy to pass through skin but not bones. Microwaves are also invisible waves, but they have enough energy to heat food. Electromagnetic waves can travel through the vacuum of space, as well as through air and water.

Maxwell's predictions were verified by the German physicist Heinrich R Hertz in 1887. By varying an electric charge, he produced electromagnetic waves that were longer than visible light waves. This discovery eventually led to the development of radio and television.

The full impact of these experiments with light was not felt until 1913, when it was discovered that the existence of line spectra held the key to the structure of the atom.

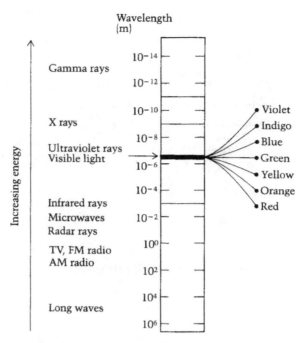

FIGURE 5.4 The electromagnetic spectrum.

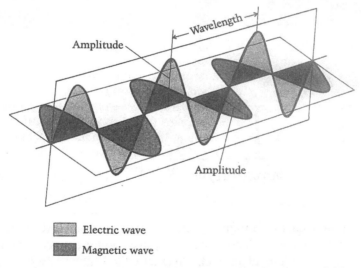

FIGURE 5.5 Light waves are one form of electromagnetic wave. They consist of an electric field and a magnetic field vibrating at right angles to each other.

5.2 Light As Energy

We have noted that light is a form of energy and that energy can be converted from one form to another. For example, a photoelectric cell converts light energy to electrical energy (in the form of electric current). Energy may also be absorbed and then re-emitted. For example, the luminous dial on a watch absorbs energy when it is exposed to a strong light. When it is removed from the light source, the dial re-emits this energy in the form of light: It glows.

In the experiments of Kirchhoff and Bunsen, each element was made to absorb energy in the form of heat. When the element got hot enough, it emitted this energy as light. Scientists still wondered why the elements emitted line spectra rather than continuous spectra, and they wondered what subatomic particle caused the line spectra.

In 1900, the German physicist Max Planck suggested that although light seems continuous, it is not. He said that electromagnetic radiation can be thought of as a stream of minute packets of energy called **quanta** (singular, *quantum*). The word quanta comes from the Latin, *quantus,* meaning "how much." According to Planck's quantum hypothesis, these bundles of energy (also referred to as *photons*) are the smallest packets of energy associated with a particular form of electromagnetic radiation, and each contains a specific, fixed amount of energy and has its own wavelength. Niels Bohr used the concept of quanta to explain line spectra of elements.

5.3 The Bohr Atom

In 1913 Bohr suggested that electrons are responsible for line spectra. He proposed the idea that electrons travel around the nucleus of the atom in **shells**, which were described as imaginary spherical surfaces roughly concentric with the nucleus. Bohr also proposed that each shell is associated with a particular **energy level** and that shells farther away from the nucleus are associated with higher energy levels. According to the Bohr model, electrons whirl around the nucleus in specific shells, just as planets travel in specific orbits around the sun. But electrons, unlike planets, are able to jump from one shell (at one energy level) to another (at a higher energy level) when they absorb enough energy from an outside source to do so (Figure 5.6). They can also fall back to their original shells by emitting this energy.

Look at the hydrogen atom in Figure 5.7. The single electron is moving around the nucleus in the shell closest to the nucleus—the shell associated with the lowest energy. In this situation, the electron is said to be in its *lowest energy level*, and the atom is in its **ground state**. As the electron moves, it neither gains nor loses

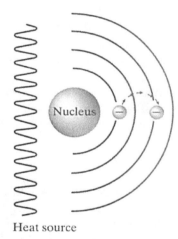

FIGURE 5.6 In the Bohr model of the atom, electrons "jump" from one shell to another when an outside source gives them enough energy to do so.

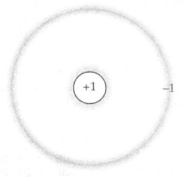

FIGURE 5.7 A hydrogen atom in its ground state.

FIGURE 5.8 An excited hydrogen atom. Note that the electron is in energy level 3, and thus the atom is in an excited state.

energy. Its energy of motion exactly counterbalances the attraction of the nucleus. For this reason, it is not pulled into the nucleus.

Now suppose we heat our hydrogen atom. According to Bohr, the electron will jump to a higher-energy shell, farther from the nucleus, when it has absorbed a certain amount of heat energy. It must absorb the entire amount before it can leave the shell it occupies. It *cannot* absorb some energy, move out toward the next shell, absorb more energy and move closer to that shell, and so on until it reaches the next shell. It must absorb all the energy it needs and then instantaneously jump to the next shell. It may then absorb enough energy at that level to jump to another level (Figure 5.8). Or it may absorb enough energy to jump two or three levels at a time. But in each case it will absorb the energy and jump instantaneously. When an electron is at an energy level above its lowest level, the atom is said to be in an **excited state**. (An electron can also absorb so much energy that it escapes completely from the atom. This process, called *ionization*, is discussed in Chapter 6.)

Eventually, the electron emits some or all of the energy that was absorbed, and it falls back to a lower energy level. The energy is emitted in quanta, the difference in energy between the higher-energy shell where the electron was and the lower-energy shell where it lands. If the electron spiraled down without a level-to-level requirement, it would emit a continuous stream of energy. But it moves only from one level to another and emits its energy in specific bursts that may be seen as light energy. It is this light, being emitted by electrons falling back to lower energy levels, that produces the unique line spectrum of hydrogen (or of any other element).

It is difficult to visualize electrons moving instantaneously from one energy level to another. Picture a marble rolling down a flight of stairs. The marble can rest on each step (shell), but it can never stop part of the way between steps. And, as the marble falls from step to step, it also loses (emits) energy in bursts.

5.4 Quantum Mechanical Model

The Bohr model of the atom contributed a great deal to our knowledge of atomic structure. The concept of energy levels for electrons, as well as the relationship between the observed spectral lines and the shifting of electrons from one energy level to another for the hydrogen atom were important to the development of atomic theory. Unfortunately, Bohr's calculations did not hold true for heavier atoms. Scientists continued to search for information to understand atomic structure.

In the decades that followed, revolutionary discoveries were made in physics. These discoveries had an impact on chemical knowledge. Experimental evidence showed that the laws of physics that applied to large objects did not apply to very small objects. In 1924, the French physicist Louis de Broglie explained that electrons in motion had properties of waves and also had mass. Two years later the Austrian physicist Erwin Schrödinger introduced a complicated mathematical equation that described electrons as having dual characteristics. He showed that in some ways the properties of electrons are best described in terms of waves, and in other ways their properties are most similar to those of particles having mass. Physicists formalized laws that could be used to characterize the motion of electrons in what we call **quantum mechanics**.

Quantum mechanical concepts have replaced the ideas Bohr presented in his model, where electrons are thought to revolve about the nucleus in planetary orbits. In this model of the atom, we think of an electron *not* as being in a carefully defined orbit, but instead as occupying a volume of space. This volume of space is called an **orbital**. The orbital represents a region of space in which the electron can be found with a 95% probability.

FIGURE 5.9 Names of the energy levels in atoms.

5.5 Energy Levels of Electrons

As we learned from both the Bohr model and the quantum mechanical model, we are most likely to find an electron at a certain specified distance from the nucleus, called an **energy level** or electron **shell**. The electrons in each shell have *approximately* the same amount of energy. We will use the terms *shell* and *energy level* interchangeably.

There are two ways to identify energy levels. One method uses the letters of the alphabet beginning with the letter *K* and continuing sequentially (Figure 5.9). The farther away a level is from the nucleus, the higher the energy of an electron at that level. The *K* energy level is associated with electrons at the lowest level, whereas the *Q* energy level holds the electrons of the highest level known. The newer method of identifying electron shells uses the *principal quantum number, n*, to designate the energy levels allowed for electrons. The shell having the lowest energy level is designated $n = 1$, then the next higher level is $n = 2$, and so on, until we reach $n = 7$. We know of no atom in its ground state having electrons in an energy level higher than the seventh. Notice that the *K* energy level and the $n = 1$ energy level are exactly the same, as are the *L* energy level and the $n = 2$ energy level, and so on.

Each energy level can hold only a certain number of electrons at any one time. Table 5.1 gives the *theoretical maximum number* of electrons for each level. The lowest energy level ($n = 1$) holds a maximum of 2 electrons, while the second, third, fourth, and fifth energy levels hold 8, 18, 32, and 50 electrons, respectively. The maximum number of electrons in each level can also be determined using a simple mathematical calculation.

Maximum number of electrons for an energy level $= 2n^2$

EXAMPLE 5.1

Calculate the maximum number of electrons permitted in the following: (a) *M* shell ($n = 3$) (b) *P* shell ($n = 6$)

TABLE 5.1 Theoretical Maximum Number of Electrons for Each Energy Level

ENERGY LEVEL		MAXIMUM NUMBER OF ELECTRONS
Letter designation	*Number*	
K	1	2
L	2	8
M	3	18
N	4	32
O	5	50
P	6	72
Q	7	98

80 ESSENTIAL CONCEPTS OF CHEMISTRY

Solution

UNDERSTAND THE PROBLEM

Using the number and letter designations, calculate the electron capacity for each shell.

DEVISE A PLAN

Use the mathematical equation $2n^2$.

CARRY OUT THE PLAN

(a) For $n = 3$, $2(3)^2 = 2(9) = 18$
(b) For $n = 6$, $2(6)^2 = 2(36) = 72$

LOOK BACK

Check the mathematics and determine if these solutions make sense.

Practice Exercise 5.1

Calculate the maximum number of electrons permitted in each of the following: (a) N shell ($n = 4$) (b) Q shell ($n = 7$)

$2(4)^2 =$ $2(7)^2 =$

5.6 Electron Subshells

Within each energy level we find electrons whose energies are very close in magnitude, but not exactly the same. The range in energies can be explained by the existence of **energy subshells** or **energy sublevels**. Electrons within each sublevel have exactly the same amount of energy.

To visualize the concept of energy sublevels, think of a tall office building with the unusual architectural feature depicted in Figure 5.10. The "building" is upside down. Each floor of the building represents an energy level [K ($n = 1$), L ($n = 2$), and so on]. Offices on each floor represent the energy sublevels. Each floor is a different size, so each has a different number of offices. In the same way, each energy level has a different number of sublevels. The sublevels are represented by the letters s, p, d, f, g, h, and i. The K energy level ($n = 1$), which is on the ground floor, has only one sublevel: the s sublevel; the L energy level ($n = 2$) has two: the s and p sublevels; and so on.

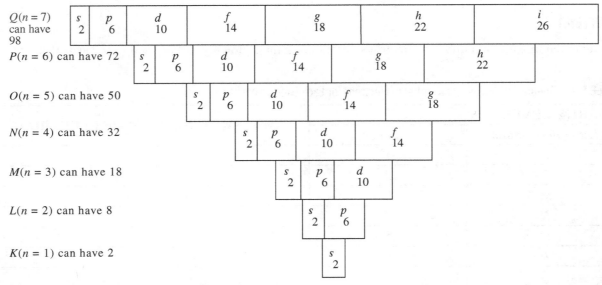

FIGURE 5.10 Energy levels and sub levels, showing maximum number of electrons in each.

Each office of our "energy level building" is a different size and can hold only a certain number of desks. That is, each sublevel is a different size and can hold only a certain number of electrons. The s sublevel can hold a maximum of two electrons, and the p sublevel a maximum of six. Figure 5.10 shows the maximum number of electrons that each sublevel can hold. We discussed the maximum number of electrons for each energy level (2, 8, 18, 32, 50, 72, 98), and now we see that the total number of electrons that a main energy level can hold is simply the sum of the numbers of electrons that all of its *sublevels* can hold.

5.7 Electron Orbitals

We have already looked at energy levels and sublevels. We will now look at the final concept that will help us understand how electrons are arranged about the nucleus of an atom. Electron *orbitals* are the regions of space around the nucleus of an atom where we are most likely to find an electron with a specific amount of energy. Each type of orbital has a specific shape and can hold a maximum of *two* electrons (Figure 5.11). For example, all s orbitals are spherical. However, the different levels of s orbitals occupy different regions of space around the nucleus. Figure 5.12 shows how the 1s, 2s, and 3s orbitals are related to each other.

The p orbitals are dumbbell shaped, as Figure 5.11 shows. There are three p orbitals in each major energy level except the first. They are called p_x, p_y, and p_z, and they correspond to three axes in space. Each of the p orbitals can hold a maximum of two electrons, so the three p orbitals combined can hold a maximum of six electrons. (This total corresponds to the maximum number of electrons that a p sublevel can hold.) The p orbitals that are associated with the different energy levels occupy different regions of space around the nucleus. Figure 5.13 shows the relationship between the $2p_x$ and $3p_x$ orbitals, and Figure 5.14 shows the relationship among the 1s, 2s, and $2p_x$ orbitals.

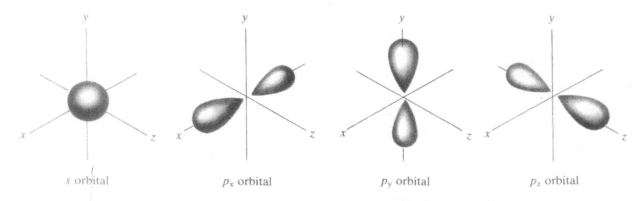

FIGURE 5.11 The shapes of the s, p_x, p_y, and p_z orbitals. The shape of the orbital defines the region of space in which an electron may be found with a probability of 95%. Darker areas are regions of higher probability.

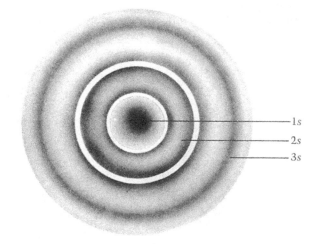

FIGURE 5.12 The relationship of the 1s, 2s, and 3s orbitals to each other.

82 ESSENTIAL CONCEPTS OF CHEMISTRY

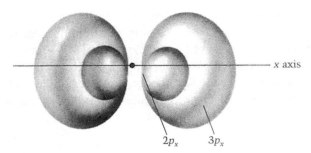

FIGURE 5.13 The spatial relationship of the $2p_x$ and $3p_x$ orbitals to each other.

Orbitals of the d and f type can hold a maximum of 10 and 14 electrons, respectively. There are five orbitals for a d sublevel and seven orbitals for an f sublevel.

To clarify the idea of an electron being found in a specific orbital with a probability of 95%, consider the following analogy. George, a bank teller, works at the first National Bank between 10:00 A.M. and 3:00 P.M. The probability of finding George anywhere in the bank between these hours on a given day is extremely high, let us say 95%. (It is not 100% because there is always the chance that he may be ill or on vacation on a given day.)

Now suppose we make a map of the inside of the bank. We mark George's position every 10 minutes during the course of the workday. For most of the day we would find him at window three. This is where we would find most of our marks on the map. However, there would also be other marks on the map where he might have walked.

This map would be very much like the orbital pictures shown in Figures 5.11 to 5.14. Where there are many marks or darker shading, we have a better chance of finding George and the electron—for example, at window three or at the edges of the orbitals in Figure 5.13. Where there are fewer marks or lighter shading, the probability of finding George or an electron is low. And the probability of finding an electron any place within the orbital (or George any place within the bank) is 95%. Just as George can take a day off, an electron can move out of its orbital and then—just as easily—move back in.

EXAMPLE 5.2

For the third energy level determine (a) the maximum number of electrons the level could hold, (b) the number of sublevels, (c) the maximum number of electrons each sublevel could hold, and (d) the number of orbitals within each sublevel.

Solution

(a) For the third energy level, $n = 3$, therefore $2(n)^2 = 2(3)^2 = 2(9) = 18$.
(b) For the third level, the sublevels are $3s$, $3p$, and $3d$.
(c) The $3s$ sublevel can hold a maximum of 2 electrons, $3p$ can hold a maximum of 6 electrons, and $3d$ can hold a maximum of 10 electrons.
(d) For $3s$ there is one orbital, three orbitals for $3p$, and five orbitals for $3d$.

Practice Exercise 5.2

Determine all of the above for the fourth energy level.

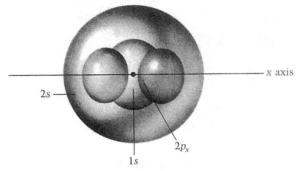

FIGURE 5.14 The relationship among the $1s$, $2s$, and $3p_x$ orbitals.

5.8 Writing Electron Configurations

The term **electron configuration** refers to the way the electrons fill the various energy levels of the atom and is what determines how an atom behaves chemically. As shown in Figure 5.7, unexcited hydrogen has its one electron in the first energy level. This seems reasonable, since the first level is associated with the lowest energy. Both of helium's electrons are also in the first energy level (Figure 5.15). Lithium, however, has three electrons. Two of them go into the first energy level, but one must go in the second level, the next-highest level. This is consistent with the fact that the first level can hold only two electrons. Atoms with 4 to 10 electrons have 2 electrons in the first level and the remainder in the second level. Neon, with 10 electrons, fills both the first and second levels (Table 5.2). Sodium, with 11 electrons, has full first and second levels and 1 electron in the third level.

Table 5.2 shows that (up to element 18) the electrons fill the energy levels in the order that we might predict. The lowest-numbered level is filled before any electrons are positioned in the next-higher-numbered level.

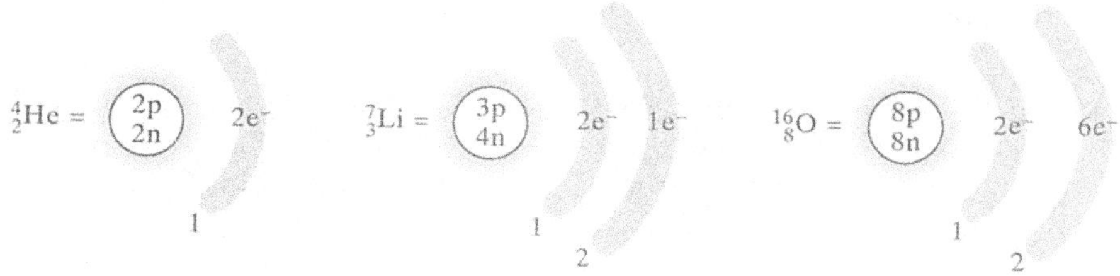

FIGURE 5.15 Some ground state-electron configurations.

TABLE 5.2 Ground-State Electron Configurations of the First 21 Elements

ELEMENT	ATOMIC NUMBER	ELECTRON CONFIGURATION			
		K (1)	L (2)	M (3)	N (4)
Hydrogen	1	1			
Helium	2	2			
Lithium	3	2	1		
Beryllium	4	2	2		
Boron	5	2	3		
Carbon	6	2	4		
Nitrogen	7	2	5		
Oxygen	8	2	6		
Fluorine	9	2	7		
Neon	10	2	8		
Sodium	11	2	8	1	
Magnesium	12	2	8	2	
Aluminum	13	2	8	3	
Silicon	14	2	8	4	
Phosphorus	15	2	8	5	
Sulfur	16	2	8	6	
Chlorine	17	2	8	7	
Argon	18	2	8	8	
Potassium	19	2	8	8	1
Calcium	20	2	8	8	2
Scandium	21	2	8	9	2

84 ESSENTIAL CONCEPTS OF CHEMISTRY

But when we write the electron configuration for $_{19}$K the pattern changes. we would expect the ground-state configuration for potassium to include two electrons in the first level, eight electrons in the second level, and nine electrons in the third level. Instead we find that potassium includes two electrons in the first level, eight electrons in the second level, eight electrons in the third level, and one electron in the fourth energy level. Why should this last electron go into the fourth level when the third level hasn't been filled?

To understand electron configuration more fully we must look further into energy sublevels. These sublevels group electrons according to energy, and the electron configuration shows how many electrons an atom has at the various energies. Since electrons do not occupy orbitals in a random fashion, we must know in exactly what order they do occupy the orbitals. Up to element number 18, we can write electron configurations by filling the energy levels closest to the nucleus first. For elements above number 18, this method does not produce the correct configuration. We can then use the **Aufbau principle** to guide us in determining electron configuration. Aufbau comes from the German word *aufbauen*, which means to build. We must consider the overlapping of energy levels, shown in Figure 5.16. For example, the 4s sublevel has a lower energy than the 3d sublevel. Therefore the 4s sublevel fills with electrons before the 3d sublevel does, even though the 4s sublevel is farther away from the nucleus than the 3d sublevel. And the 5s sublevel fills with electrons before the 4d and 4f sublevels.

Now you can write the electron configurations for any element with any atomic number—by simply filling the lowest energy *sublevel* first. The Aufbau diagram in Figure 5.17 will help you do this. All you have to do is follow the diagonal arrows, starting at the upper left. That is, the 1s sublevel is filled first, then the 2s sublevel,

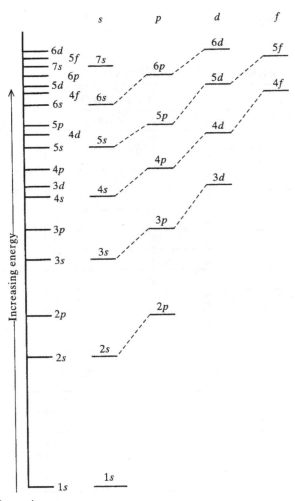

FIGURE 5.16 How energy levels overlap.

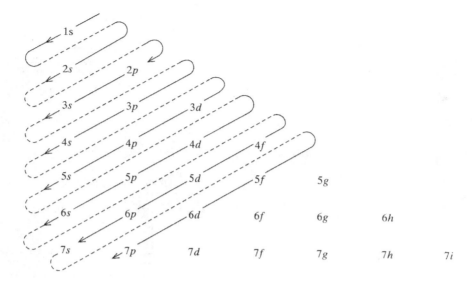

FIGURE 5.17 The order in which electrons fill energy sublevels.

followed by 2p, 3s, 3p, 4s, 3d, and 4p sublevels, and so on. Unfortunately, even using the Aufbau diagram you will not be able to predict the electron configurations for *all* elements. There are a few exceptions.

When writing electron configurations we do not use words, we use a system involving shorthand. We represent the K energy level by the number 1, the L energy level by the number 2, the M energy level by the number 3, and so on. Then, for example, a 2p electron is an electron in the p sublevel of level 2.

Here are two examples of the notation we will use to write electron configurations:

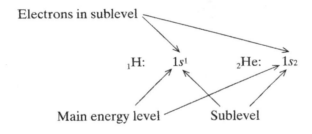

Note that the superscript to the right of the sublevel letter tells us the number of electrons in that sublevel. Here are some more examples:

$_3$Li: $1s^2 2s^1$ $_{18}$Ar: $1s^2 2s^2 2p^6 3s^2 3p^6$
$_4$Be: $1s^2 2s^2$ $_{19}$K: $1s^2 2s^2 2p^6 3p^6 4s^1$
$_5$B: $1s^2 2s^2 2p^1$ $_{20}$Ca: $1s^2 2s^2 2p^6 3s^2 3p^6 4s^2$
$_6$C: $1s^2 2s^2 2p^2$ $_{21}$Sc: $1s^2 2s^2 2p^6 3s^2 3p^6 4s^2 3d^1$

Because the 4s sublevel is lower in energy than the 3d sublevel, the last electron in potassium (K) is in the 4s sublevel and not the 3d sublevel (check Figure 5.17 again). The same is true of calcium (Ca). But beginning will scandium (Sc), the 3d sublevel starts to fill. This is because the 4s sublevel is now completely filled, and the 3d sublevel is now the lowest unfilled energy sublevel. Sublevels continue to fill in this manner. Table 5.3 gives the resulting electron configurations for all elements, and Figure 5.18 shows the pattern of filling in the periodic table.

EXAMPLE 5.3

Using Figure 5.17 write the electron configuration for each of the following elements: (a) $_8$O (b) $_{30}$Zn

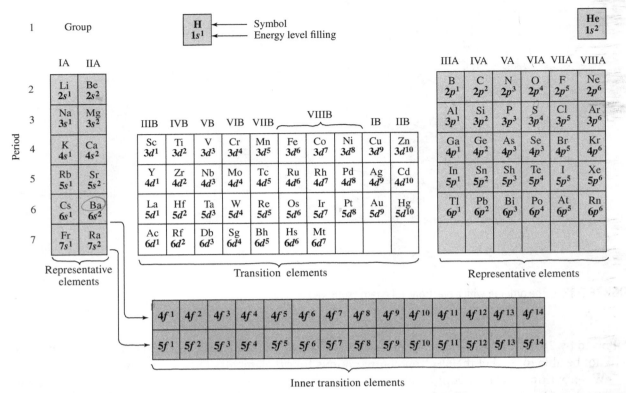

FIGURE 5.18 The general pattern for filling energy levels of the atoms of each element. (*Note:* There are some exceptions to this general pattern for certain transition and inner-transition elements, as shown in Table 5.3.)

Solution

Follow the arrows in Figure 5.17. Remember not to place more than two electrons in an *s* sublevel, six in a *p* sublevel, ten in a *d*, and so on.

(a) $_8$O: $1s^2 2s^2 2p^4$
(b) $_{30}$Zn: $1s^2 2s^2 2p^6 3s^2 3p^6 4s^2 3d^{10}$

Practice Exercise 5.3

Using Figure 5.17 write the electron configuration for each of the following elements:

(a) $_{12}$Mg   (b) $_{34}$Se

5.9 The Importance of Electron Configuration

When we know the electron configurations of elements, we can predict how elements will react with each other. The number of electrons in the **outermost shell** (the shell farthest away from the nucleus, by position) controls the chemical properties of an element. Elements with similar outer electron configurations behave in very similar ways. For instance, the elements Ne, Ar, Kr, Xe, and Rn are unreactive; in other words, they are almost chemically inert. Their atoms all have *eight* electrons in their outermost energy levels. On the other hand, the elements Li, Na, K, Rb, Cs, and Fr all react violently with water; their atoms all have *one* electron in their outermost energy levels.

The electron configurations of the elements also help us predict how the atoms of some elements will bond to each other. We will discuss that in Chapter 7.

TABLE 5.3 Electron Configuration of the Elements

ATOMIC NUMBER	ELEMENT	1	2		3			4				5				6				7
		s	s	p	s	p	d	s	p	d	f	s	p	d	f	s	p	d	f	s
1	H	1																		
2	He	2																		
3	Li	2	1																	
4	Be	2	2																	
5	B	2	2	1																
6	C	2	2	2																
7	N	2	2	3																
8	O	2	2	4																
9	F	2	2	5																
10	Ne	2	2	6																
11	Na	2	2	6	1															
12	Mg	2	2	6	2															
13	Al	2	2	6	2	1														
14	Si	2	2	6	2	2														
15	P	2	2	6	2	3														
16	S	2	2	6	2	4														
17	Cl	2	2	6	2	5														
18	Ar	2	2	6	2	6														
19	K	2	2	6	2	6		1												
20	Ca	2	2	6	2	6		2												
21	Sc	2	2	6	2	6	1	2												
22	Ti	2	2	6	2	6	2	2												
23	V	2	2	6	2	6	3	2												
24	Cr	2	2	6	2	6	5	1												
25	Mn	2	2	6	2	6	5	2												
26	Fe	2	2	6	2	6	6	2												
27	Co	2	2	6	2	6	7	2												
28	Ni	2	2	6	2	6	8	2												
29	Cu	2	2	6	2	6	10	1												
30	Zn	2	2	6	2	6	10	2												
31	Ga	2	2	6	2	6	10	2	1											
32	Ge	2	2	6	2	6	10	2	2											
33	As	2	2	6	2	6	10	2	3											
34	Se	2	2	6	2	6	10	2	4											
35	Br	2	2	6	2	6	10	2	5											
36	Kr	2	2	6	2	6	10	2	6											
37	Rb	2	2	6	2	6	10	2	6			1								
38	Sr	2	2	6	2	6	10	2	6			2								
39	Y	2	2	6	2	6	10	2	6	1		2								

(Continued)

TABLE 5.3 Electron Configuration of the Elements (*Continued*)

ATOMIC NUMBER	ELEMENT	1	2		3			4				5				6				7
		s	s	p	s	p	d	s	p	d	f	s	p	d	f	s	p	d	f	s
40	Zr	2	2	6	2	6	10	2	6	2		2								
41	Nb	2	2	6	2	6	10	2	6	4		1								
42	Mo	2	2	6	2	6	10	2	6	5		1								
43	Tc	2	2	6	2	6	10	2	6	6		1								
44	Ru	2	2	6	2	6	10	2	6	7		1								
45	Rh	2	2	6	2	6	10	2	6	8		1								
46	Pd	2	2	6	2	6	10	2	6	10										
47	Ag	2	2	6	2	6	10	2	6	10		1								
48	Cd	2	2	6	2	6	10	2	6	10		2								
49	In	2	2	6	2	6	10	2	6	10		2	1							
50	Sn	2	2	6	2	6	10	2	6	10		2	2							
51	Sb	2	2	6	2	6	10	2	6	10		2	3							
52	Te	2	2	6	2	6	10	2	6	10		2	4							
53	I	2	2	6	2	6	10	2	6	10		2	5							
54	Xe	2	2	6	2	6	10	2	6	10		2	6							
55	Cs	2	2	6	2	6	10	2	6	10		2	6			1				
56	Ba	2	2	6	2	6	10	2	6	10		2	6			2				
57	La	2	2	6	2	6	10	2	6	10		2	6	1		2				
58	Ce	2	2	6	2	6	10	2	6	10	2	2	6			2				
59	Pr	2	2	6	2	6	10	2	6	10	3	2	6			2				
60	Nd	2	2	6	2	6	10	2	6	10	4	2	6			2				
61	Pm	2	2	6	2	6	10	2	6	10	5	2	6			2				
62	Sm	2	2	6	2	6	10	2	6	10	6	2	6			2				
63	Eu	2	2	6	2	6	10	2	6	10	7	2	6			2				
64	Gd	2	2	6	2	6	10	2	6	10	7	2	6	1		2				
65	Tb	2	2	6	2	6	10	2	6	10	9	2	6			2				
66	Dy	2	2	6	2	6	10	2	6	10	10	2	6			2				
67	Ho	2	2	6	2	6	10	2	6	10	11	2	6			2				
68	Er	2	2	6	2	6	10	2	6	10	12	2	6			2				
69	Tm	2	2	6	2	6	10	2	6	10	13	2	6			2				
70	Yb	2	2	6	2	6	10	2	6	10	14	2	6			2				
71	Lu	2	2	6	2	6	10	2	6	10	14	2	6	1		2				
72	Hf	2	2	6	2	6	10	2	6	10	14	2	6	2		2				
73	Ta	2	2	6	2	6	10	2	6	10	14	2	6	3		2				
74	W	2	2	6	2	6	10	2	6	10	14	2	6	4		2				
75	Re	2	2	6	2	6	10	2	6	10	14	2	6	5		2				
76	Os	2	2	6	2	6	10	2	6	10	14	2	6	6		2				
77	Ir	2	2	6	2	6	10	2	6	10	14	2	6	7		2				
78	Pt	2	2	6	2	6	10	2	6	10	14	2	6	9		1				
79	Au	2	2	6	2	6	10	2	6	10	14	2	6	10		1				

ATOMIC NUMBER	ELEMENT	1	2		3			4				5				6				7
		s	s	p	s	p	d	s	p	d	f	s	p	d	f	s	p	d	f	s
80	Hg	2	2	6	2	6	10	2	6	10	14	2	6	10		2				
81	Tl	2	2	6	2	6	10	2	6	10	14	2	6	10		2	1			
82	Pb	2	2	6	2	6	10	2	6	10	14	2	6	10		2	2			
83	Bi	2	2	6	2	6	10	2	6	10	14	2	6	10		2	3			
84	Po	2	2	6	2	6	10	2	6	10	14	2	6	10		2	4			
85	At	2	2	6	2	6	10	2	6	10	14	2	6	10		2	5			
86	Rn	2	2	6	2	6	10	2	6	10	14	2	6	10		2	6			
87	Fr	2	2	6	2	6	10	2	6	10	14	2	6	10		2	6			1
88	Ra	2	2	6	2	6	10	2	6	10	14	2	6	10		2	6			2
89	Ac	2	2	6	2	6	10	2	6	10	14	2	6	10		2	6	1		2
90	Th	2	2	6	2	6	10	2	6	10	14	2	6	10		2	6	2		2
91	Pa	2	2	6	2	6	10	2	6	10	14	2	6	10	2	2	6	1		2
92	U	2	2	6	2	6	10	2	6	10	14	2	6	10	3	2	6	1		2
93	Np	2	2	6	2	6	10	2	6	10	14	2	6	10	4	2	6	1		2
94	Pu	2	2	6	2	6	10	2	6	10	14	2	6	10	6	2	6			2
95	Am	2	2	6	2	6	10	2	6	10	14	2	6	10	7	2	6			2
96	Cm	2	2	6	2	6	10	2	6	10	14	2	6	10	7	2	6	1		2
97	Bk	2	2	6	2	6	10	2	6	10	14	2	6	10	8	2	6	1		2
98	Cf	2	2	6	2	6	10	2	6	10	14	2	6	10	10	2	6			2
99	Es	2	2	6	2	6	10	2	6	10	14	2	6	10	11	2	6			2
100	Fm	2	2	6	2	6	10	2	6	10	14	2	6	10	12	2	6			2
101	Md	2	2	6	2	6	10	2	6	10	14	2	6	10	13	2	6			2
102	No	2	2	6	2	6	10	2	6	10	14	2	6	10	14	2	6			2
103	Lr	2	2	6	2	6	10	2	6	10	14	2	6	10	14	2	6	1		2
104	Db	2	2	6	2	6	10	2	6	10	14	2	6	10	14	2	6	2		2
105	Jl	2	2	6	2	6	10	2	6	10	14	2	6	10	14	2	6	3		2
106	Rf	2	2	6	2	6	10	2	6	10	14	2	6	10	14	2	6	4		2
107	Bh	2	2	6	2	6	10	2	6	10	14	2	6	10	14	2	6	5		2
108	Hn	2	2	6	2	6	10	2	6	10	14	2	6	10	14	2	6	6		2
109	Mt	2	2	6	2	6	10	2	6	10	14	2	6	10	14	2	6	7		2

SUMMARY

When a beam of sunlight passes through a glass prism, the sunlight separates into a rainbow of colors called the visible spectrum. Each color of light is associated with a specific amount of energy. The visible spectrum is part of the electromagnetic spectrum, and both are continuous spectra—their various parts merge into each other. In the 1850s, Kirchhoff and Bunsen showed that when a sample of an element is heated until it glows, it produces a line spectrum of visible light. Each element produces its own characteristic line spectrum.

In 1913 Niels Bohr proposed a model of the atom to explain these line spectra. The model consisted of shells, or imaginary spherical surfaces, roughly concentric with the nucleus. Each shell was associated with a particular energy level. Bohr postulated that electrons travel around the nucleus in their orbits. However, an electron can jump from one shell to another at a higher energy level when it absorbs enough energy. It can fall back to the original shell by emitting this energy in bursts. These separate bursts show up as line spectra.

Revolutionary discoveries occurred in physics in the decades that followed. These discoveries gave rise to the quantum mechanical model of the atom, which is based on a series of mathematical equations. The electrons in an atom were now thought of as occupying energy levels, with each energy level being composed of sublevels. The probability of finding an electron in an orbital, or region of space about the nucleus, was postulated.

Using this information, the electron configuration, or arrangement of electrons about the nucleus, could be determined. The various energy levels of the atom have been given the letter designations K through Q and the numbers 1 through 7. (Level K or 1 is the lowest energy level.) Associated with each level is a theoretical maximum electron population. Lower energy levels are usually filled before electrons are positioned in higher energy levels. Using the Aufbau principle, electron configurations can be determined and represented using a shorthand notation. The electron configurations endow the elements with many of their properties; elements with similar outer electron configurations exhibit similar chemical behavior.

KEY TERMS

Aufbau principle (**5.8**)
continuous spectrum (**5.1**)
electromagnetic spectrum (**5.1**)
electron configuration (**5.8**)
Energy level (**5.3**)
energy sublevel (**5.6**)
energy subshell (**5.6**)
excited state (**5.3**)

ground state (**5.3**)
line spectrum (**5.1**)
orbital (**5.4**)
outermost shell (**5.9**)
quanta (**5.2**)
quanta (**5.2**)
quantum mechanics (**5.4**)
shell (**5.5**)

SELF-TEST EXERCISES

LEARNING GOAL 1

Line Spectra and Continuous Spectra

1. How does a line spectrum differ from a continuous spectrum?
◀ 2. What is a spectrum? What is the difference between the electromagnetic spectrum and the visible spectrum?
◀ <u>3</u>. How could a spectroscope help scientists analyze the sun and other extraterrestrial bodies?
4. Which end of the visible spectrum has the highest energy, red or violet? Which end has the longest wavelength?
5. Can you discover the origin of the word *helium*?
(*Hint:* See the *Handbook of Chemistry and Physics*.)
6. Do you think line spectra could be used to identify compounds? Explain.

LEARNING GOAL 2

Bohr Model of the Atom

7. Explain the major differences among the Thomson, Rutherford, and Bohr models of the atom.
8. Explain how the *ground state* of an atom differs from its *excited state*.
<u>9</u>. What is Bohr's most important contribution to our ideas about the structure of the atom?
10. According to the Bohr model of the atom, do electrons "spiral" or "jump" from one energy level to another? What type of spectrum would be produced by each of these types of movements by electrons?

CHAPTER 5: ATOMIC THEORY, PART 2

LEARNING GOAL 3

Shells and Energy Levels

11. What subatomic particles produce the line spectrum of an element?

12. Complete the following sentence with the word *higher* or *lower:* The farther away an energy level is from the nucleus, the _____ the energy of an electron in that level.

◀ *13*. What is meant by shells and energy levels? How many shells are known to exist in atoms? State their letter and number designations.

14. Using the $2n^2$ rule, calculate the maximum number of electrons that can populate each of the first three energy levels.

LEARNING GOAL 4

Difference Between Ground State and Excited State of an Atom

15. What do we mean by the ground state of an atom? What do we mean by the excited state of an atom?

16. Describe the ground-state configuration of a lithium atom. What changes could take place if the lithium atom were in an excited state?

17. If a neutral hydrogen atom has an electron in the second energy level, is the atom in its ground state or an excited state?

◀ 18. A particular neutral magnesium atom has two electrons in the first shell, eight electrons in the second shell, and two electrons in the third shell. Would you characterize this atom as being in the ground state or in an excited state?

LEARNING GOAL 5

Quantum Mechanical Model of the Atom

19. Why was the Bohr model of the atom inadequate to explain the behavior of the atom?
20. What new knowledge regarding the laws of physics was discovered in the early 1900s?
21. Discuss Louis de Broglie's contributions to the field of atomic theory.
22. Discuss Erwin Schrodinger's contributions to the knowledge of the behavior of electrons.

LEARNING GOAL 6

Energy Sublevels

◀ *23*. How many energy sublevels does each of the following energy levels contain?

   (a) energy level 1
   (b) energy level 2
   (c) energy level 3
   (d) energy level 4

24. How many energy sublevels does energy level 7 contain? Name them.

25. How is the total number of electrons an energy level contains related to the number of electrons a sublevel can hold?

26. The third energy level can hold a maximum of 18 electrons. How many sublevels does this energy level contain? How many electrons does each sublevel contain?

◀ *27*. Determine the following for the sixth energy level of an atom:

   (a) number of subshells it contains
   (b) maximum number of electrons this energy level can hold

92 ESSENTIAL CONCEPTS OF CHEMISTRY

28. Determine the following for the fourth energy level of an atom:
    (a) number of subshells it contains
    (b) maximum number of electrons this energy level can hold

LEARNING GOAL 7

Electron Orbitals

◀ 29. Using an analogy similar to the one in the text, explain what is meant by 95% probability and describe how it relates to the quantum mechanical model of an atom.
30. What is meant by the "region of highest probability of finding an electron"?
<u>31</u>. For the fifth energy level, determine the number of orbitals in the first two sublevels.
32. For the sixth energy level, determine the number of orbitals in the first three sublevels.
33. Describe the shape of the $2s$, $3s$, and $4s$ electron orbitals.
34. Describe the shape of the $4p$ orbitals.
<u>35</u>. Which of the following does not exist?

  (a) $2s$ (b) $2p$ (c) $2d$ (d) $2f$

36. Which of the following does not exist?

  (a) $3s$ (b) $3p$ (c) $3d$ (d) $3f$

LEARNING GOAL 8

Writing Electron Configurations

<u>37</u>. Using Figure 5.17 write the electron configurations for the following elements:
   (a) $_7N$ (b) $_{23}V$
   (c) $_{54}Xe$ (d) $_{51}Sb$

38. Using Figure 5.17 write the electron configurations for the following elements:
   (a) $_{50}Sn$ (b) $_{27}Co$
   (c) $_{86}Rn$ (d) $_{35}Br$

<u>39</u>. Name the neutral elements with the following electron configurations. (Use Table 5.3.)
   (a) $1s^2 2s^2 2p^6 3s^2 3p^6 3d^{10} 4s^2 4p^6 4d^{10} 5s^2 5p^6 6s^2$
   (b) $1s^2 2s^2 2p^6 3s^2 3p^6 4s^2$
   (c) $1s^2 2s^2 2p^6 3s^2 3p^6 3d^{10} 4s^2 4p^6 5s^2$
   (d) $1s^2 2s^2$

40. Name the neutral elements with the following electron configurations. (Use Table 5.3.)
   (a) $1s^2$
   (b) $1s^2 2s^2 2p^3$
   (c) $1s^2 2s^2 2p^6 3s^2 3p^6 3d^{10} 4s^1$
   (d) $1s^2 2s^2 2p^6 3s^2 3p^6 3d^{10} 4s^2 4p^4$

◀ <u>41</u>. A certain neutral element has 2 electrons in the first level, 8 electrons in the second level, 18 electrons in the third level, and 3 electrons in the fourth level. List the following information for this element:
   (a) atomic number
   (b) total number of s electrons
   (c) total number of p electrons
   (d) total number of d electrons
   (e) number of protons

42. A certain neutral element has 2 electrons in the first level, 8 electrons in the second level, 18 electrons in the third level, and 2 electrons in the fourth level. List the following information for this element:

    (a) atomic number
    (b) total number of s electrons
    (c) total number of p electrons
    (d) total number of d electrons
    (e) number of protons

43. Write electron configurations to determine which of the following elements belong to the same groups: elements with atomic numbers 6, 19, 17, 14, 3, and 35.

44. Write electron configurations to determine which of the following elements belong to the same groups: elements with atomic numbers 5, 18, 16, 36, 10, and 13.

45. Use the $2n^2$ rule to calculate the maximum number of electrons that the seventh energy level can hold.

46. Use the $2n^2$ rule to calculate the maximum number of electrons that the fourth energy level can hold.

47. Using Figure 5.17, write the electron configurations for each of the following elements:

    (a) $_{31}$Ga (b) $_{51}$Sb (c) $_{82}$Pb (d) $_{88}$Ra

48. Using Figure 5.17 write the electron configurations for the following elements:

    (a) $_3$Li (b) $_{12}$Mg (c) $_{56}$Ba

49. Name the neutral elements with the following electron configurations. (Use Table 5.3.)

    (a) $1s^2 2s^2 2p^6 3s^2$
    (b) $1s^2 2s^2 2p^6 3s^2 3p^6 3d^{10} 4s^2 4p^6 5s^2$
    (c) $1s^2 2s^2 2p^6 3s^2 3p^6 3d^{10} 4s^2 4p^6 4d^{10} 4f^{14} 5s^2 5p^6 6s^2$
    (d) $1s^2 2s^2 2p^6 3s^2 3p^6 3d^{10} 4s^2 4p^6 4d^{10} 4f^{14} 5s^2 5p^6 5d^{10} 6s^2 6p^1$

50. Name the neutral elements with the following electron configurations. (Use Table 5.3.)

    (a) $1s^2 2s^2 2p^6 3s^1$
    (b) $1s^2 2s^2 2p^6 3s^2 3p^6 3d^{10} 4s^2 4p^6 4d^{10} 4f^{12} 5s^2 5p^6 6s^2$
    (c) $1s^2 2s^2 2p^6 3s^2 3p^6 3d^{10} 4s^2 4p^6 4d^{10} 4f^{14} 5s^2 5p^6 6s^2$
    (d) $1s^2 2s^2 2p^6 3s^2 3p^6 3d^{10} 4s^2 4p^6 4d^{10} 4f^{14} 5s^2 5p^6 5d^{10} 6s^2 6p^6$

51. A given neutral element has 2 electrons in the first level, 8 electrons in the second level, 18 electrons in the third level, 22 electrons in the fourth level, 8 electrons in the fifth level, and 2 electrons in the sixth level. List the following information for this element:

    (a) atomic number
    (b) total number of s electrons
    (c) total number of p electrons
    (d) total number of d electrons
    (e) total number of f electrons
    (f) number of protons

52. A given neutral element has 2 electrons in the first level, 8 electrons in the second level, 18 electrons in the third level, 32 electrons in the fourth level, 9 electrons in the fifth level, and 2 electrons in the sixth level. List the following information for this element:

    (a) atomic number
    (b) total number of s electrons
    (c) total number of p electrons
    (d) total number of d electrons
    (e) total number of f electrons
    (f) number of protons

LEARNING GOAL 9

Importance of Electron Configuration

53. Explain the importance of electron configuration with regard to the chemical behavior of atoms.
54. Complete the following sentence: The number of electrons in the _____ shell determines the chemical properties of an element.
55. The elements in Group VIIIA of the periodic table (this group is the vertical column of elements farthest to the right in the periodic table) have eight electrons in their outermost energy level. (The exception to this is helium, which has only two electrons in its outermost energy level.) What is chemically unique about the Group VIIIA elements?
56. Separate the following into groups (vertical columns) of elements that exhibit similar chemical behavior:

    (a) Li   (b) Ne   (c) K   (d) Rb   (e) Kr   (f) Na   (g) Cs   (h) Rn   (i) Xe

EXTRA EXERCISES

57. Write electron configurations for all elements with atomic numbers between 1 and 20.
58. Write electron configurations for all elements with atomic numbers between 1 and 20 that have the same number of electrons in the outermost energy level as a magnesium atom.
◀ 59. What was the reasoning behind replacement of the Rutherford model of the atom with Bohr's model?
60. Explain why the line spectrum of an element is sometimes referred to as its fingerprint.
61. Explain briefly how electrons produce a line spectrum.
62. In Chapter 4 you learned that not all atoms of an element need be identical and that these different atoms are called isotopes. Do isotopes of the same element produce the same line spectrum? Explain.
63. In this chapter we stated that a marble rolling down a flight of stairs is similar to an electron jumping from one energy level to the next. Can you think of other analogous phenomena?
64. What can happen to an electron in an atom when it absorbs too much energy while being excited?
65. What would be the atomic number of an element whose energy levels 1, 2, 3, 4, 5, 6, and 7 were filled with the maximum number of electrons?

CHAPTER 6

THE PERIODIC TABLE
Keeping Track of the Elements

This is the story of a physician who tracked down the cause of death one of his patients in order to try to save other people who might also be at risk. Our story begins in 1955. A 49-year-old woman, Mary Brown, visited her doctor complaining of pain in her jaw and a severe lack of energy. Routine blood tests showed that she was anemic. Anemia may be caused by bleeding, poor nutrition, drugs, or inadequate red blood cell production caused by tumors in the bone marrow.

After elimination of the first three possibilities, her physician, Dr. Ingram, suspected a problem with the bone marrow. A bone marrow biopsy was performed and Mary was diagnosed as having a malignant tumor of the bone marrow. The pain in her jaw was caused by inflammation of the bone marrow of the jaw.

Malignant tumors invade surrounding tissues and send out cells that form new tumors at distant sites. The formation of new tumors at distant sites is called *metastasis*. One year after diagnosis, Mary Brown, then 50 years old, died due to widespread metastasis of her cancer.

Dr. Ingram, upset and puzzled about how his patient of ten years, who seemed to have lived a healthy lifestyle, developed bone marrow cancer, decided to investigate. Dr. Ingram's investigation began with Mary's surviving family members, who gave the doctor a more complete history of Mary's life. The mystery was soon solved. Dr. Ingram determined that the cause of Mary's illness and subsequent death was her previous occupation. In the early 1920s, Mary was employed by a company that applied a radium-based luminous paint on watch and instrument dials. Workers who performed this task would keep a fine point on their paint brushes by turning the bristles on their tongues. When the dials were painted with this material, they glowed in the dark. Such items became very popular with consumers.

Although radium-based paint was very popular with consumers, it proved to be toxic to people like Mary who worked with it. Dr. Ingram knew that radium produces ionizing radiation that damages the body and can cause cancer. Unfortunately, working with radium for five years and ingesting small amounts of it took its toll on Mary and many of her coworkers.

Dr. Ingram recalled that radium has chemical properties that resemble those of calcium because the arrangements of electrons (which determine the properties of elements) are very similar in radium and calcium atoms. These two elements are so close in electron arrangement that the body is unable to distinguish them from one another. So radium can be incorporated into the bones and teeth just as easily as calcium is. Once inside the body, however, radium produces damaging ionizing radiation.

Working with radium is an occupational hazard. Unfortunately, classifying occupational carcinogens is not always easy because there is often a long period between exposure to the carcinogen and development of cancer. This was the case with radium. Many workers were exposed to radium-based paint while manufacturing watches and instrument dials, and only years later were the dangers of radium recognized. Dr. Ingram hoped that his research would help those people who were exposed to the radium so that they could be told of the risks that they face and could seek proper medical attention.

LEARNING GOALS

After you've studied this chapter, you should be able to:

1. Discuss the historical basis and significance of the periodic table.
2. Distinguish between a period and a group in the periodic table.
3. State the difference between an A-group (representative) element and a B-group element (transition metal).
4. Predict trends for properties such as atomic radius, ionization potential, and electron affinity.

INTRODUCTION

In preceding chapters, we referred to the periodic table for information about the elements, such as atomic number and mass. The periodic table can be used to predict many properties that can be useful to us. In this chapter, we trace the development of the periodic table and see what kinds of additional information it holds.

6.1 History

From the time of the ancient Greeks up to 1800, only 14 elements were known. Then, in a short span of 10 years, from 1800 to 1810, 14 more elements were discovered. By 1830, 45 elements were known. As new elements were discovered, chemists probably began to feel insecure. All these elements had different properties, and there didn't seem to be any relationship among them. And chemists wondered how many elements actually existed. To answer this question they had to find some relationship among the known elements.

One of the first chemists to notice some order among the elements was the German scientist Johann Döbereiner. He published an account of his observations in 1829. It occurred to Döbereiner that bromine had chemical and physical properties somewhere between those of chlorine and iodine and that bromine's atomic mass was almost midway between those of chlorine and iodine (Figure 6.1). Could this be nothing more than coincidence?

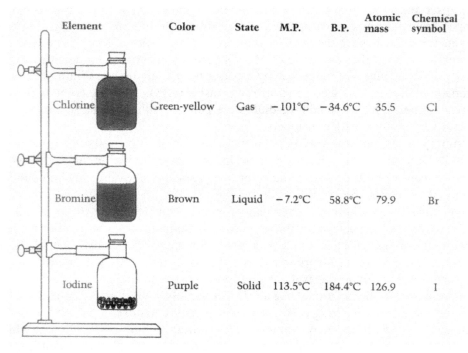

FIGURE 6.1 Some similar elements.

Element	Color	State	M.P. (°C)	B.P. (°C)	Atomic mass (amu)
Ca	Silver-white	Solid	842.8	1487	40.08
Sr	Silver white to pale yellow	Solid	769	1384	87.62
Ba	Yellow-silver	Solid	725	1140	137.34
S	Yellow	Solid	113	444	32.06
Se	Blue-gray	Solid	217	685	98.96
Te	Silver-white	Solid	452	1390	127.6

FIGURE 6.2 More similar elements.

Döbereiner searched and found two more groups of similar elements. The first was made up of calcium (Ca), stronium (Sr), and barium (Ba); the second included sulfur (S), selenium (Se), and tellurium (Te) (Figure 6.2). He called these groups *triads* (sets of three).

In 1864 the English chemist John Newlands arranged the known elements in order of increasing atomic mass. He came upon the idea of arranging them in vertical columns. Because he noticed that the eighth element had chemical and physical properties that were similar to those of the first, he let the eighth element start a new column (Figure 6.3). Newlands called his arrangement the *Law of Octaves*. But there were many places in his arrangement where dissimilar elements were next to each other.

In 1869 the German chemist Julius Lothar Meyer devised an incomplete periodic table consisting of 56 elements. In the same year the Russian chemist Dmitri Mendeleev (pronounced "Menduh-LAY-eff ") arranged the elements in order of increasing atomic mass. In cases where discrepancies arose, *the chemical properties were used to place the elements*. Mendeleev established horizontal rows, or *periods*. Hydrogen by itself made up the first period. Each of the next two periods contained seven elements. (The noble gases had not yet been discovered.) The periods after that contained more than seven elements. This is the point at which Mendeleev's table differed from Newlands's table.

In arranging the table, Mendeleev occasionally put heavier elements before lighter ones to keep elements with the same chemical properties in the same vertical column, or *group* (Figure 6.4). Sometimes he left open spaces in the table, where he reasoned that unknown elements should go. This was the case with gallium. In 1869, when he made up his table, the element gallium was unknown; however, Mendeleev *predicted* its existence. He based his prediction on the properties of aluminum (which appeared directly above gallium in the table). Mendeleev even went so far as to predict the melting point, boiling point, and atomic mass of the then-unknown gallium, which he called eka-aluminum. Six years later, while analyzing zinc ore, the French chemist Lecoq de Boisbaudran discovered the element gallium. Its properties were almost identical to those Mendeleev had predicted (Figure 6.5).

The periodic tables proposed by both Mendeleev and Meyer were based on increasing atomic masses. Mendeleev used similarities among chemical properties to account for discrepancies among elements. Certain irregularities in these tables were corrected by the work of Henry G. J. Moseley, a British physicist. Moseley

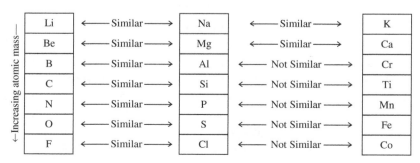

FIGURE 6.3 Newlands's Law of Octaves.

98 ESSENTIAL CONCEPTS OF CHEMISTRY

Period	I	II	III	IV	V	VI	VII	VIII		
1	1 H							2		
2	3 Li	4 Be	5 B	6 C	7 N	8 O	9 F	10		
3	11 Na	12 Mg	13 Al	14 Si	15 P	16 S	17 Cl	18		
4	19 K	20 Ca	21	22 Ti	23 V	24 Cr	25 Mn	26 Fe	27 Co	28 Ni
	29 Cu	30 Zn	31	32	33 As	34 Se	35 Br	36		
5	37 Rb	38 Sr	39 Y	40 Zr	41 Nb	42 Mo	43	44 Ru	45 Rh	46 Pd
	47 Ag	48 Cd	49 In	50 Sn	51 Sb	52 Te	53 I	54		
6	55 Cs	56 Ba	57 La	72	73 Ta	74 W	75	76 Os	77 Ir	78 Pt
	79 Au	80 Hg	81 Tl	82 Pb	83 Bi	84	85	86		
7	87	88	89	104	105					

58 Ce	59	60	61	62	63	64	65 Tb	66	67	68 Er	69	70	71
90 Th	91	92 U	93	94	95	96	97	98	99	100	101	102	103

FIGURE 6.4 Mendeleev's original periodic table arranged in modern form.

was able to determine the nuclear charge on the atoms of known elements. He concluded that elements should be arranged by increasing atomic number rather than by atomic mass. This research led to the development of the *periodic* law, which states that the chemical properties of the elements are periodic functions of their atomic numbers. Elements with similar chemical properties recur at regular intervals, and they are placed accordingly on the periodic table.

		Eka-aluminum	Gallium
Atomic mass	→	Predicted: 68	Observed: 69.9
Oxide formula	→	Predicted: E_2O_3	Observed: Ga_2O_3
Oxide in acid	→	Predicted: should dissolve in acid	Observed: dissolves in acid
Salt formation	→	Predicted: should form salts as EX_3	Observed: forms salts as GaX_3

FIGURE 6.5 Predicted and actual properties of eka-aluminum, which was named gallium.

6.2 The Modern Periodic Table

The periodic table of today is similar to Mendeleev's but has many more elements—those that have been discovered since 1869. Today's table consists of seven horizontal rows called **periods** and a number of vertical columns called **groups** (or **families**) (Table 6.1).

The groups are numbered with Roman numerals. All the elements in each group have the same number of electrons in their outermost shells, so they all behave similarly. For example, the Group IA elements react violently when they come into contact with water. And all elements in Group IA have one electron in their outermost shell.

Some of the groups in the periodic table are labeled with a Roman numeral followed by A, others with a Roman numeral followed by B. The A groups are called the **representative elements**. The B groups are called the **transition metals**. You can also see that as we move from the top to the bottom of the periodic table (in other words, from period 1 to period 7), the periods get larger—they have more elements in them. In fact, periods 6 and 7 are so large that to fit the table on one page, we have to write part of each period *below* the rest of the table (the 15 lanthanides and the 15 actinides).

The most significant changes to the periodic table in the last few decades have been the names of the newest elements. Traditionally, the creator or discoverer of an element has the privilege of proposing its name. And choosing a name is no trivial matter! It may be influenced by issues ranging from national pride to professional rivalry. But skepticism about which labs were actually the first to create specific elements led to worldwide disagreement about the names of these newly created elements 104 through 117. The American Chemical Society proposed one set of names, but the International Union of Pure and Applied Chemistry (IUPAC) proposed a

TABLE 6.1 Periodic Table of the Elements

different set. After a period of time, however, the IUPAC developed what it considered a "fair compromise between the various claims and suggestions" of the American, German, and Russian teams that discovered the elements. Finally, publishers of chemistry textbooks and scientific papers could avoid the confusion caused by five different versions of the periodic table. Work continues by the IUPAC in naming the most recently discovered elements, atomic numbers 113 through 118.

6.3 Periodic Trends

When we examine the periodic table, we can see certain trends in the properties of the elements. For example, the elements in Group IA (at the left) are all metals. As we move across to the right, we continue to encounter metals through more than half the width of the table. Then we reach elements that have properties intermediate between metals and nonmetals; these are the metalloids. As we move still further to the right, we finally reach the nonmetals. (See Section 3.9 for a review of these terms.)

Other trends in properties such as atomic radius, the tendency to lose or gain electrons, and some physical properties are related to the elements' position in the periodic table because of their electron configuration. In the remainder of this chapter, we discuss these trends as well as the properties themselves and their structural basis.

6.4 Periodicity and Electron Configuration

All the elements within each A group have the same number of electrons in their outermost shell. Moreover, for the A-group elements, *the group number tells us how many outer electrons there are*. For example, all the elements in Group VIA have six electrons in their outermost energy level. All the elements in Group IA have one electron in their outermost energy level.

For the B-group elements, there is no clear connection between group number and the number of electrons in the outermost energy level. For example, Group IVB elements *do not* have four electrons in their outermost energy level.

6.5 Similarities Among Elements in a Group and Period

The configuration of the outermost energy level determines the chemical behavior of an element. Thus all elements in the same group should behave (react) similarly, because they all have the same outer-shell electron configuration. In other words, the properties that these elements exhibit as a result of their interaction with other substances are similar. But what about elements in the same period? In general, the elements in any period have different outer-shell electron configurations, so these elements have different chemical properties.

For example, in period 3, Al (with two $3s$ electrons and one $3p$ electron) is a metal, whereas Cl (with two $3s$ electrons and five $3p$ electrons) is a nonmetal. However, this is not so for the transition metals in periods 4 through 7. Consider the elements Mn, Fe, Co, and Ni in period 4. Their electron configurations differ because they have different numbers of electrons in their $3d$ sublevels. However, level 3 is *not* their outermost energy level. The outermost level for these transition metals is the $4s$ sublevel. We would therefore expect these elements to exhibit somewhat similar behavior, and they do. Moreover, the same is true of the transition elements in periods 5 through 7.

To summarize:

1. Elements in any one group behave in similar ways, because they have (with few exceptions) *identical electron configurations in the outermost shell.*

2. Elements in any one period generally do not behave similarly.

3. However, because of the manner in which electrons fill their shells (filling some inner shells after outer shells), the transition metals—that is, the B groups—show somewhat similar chemical behavior.

6.6 Atomic Radius

The arrangement of the periodic table is helpful in predicting the relative sizes of atoms within any period or group. We shall consider the **atomic radius** (radius of an atom) to be the distance from the center of the nucleus to the outermost electron. Atomic radii are measured with a unit called the *angstrom*, named after a Swedish physicist. One angstrom (1 Å) is equal to 10^{-10} m, or one ten-billionth of a meter.

In any group in the periodic table, each element has one more energy level than the element above it (Figure 6.6). Therefore, as we move down through a particular group, the atomic radii of the elements should increase.

But what happens to the atomic radii of the elements as we move from left to right across a period? You might assume that because the number of electrons in the atom is increasing, the size of the atom should also increase. This is not the case. Remember that as we move across a period, the number of protons increases along with the number of electrons. Therefore there is a greater pull on the electrons of the inner energy levels, which usually causes a shrinking of the atomic radius (Figure 6.7).

Using the elements in period 2, we can explain what happens. In lithium (atomic number 3), three protons attract two K electrons and one L electron. In beryllium (atomic number 4), four protons attract two K electrons and two L electrons. In boron, five protons attract two K electrons and three L electrons. This increased positive charge acting on the K electrons makes the K energy level shrink. The L energy level also shrinks, and this results in a decreased atomic radius (Figure 6.8).

Also keep in mind that for the transition metals, atomic radii change very little because electrons are filling the d and f subshells. Table 6.2 shows the atomic radii of all the elements.

EXAMPLE 6.1

List the following elements in order of *increasing* atomic radius: Mg, K, and Ca. Use only the periodic table (Table 6.1); don't look up the atomic radii.

Solution

UNDERSTAND THE PROBLEM

We must order the elements in terms of atomic radius without knowing the specific atomic radius of each element. In other words, we must look at periodic trends.

DEVISE A PLAN

Locate the position of each element in the periodic table, remembering that the atomic radius increases from the top to the bottom of a group of elements and decreases from left to right across a period of elements.

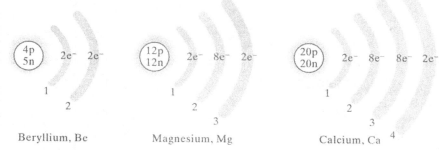

FIGURE 6.6   Relative sizes of some Group IIA elements.

102 ESSENTIAL CONCEPTS OF CHEMISTRY

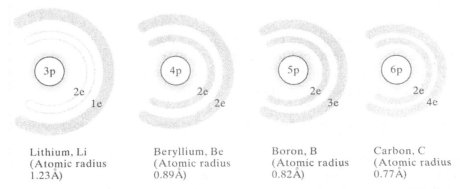

FIGURE 6.7 The atomic radii of the elements decrease as we move from left to right across a period. The increased nuclear charge results in a greater pull on the electrons of the inner energy levels, so the atom shrinks.

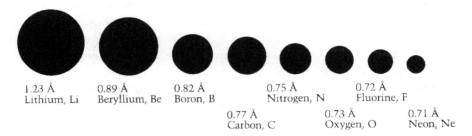

FIGURE 6.8 The decrease of atomic radii within a period.

Table 6.2 Periodic Table of Atomic Radii (in Angstroms)

Period	Group IA	IIA	IIIB	IVB	VB	VIB	VIIB	VIIIB			IB	IIB	IIIA	IVA	VA	VIA	VIIA	VIIIA
1	1 H 0.32																	2 He 0.31
2	3 Li 1.23	4 Be 0.89			Transition Elements								5 B 0.82	6 C 0.77	7 N 0.75	8 O 0.73	9 F 0.72	10 Ne 0.71
3	11 Na 1.54	12 Mg 1.36											13 Al 1.18	14 Si 1.11	15 P 1.06	16 S 1.02	17 Cl 0.99	18 Ar 0.98
4	19 K 2.03	20 Ca 1.74	21 Sc 1.44	22 Ti 1.32	23 V 1.22	24 Cr 1.18	25 Mn 1.17	26 Fe 1.17	27 Co 1.16	28 Ni 1.15	29 Cu 1.17	30 Zn 1.25	31 Ga 1.26	32 Ge 1.22	33 As 1.20	34 Se 1.17	35 Br 1.14	36 Kr 1.12
5	37 Rb 2.16	38 Sr 1.91	39 Y 1.62	40 Zr 1.45	41 Nb 1.34	42 Mo 1.30	43 Tc 1.27	44 Ru 1.25	45 Rh 1.25	46 Pd 1.28	47 Ag 1.34	48 Cd 1.48	49 In 1.44	50 Sn 1.40	51 Sb 1.40	52 Te 1.36	53 I 1.33	54 Xe 1.31
6	55 Cs 2.35	56 Ba 1.98	57–71 La–Lu	72 Hf 1.44	73 Ta 1.34	74 W 1.30	75 Re 1.28	76 Os 1.26	77 Ir 1.27	78 Pt 1.30	79 Au 1.34	80 Hg 1.49	81 Tl 1.48	82 Pb 1.47	83 Bi 1.46	84 Po 1.46	85 At 1.45	86 Rn
7	87 Fr	88 Ra 2.20	89–103 Ac–Lr	104 Rf	105 Db	106 Sg	107 Bh	108 Hs	109 Mt									

57 La 1.69	58 Ce 1.65	59 Pr 1.64	60 Nd 1.64	61 Pm 1.63	62 Sm 1.62	63 Eu 1.85	64 Gd 1.62	65 Tb 1.61	66 Dy 1.60	67 Ho 1.58	68 Er 1.58	69 Tm 1.58	70 Yb 1.70	71 Lu 1.56
89 Ac 2.0	90 Th 1.65	91 Pa	92 U 1.42	93 Np	94 Pu	95 Am	96 Cm	97 Bk	98 Cf	99 Es	100 Fm	101 Md	102 No	103 Lr

```
  6        ← Atomic number
  C
 0.77      ← Atomic radius
```

CARRY OUT THE PLAN

Remember that magnesium (Mg) is in period 3, whereas potassium (K) and calcium (Ca) are in period 4. Therefore, an atom of Mg has the smallest atomic radius of the three atoms in question. Next we look at K and Ca, the two elements that are in the same period. Since K is in Group IA and Ca is in Group IIA, a K atom has a larger atomic radius than a Ca atom. Therefore, of the three atoms, Mg is the smallest, Ca is larger, and K is the largest.

LOOK BACK

We glance at the periodic table to confirm our reasoning.

Practice Exercise 6.1

List the following elements in order of *decreasing* atomic radius: S, O, N, and B. Use only the periodic table (Table 6.1); don't look up the atomic radii.

6.7 Ionization Potential

When an electron is pulled completely away from a neutral atom, what remains behind is a positively charged particle called an *ion*. **A positive ion** (also called a **cation**) is *an atom (or group of atoms) that has lost one or more electrons*. **A negative ion** (also called an **anion**) is *an atom (or group of atoms) that has acquired one or more electrons*.

The process of removing an electron from a neutral atom is called *ionization*. Here is what happens when a neutral atom of potassium becomes ionized:

$$K \rightarrow K^{1+} + 1e^{1-}$$

The equation shows the neutral potassium atom (on the left) losing one electron to become a potassium ion (on the right). The superscript 1+ on the potassium shows that it has become a *positively charged particle*. That is, it has lost an electron ($1e^{1-}$), which is negatively charged. Thus the potassium is left with one more positive charge (proton) than negative charge (electron). It is no longer electrically neutral.

The **ionization potential** (also called *ionization energy*) of an element is *the energy needed to pull an electron away from an isolated ground-state atom of that element*. The farther away from the nucleus an electron is, the less it is attracted by the positive charge of the nucleus and the less energy is needed to pull that electron away from the atom. For this reason, the electron is always pulled away from the outermost energy level. In any group of the periodic table, the ionization potential decreases as we move down the group (Figure 6.9).

Ionization energy is usually measured in units called calories. A calorie is the amount of heat needed to raise the temperature of one gram of water one degree Celsius.

Within each period, ionization potential generally increases from left to right (Table 6.3). This results from the decrease in atomic radius (and thus the stronger attraction between nucleus and outer electrons) as we move

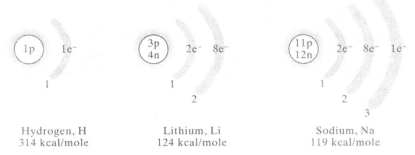

FIGURE 6.9 The decrease of ionization potential within a group.

TABLE 6.3 Periodic Table of Ionization Potentials (in Kilocalories per Mole)

Period	IA	IIA	IIIB	IVB	VB	VIB	VIIB		VIIIB		IB	IIB	IIIA	IVA	VA	VIA	VIIA	VIIIA
1	1 H 314																	2 He 566
2	3 Li 124	4 Be 215			Transition Elements								5 B 191	6 C 260	7 N 335	8 O 312	9 F 402	10 Ne 498
3	11 Na 119	12 Mg 176											13 Al 138	14 Si 188	15 P 254	16 S 239	17 Cl 300	18 Ar 363
4	19 K 100	20 Ca 141	21 Sc 151	22 Ti 158	23 V 155	24 Cr 156	25 Mn 171	26 Fe 182	27 Co 181	28 Ni 176	29 Cu 178	30 Zn 216	31 Ga 138	32 Ge 187	33 As 242	34 Se 225	35 Br 273	36 Kr 323
5	37 Rb 96	38 Sr 131	39 Y 152	40 Zr 160	41 Nb 156	42 Mo 166	43 Tc 172	44 Ru 173	45 Rh 178	46 Pd 192	47 Ag 174	48 Cd 207	49 In 133	50 Sn 169	51 Sb 199	52 Te 208	53 I 241	54 Xe 280
6	55 Cs 90	56 Ba 120	57–71 La–Lu	72 Hf 127	73 Ta 140	74 W 184	75 Re 181	76 Os 201	77 Ir 212	78 Pt 208	79 Au 212	80 Hg 241	81 Tl 141	82 Pb 1.47	83 Bi 171	84 Po 184	85 At 196	86 Rn 248
7	87 Fr	88 Ra 122	89–103 Ac–Lr	104 Rf	105 Db	106 Sg	107 Bh	108 Hs	109 Mt									

57 La 129	58 Ce 159	59 Pr 133	60 Nd 145	61 Pm	62 Sm 129	63 Eu 131	64 Gd 142	65 Tb 155	66 Dy 157	67 Ho	68 Er	69 Tm	70 Yb 143	71 Lu 115
89 Ac 162	90 Th	91 Pa	92 U 92	93 Np	94 Pu	95 Am	96 Cm	97 Bk	98 Cf	99 Es	100 Fm	101 Md	102 No	103 Lr

6 C 260 ← Atomic number / Ionization potential

from left to right in the periodic table. The ionization potentials of the transition metals in each period do not vary much, because their atomic radii do not vary much (see Section 6.6).

EXAMPLE 6.2

List the following elements in order of *increasing* ionization potential: Ca, Mg, and K. Use only the periodic table (Table 6.1); don't look up the atomic radii.

Solution

Locate the position of each element in the periodic table. Remember that the ionization potential decreases from the top to the bottom of a group of elements and increases from left to right across a period of elements. Magnesium (Mg) is in period 3, whereas potassium (K) and calcium (Ca) are in period 4. Therefore, an atom of Mg has the highest ionization potential of the three atoms in question.

Now consider K and Ca, the two elements that are in the same period. Because K is in Group IA and Ca is in Group IIA, a Ca atom has a higher ionization potential than a K atom.

Therefore, of the three atoms, K has the lowest ionization potential, Ca has a higher ionization potential, and Mg has the highest ionization potential. (*Note*: Look back at the solution of Example 6.1. Notice how the ionization potential varies inversely with the atomic radius. In other words, as the atomic radius of an atom gets larger, the ionization potential gets smaller, and vice versa.)

Practice Exercise 6.2

List the following elements in order of decreasing ionization potential: S, O, N, and B. Use only the periodic table (Table 6.1); don't look up the atomic radii.

TABLE 6.4 Electron Affinities of Group VIIA Elements

ELEMENT	ELECTRON AFFINITY (IN ELECTRON VOLTS)	OUTER ELECTRON CONFIGURATION
F	3.6	$2s^2\ 2p^5$
Cl	3.75	$3s^2\ 3p^5$
Br	3.53	$4s^2\ 4p^5$
I	3.2	$5s^2\ 5p^5$

6.8 Electron Affinity

An atom's ability to acquire an additional electron is called its **electron affinity** and is another important factor that determines its chemical properties. The process in which a neutral atom acquires an electron can be represented in this way:

$$Cl + e^{1-} \rightarrow Cl^{1-}$$

The equation shows that when a neutral chlorine atom picks up an electron, it becomes a negatively charged chloride ion. When the chlorine atom picks up an electron to form a chloride ion, it gives up something in return: *energy*. The amount of energy the chlorine atom releases is a measure of its electron affinity. Electron affinity depends on the attraction between an electron and the nucleus of an atom.

As we move down a group, the electron affinity of the elements generally decreases. If we look at some Group VIIA elements, we can see a decrease in electron affinity as we move from chlorine to iodine (Table 6.4). This is because in iodine, the additional electron is added to the fifth energy level, whereas in chlorine the additional electron is added to the third energy level. (Remember that the closer the electron is to the nucleus, the more it is attracted to the nucleus, and electron affinity depends on how strongly an electron is attracted to the nucleus.) Fluorine's unexpectedly low electron affinity is caused by the special nature of this element. (Some scientists feel that the electron affinity value for fluorine is reasonable but that the electron affinity value for chlorine is too high.)

If we examine the trend of electron affinity within a period, we find that it *increases* as we go from left to right in a period, and the attraction between the nucleus and the electron becomes greater.

Table 6.5 summarizes what we have discussed about trends in the periodic table. As your study of chemistry progresses, you will find that you rely more and more on information conveyed by the periodic table.

SUMMARY

As more and more elements were discovered during the eighteenth and nineteenth centuries, scientists attempted to find relationships among the properties of the various known elements. Finally, in 1869, Dmitri Mendeleev arranged the elements in a grid or table form, with horizontal rows called periods and vertical columns called groups. Elements with similar chemical properties were positioned in the same group. The table ordered the known elements and their properties so well that Mendeleev was able to use it to predict the properties of yet undiscovered elements. The periodic table of today is similar to Mendeleev's. It consists of seven periods and sixteen groups. For elements in the A groups (representative elements), the group number is the same as the number of electrons in the outermost shell. For elements in the B groups (transition metals), there is not a

TABLE 6.5 Summary of Trends in the Periodic Table

TREND OF	TOP TO BOTTOM IN A GROUP	LEFT TO RIGHT IN A PERIOD
Atomic radius	Increases	Decreases
Ionization potential	Decreases	Increases
Electron affinity	Decreases	Increases

simple connection between group number and number of electrons in the outermost energy level. However, because the transition metals fill inner shells after outer shells, they show quite similar chemical behavior.

The atomic radius of an atom is the distance from the center of the nucleus to the outermost electron. As we move from the top to the bottom of any group in the periodic table, atomic radius generally increases. And as we move from left to right across a period, atomic radius generally decreases. The ionization potential for an element is the energy needed to pull an outershell electron away from an isolated atom of that element. As we move from the top to the bottom of a group, the ionization potential decreases. And as we move from left to right across a period, the ionization potential generally increases. Electron affinity, a measure of an atom's ability to acquire additional electrons, is related to the energy released when an additional electron is added to a neutral atom. As we move down a group in the periodic table, the electron affinity of the elements generally decreases, whereas as we move from left to right across a period, the electron affinity increases.

KEY TERMS

anion (**6.7**)
atomic radius (**6.6**)
cation (**6.7**)
electron affinity (**6.8**)
group (family) (**6.2**)
ionization potential (**6.7**)

negative ion (**6.7**)
period (**6.2**)
positive ion (**6.7**)
representative element (**6.2**)
transition metal (**6.2**)

SELF-TEST EXERCISES

LEARNING GOAL 1

Historical Basis and Significance of Periodic Table

◀ 1. Julius Lothar Meyer published a periodic table at about the same time that Dmitri Mendeleev published his table. Why do you think that Mendeleev received the major portion of the credit for discovering the periodic law?
2. How did knowledge of the periodic law help future scientists discover new elements?
◀ 3. Locate at least two instances in the periodic table (Table 6.1) in which a heavier element appears before a lighter one.
4. What was Mendeleev's reason for placing some heavier elements before lighter ones in his table? Why didn't this upset the "logic" of his table?
◀ 5. Why aren't elements 57 to 71 and 89 to 103 shown with the other elements in the body of Table 6.1?
6. Discuss the contributions to the development of the periodic table made by (a) Dobereiner, (b) Newlands, (c) Mendeleev.

LEARNING GOAL 2

Periods and Groups in the Periodic Table

7. Give the name and symbol of the element from the information that follows.

   (a) The element is in period 4 and group IIIA.
   (b) The element is in period 3 and group VIA.
   (c) The element is in period 6 and group IIB.
   (d) The element is in period 7 and group VIIB.

8. In the periodic table, a horizontal row is called a _____. A vertical column is called a _____.

◀ 9. What similarity is there among the outermost energy levels of elements within the same group?
10. "Elements in the B groups (the transition metals) show somewhat similar chemical behavior." Explain why this statement is true.

CHAPTER 6: THE PERIODIC TABLE

LEARNING GOAL 3

A-Group and B-Group Elements

11. What does the group number tell us about the A-group elements?

◀ *12*. How many elements are in the following periods?

(a) period 2
(b) period 3
(c) period 4
(d) period 5
(e) period 6
(f) period 7

13. The *representative* elements are also known as the _____ (A/B) group elements. The *transition metals* are also known as the _____ (A/B) group elements.

14. "Elements within any one group have the same number of electrons in their outermost energy levels." Is this statement true for the A-group or the B-group elements?

LEARNING GOAL 4

Periodic Trends

◀ *15*. List the following elements in order of increasing ionization potential: Br, Sr, I, Te, and Ba. Use only the periodic table (Table 6.1); don't look up the ionization potentials.

16. What is an angstrom, and how big is it?

17. List the following elements in order of increasing atomic radius: In, P, I, Sb, and As. Once again, use only the periodic table.

18. Does atomic radius increase or decrease as we move from the top to the bottom of a group in the periodic table? Why?

19. Define *electron affinity*.

20. As we move from left to right across a period in the periodic table, does atomic radius increase or decrease? Why?

◀ 21. After you've removed one electron from a neutral atom, would you need more or less energy to remove a second electron? Why?

22. List the following elements in order of increasing atomic radius: In, Tl, H, F, and S. (Use the periodic table.)

23. Where in the periodic table do we find the following?

(a) elements with the highest electron affinity
(b) elements with the lowest ionization potential
(c) the most unreactive elements

24. What name is given to the energy needed to pull an electron away from an isolated atom?

25. Complete the following statements:

(a) The ionization potential _____ (increases/decreases) as one moves down a *group* of elements.

(b) The ionization potential _____ (increases/decreases) as one moves from left to right across a *period* of elements.

(c) The atomic radius _____ (increases/decreases) as one moves from left to right across a *period* of elements.

◀ 26. Explain why it is more difficult to remove an electron from an atom that is located at the top of a group in the periodic table than to remove an electron from an atom that is located at the bottom of a group.

◂ 27. List the following elements in order of increasing atomic radius: Cl, Ca, Sr, Rb, and S. Use only the periodic table (Table 6.1); don't look up the atomic radii.

◂ 28. The elements sodium and chlorine are both in period 3 of the periodic table. Sodium has an ionization potential of 119 kcal/mole, and chlorine has an ionization potential of $30\overline{0}$ kcal/mole. Why is the ionization potential of chlorine so much greater?

29. List the following elements in order of increasing ionization potential: P, I, Sb, and As. Once again, use only the periodic table.

30. List the following elements in order of increasing ionization potential: Br, Ge, Se, Cs, and F. Use only the periodic table.

31. A positive ion is an atom (or group of atoms) that has (a) gained a proton, (b) lost a proton, (c) gained an electron, (d) lost an electron.

32. The quantity that depends on the attraction between an electron and the nucleus of an atom is called _____.

33. A Cl^{1-} ion is a Cl atom that has (a) gained a proton, (b) lost a proton, (c) gained an electron, (d) lost an electron.

34. Electron affinity _____ (increases/decreases) as one moves down a *group* of elements.

35. An Mg^{2+} ion is an Mg atom that has (a) gained two protons, (b) lost two protons, (c) gained two electrons, (d) lost two electrons.

36. Electron affinity _____ (increases/decreases) as one moves from left to right across a *period* of elements.

EXTRA EXERCISES

37. (a) Explain why a bromine atom is smaller than a potassium atom, even though a bromine atom contains more subatomic particles.

    (b) Explain why the ionization potential for a sodium atom is greater than the ionization potential for a rubidium atom.

38. What relationship is there with regard to chemical properties among

    (a) all elements in a particular period?

    (b) all elements in a particular group?

39. Without using the periodic table, find the electron configurations of the following elements. State which ones are chemically similar.

    (a) $_{11}Na$ (b) $_{13}Al$ (c) $_8O$ (d) $_{16}S$ (e) $_5B$ (f) $_3Li$

40. Without using the periodic table, find the electron configurations of the following elements. State which ones are chemically similar.

    (a) $_1H$ (b) $_3Li$ (c) $_4Be$ (d) $_{12}Mg$ (e) $_{20}Ca$

41. What would you expect the atomic number to be for the noble gas that follows radon?

42. Based on periodic trends, which group of atoms is most likely to form (a) 1+ ions? (b) 1− ions?

43. Explain the difference between a chlorine atom and a chloride ion.

44. Which member of each of the following pairs should have the larger radius? Why?

    (a) K, Br (b) K, Na (c) K, K^{1+}

45. Determine how many electrons are in the outermost energy level of a neutral atom of the following elements:

    (a) sodium (b) carbon (c) nitrogen

46. Which has the higher ionization potential, calcium or magnesium? Why?

47. How many elements are in the (a) first period? (b) second period? (c) third period? (d) fourth period? (e) fifth period? (f) sixth period?
48. Which element of each pair has the smaller ionization potential?

    (a) K, Ca   (b) O, Se

49. Which element of each pair has the larger atomic radius?

    (a) K, Ca   (b) O, Se

50. What would be the group number for element 114?
51. Why are the atomic masses of elements 94 through 109 listed in parentheses?
52. If you had the responsibility of naming the next discovered element, what would you name it?
53. We know that the number of electrons in the outermost energy level of the Group IIIB elements is 3 because the group number reflects the number of electrons in the outermost energy level. True or false? Why?
54. We know that the number of electrons in the outermost energy level of the Group IIIA elements is 3 because the group number reflects the number of electrons in the outermost energy level. True or false? Why?
55. By writing electron configurations, determine which of the following elements belong to the same groups: elements with atomic number 15, 38, 50, 56, 14, and 33.
56. By writing electron configurations, determine which of the following elements belong to the same groups: elements with atomic number 16, 36, 39, 8, and 34.

CUMULATIVE REVIEW Chapters 4–6

1. How are protons, neutrons, and electrons distributed in the atom?
2. How did Eugen Goldstein choose the term *cathode rays*?
3. What experiment did Sir William Crookes perform to determine whether cathode rays were composed of particles or waves?
4. What is the photoelectric effect?
5. What reasoning led physicist J. J. Thomson to develop the "plum-pudding" theory of the atom?
6. When Rutherford tested Thomson's theory of the atom with his own experiments, what results led him to devise a new model of the atom?
7. Why did Rutherford's model of the atom render Thomson's atomic model obsolete?
8. What type of particle adds to the mass of the atom without affecting its charge?
9. John Dalton said that atoms of an element must be identical in size, mass, and shape. How does the existence of isotopes disprove this theory?
◀ 10. Give the number of protons that are found in each of the following isotopes:

    (a) $^{59}_{27}Co$ (b) $^{90}_{40}Zr$ (c) $^{74}_{34}Se$

11. Using Table 4.2, calculate the atomic mass of neon.
12. Using Table 4.2, determine how many ^{15}N atoms there are in a sample containing 100,000 nitrogen atoms.
◀ 13. Write the standard isotopic notation for the following elements. (Use the periodic table to get the symbols of the elements.)

    (a) 17p, 17e, 18n   (b) 14p, 14e, 14n

14. The hypothetical element Z exists in two isotopic forms, ^{50}Z and ^{52}Z. The atomic mass of Z is 50.5. What is the percentage abundance of each isotope?
15. State the number of protons, electrons, and neutrons in $^{73}_{32}Ge$.
16. How many electrons are there in a neutral $^{60}_{26}Fe$ atom?

110 ESSENTIAL CONCEPTS OF CHEMISTRY

17. Of the three isotopes of oxygen, which is the most abundant: oxygen-16, oxygen-17, or oxygen-18?
18. Using Table 4.2, determine how many ^{7}Li atoms there are in a sample containing 2.0×10^3 lithium atoms.
◀ 19. Using Table 4.2, determine how many atoms of ^{7}Li you would find in a sample of lithium that has 1.000×10^6 atoms of Li.
20. Using Table 4.2, determine how many atoms of Si you would need to get exactly 470,000 atoms of ^{29}Si.
21. Do electrons, protons, or neutrons produce the line spectrum of an element?
22. As an electron travels to an energy level farther from the nucleus, the energy of that electron _____.

   (a) increases   (b) decreases   (c) remains the same

23. How many electrons are in the outermost energy level of an oxygen atom?
24. Without using the periodic table, determine the electron configurations of the following elements:

   (a) $_1$H   (b) $_3$Li   (c) $_{11}$Na   (d) $_{19}$K

25. A neutral atom of magnesium has _____ electrons in its outermost energy level. (You may consult the periodic table.)
26. Name the element that has the following groundstate configuration. You may consult the periodic table.

   K, 2 electrons; L, 8 electrons; M, 6 electrons

◀ 27. The chemical properties of an element are determined by the number of electrons in the _____.

   (a) energy level closest to the nucleus
   (b) outermost energy level
   (c) second energy level
   (d) none of these

28. Why is phosphorus in Group VA?
◀ 29. Write electron configurations for all elements with atomic numbers between 1 and 20 that have the same number of electrons in the outermost energy level as a lithium atom.
30. What is the atomic number of an element whose energy levels K, L, M, and N are filled to capacity?
31. When a rainbow is produced naturally, what acts as the prism?
32. How are line spectra used in chemical laboratories today?
33. In a detective story it was discovered that the victim died of arsenic poisoning. This information was made available by analysis of samples of the victim's hair. How could spectral analysis have made this discovery possible?
34. What is the major source of radiant energy?
35. Waves on a lake travel through water, and sound waves travel through air. Through what medium do electromagnetic waves travel?
36. What types of waves are included in the electromagnetic spectrum?

   (a) visible   (b) invisible   (c) both of these   (d) neither of these

37. How did Heinrich R. Hertz verify the theory of electromagnetic radiation proposed by James Clerk Maxwell?
38. Give an example of light energy being absorbed and re-emitted.
39. Give an example of heat energy being absorbed and re-emitted as light energy.
◀ 40. Bundles of energy absorbed or emitted by electrons are called _____.
41. State the periodic law and explain how it is the basis of the modern-day periodic table.
◀ 42. What was the contribution of Henry G. J. Moseley in terms of the periodic table?
43. The element with atomic number 107 is an artificially produced radioactive element. Both Soviet and German scientists claim to have discovered it, but neither claim has been officially accepted. To produce an isotope of the element with atomic number 107, a sample of bismuth-209 was bombarded with chromium-54. Write the electron configuration of the element with atomic number 107. What is its group number?

44. In 1974 a team of U.S. scientists working under Albert Ghiorso at the Lawrence Berkeley Laboratory in Berkeley, California, bombarded the element californium-249 with a beam of oxygen-18 to produce an isotope of the element with atomic number 106. Write the electron configuration for the element with atomic number 106. What is its group number?
45. Our modern periodic table contains cases where elements are not in proper order according to atomic mass. Where do these cases occur?
46. Ionization potential increases as one moves down a group of elements. True or false?
47. Atomic radius decreases as one moves from left to right across a period of elements. True or false?
◀ 48. A Ca^{2+} ion is a calcium atom that has gained two protons. True or false?
49. The element with atomic number 108 should behave like what other elements in the periodic table?
50. Name the neutral element with the electron configuration $1s^2 2s^2 2p^6 3s^1$.
51. The elements in the periodic table show properties that are related to their atomic numbers. True or false?
◀ 52. The element with the electron configuration $1s^2 2s^2 2p^6$ belongs in group VIA. True or false?
53. Lithium and fluorine are both in period 2 of the periodic table. Which element has the higher ionization potential? Why?
54. Write the electron configuration for $_{14}Si$.
55. The volume of space occupied by an electron is called an orbital. True or false?
◀ 56. On the basis of periodic trends, which group of atoms is most likely to form (a) 2+ ions? (b) 2-ions?
57. Which member of each of the following pairs should have the larger radius?

(a) Li, Na (b) Ca, Ca^{2+}

58. Which element has the lower ionization potential, Mg or Cl?
59. Which element has the larger atomic radius, Li or Cs?
60. Which element has the larger atomic radius, Li or F?

CHAPTER 7

CHEMICAL BONDING
How Atoms Combine

The time is about five billion years ago. You are suspended in the void of space. But look, our sun is not there. Neither is the Earth or any of the other planets in our present solar system. In the distance we see what looks like a giant star. Its light flickers, and it is about to explode.

There is no noise in space, but we see a tremendous explosion that seems to envelop us. The black void is filled with blinding light. It is our solar system forming before our eyes, out of the remains of an exploding star! All of the basic elements that we are familiar with in the periodic table are present. The most abundant element is hydrogen. A tremendous amount of hydrogen forms our sun. Other elements that are present begin to coalesce and form planets. Many of these elements combine to form compounds.

On one particular planet the conditions are just right so that, over the course of the next five billion years, the simple compounds that are formed react with each other to form more complex compounds. Eventually these compounds form living organisms of a very complex nature. This planet is the third planet from the sun, and we are just some of those complex organisms found living on it.

LEARNING GOALS

After you've studied this chapter, you should be able to:

1. Write Lewis (electron-dot) structures for the A-group elements.
2. Define and give an example of a covalent bond and a coordinate covalent bond.
3. Name the diatomic elements from memory.
4. Write electron-dot structures for various covalent compounds.
5. Define and give an example of an ionic bond.
6. Use the concept of electronegativity to find the ionic percentage and covalent percentage of a chemical bond.
7. Distinguish between a polar and a nonpolar bond.
8. Distinguish between a polar and a nonpolar molecule.

114 ESSENTIAL CONCEPTS OF CHEMISTRY

INTRODUCTION

Matter is held together by forces acting between atoms. When chemical reactions occur, atoms can combine to form compounds. In doing so, they join together in such a way as to attain more stable configurations that have lower levels of chemical potential energy.

Atoms have the ability to gain, lose, or share electrons. This enables them to form compounds in two different ways. When electrons are gained and lost, the compounds formed are held together by forces that bind oppositely charged ions. When electrons are shared, the compounds formed are held together by forces that bind atoms together to form molecules.

Whether the binding forces are due to electron loss and gain or to sharing, when atoms combine they are said to be joined together by a **chemical bond**. They have attained a more stable arrangement—usually eight electrons in their outermost shell. In this chapter, we examine the types of bonds that tie atoms together in molecules and unite elements in compounds.

7.1 Lewis (Electron-Dot) Structures

Because it is the electrons in the outermost energy levels that form the bonds, we will be concerned only with these electrons in the elements we study. A very useful notation for showing the outermost electrons of an atom (and the bonds they form) was devised in 1916 by G. N. Lewis, an American physical chemist. In this *electron-dot* notation, the outer electrons are shown as dots on the sides of the symbol for the element. (Figure 7.1 provides several examples.) Each dot corresponds to an outer-shell electron. Although dot placement isn't critical, one method is to place the dots one to a side, beginning at the top and moving clockwise. A second dot is then placed on a side only after there is one dot on each of the four sides of the symbol. If you move from left to right in Figure 7.1, you will see how each additional dot (as necessary) is placed on the diagram. Electron-dot structures make it easier to see the role outer-shell electrons play in chemical bonding.

Recall now that all the elements in the same A group have the same number of electrons in their outermost energy level. This means that all the elements in any A group have the same electron-dot diagram. Figure 7.2 shows the diagrams for fluorine and bromine, which are both in Group VIIA. Both have seven dots. And Figure 7.3 shows the diagrams for nitrogen and arsenic; both are in Group VA, so both have five dots.

Now that you know how to write the electron-dot notation for an atom of an element, we'll learn how to use this notation to draw the *electron-dot structure* of a compound. The **electron-dot structure** is *a*

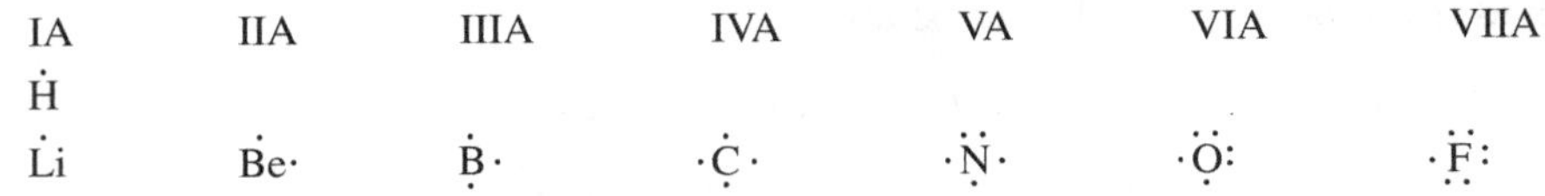

IA	IIA	IIIA	IVA	VA	VIA	VIIA
Ḣ						
Li	Be·	B·	·C·	·N·	·Ö:	·F̈:

FIGURE 7.1 Electron-dot diagrams of some A-group elements.

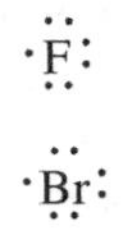

·F̈:

·B̈r:

FIGURE 7.2 The electron-dot diagrams of all Group VIIA elements are the same.

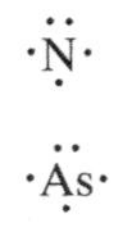

·N·

·As·

FIGURE 7.3 The electron-dot diagrams of all Group VA elements are the same.

schematic representation of the bonding in a molecule of a compound. It does not represent the physical position of the electrons in the molecule. However, the electron-dot notation is useful for understanding how bonding occurs.

7.2 The Covalent Bond: The Octet Rule

It has long been known that the Group VIIIA elements tend to be very unreactive and stable. Because of their tendency to keep to themselves, these elements are called the *noble gases*. Researchers believe that this stability is connected with the fact that each of these Group VIIIA elements has eight electrons in its outermost energy level. (The element helium is an exception because its atom has only two electrons, but these two electrons fill the outermost energy level of the helium atom.) The atoms of other elements can achieve the same type of stability by obtaining eight electrons in their outermost energy levels. We state this relationship formally as the **octet rule**, which tells us that when forming compounds, atoms of elements gain, lose, or share electrons such that a noble gas configuration is achieved for each atom. When atoms share electrons with other atoms, a *covalent bond* is formed. A **covalent bond** is *a chemical bond in which two atoms share a pair of electrons*.

To see how electrons are shared, imagine a sample of fluorine atoms in a closed container. A fluorine atom has seven electrons in its outermost energy level. It needs eight electrons in that energy level to complete its octet. One way to obtain another electron is to share an outer electron belonging to another fluorine atom (Figure 7.4). (Examine Figure 7.4, and think of the arrow as indicating that the items on the left combine to form what is on the right. We will discuss such chemical equations in more detail in Chapter 10.) When a fluorine atom shares one of its electrons with another fluorine atom, a fluorine molecule (F_2) is formed. But now each atom has a stable octet. Evidence of this stability is the fact that fluorine is always found as F_2 molecules.

The type of covalent bond formed between the two fluorine atoms is called a **single covalent bond** because each atom has shared a *single* electron with the other to form one bond. The single bond is the most common type of covalent bond.

The Diatomic Elements

Diatomic elements are elements that are found naturally in molecules with two atoms each. Fluorine is thus a *diatomic element*. In fact, all the Group VIIA elements are diatomic. They all form single covalent bonds (Figure 7.5). Hydrogen, oxygen, and nitrogen also exist in diatomic forms (Figure 7.6) because they share each other's electrons.

An easy way to remember the diatomic elements is with the memory aid HONClBrIF (pronounced "HON kel brif"). Remember that these atoms exist as diatomic molecules because doing so yields a stable octet of electrons or, in the case of hydrogen, a stable duet (two electrons, like helium).

FIGURE 7.4 Bonding between fluorine atoms.

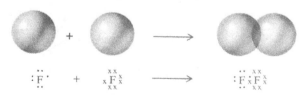

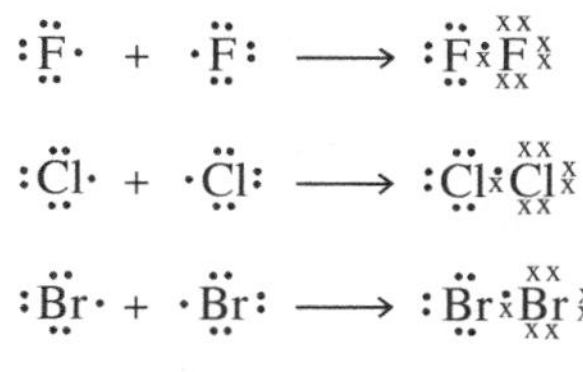

FIGURE 7.5 Bonding diagrams of Group VIIA diatomic elements.

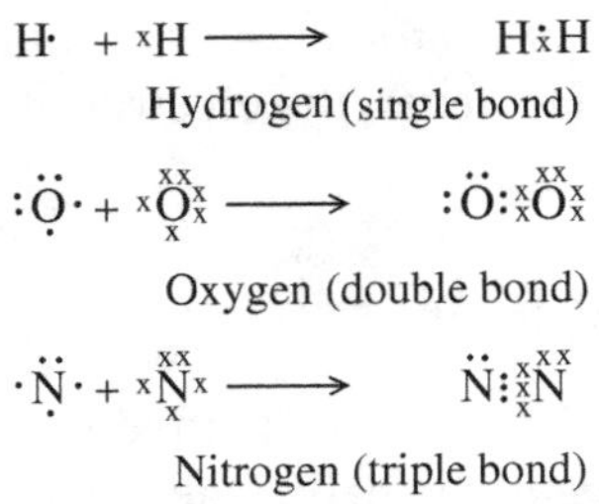

116 ESSENTIAL CONCEPTS OF CHEMISTRY

$$H\cdot + {}_xH \longrightarrow H{:}H$$
Hydrogen (single bond)

$$:\ddot{O}\cdot + {}_x\ddot{O}{}_x^x \longrightarrow :\ddot{O}{:}\ddot{O}{}_x^x$$
Oxygen (double bond)

$$\cdot\ddot{N}\cdot + {}_x\ddot{N}{}_x^x \longrightarrow \ddot{N}{:}\ddot{N}$$
Nitrogen (triple bond)

FIGURE 7.6 Bonding diagrams for other diatomic elements. Hydrogen needs only two electrons to complete its outer energy level. In O_2, two electrons from each contribute to the bond. Three electrons from each nitrogen atom contribute to the bond.

Other Molecules Having Single Covalent Bonds

Now let's see how some elements obtain eight electrons (or, in the case of hydrogen, two electrons) in their outermost energy levels by forming compounds. We will use the electron-dot notation to show the bonding.

Water, which has the formula H_2O, bonds in this way:

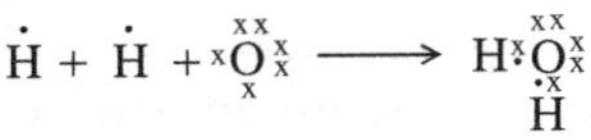

$$\dot{H} + \dot{H} + {}_x\ddot{O}{}_x^x \longrightarrow H{:}\ddot{O}{}_x^x$$
$$H$$

In this diagram an X is used to represent the oxygen's electrons. The two hydrogens share their single electrons with the oxygen. So the oxygen obtains eight electrons in its outermost energy level, and each hydrogen now has its two electrons.

In ammonia (NH_3), methane (CH_4), and hydrogen chloride (HCl), the bonding looks like this:

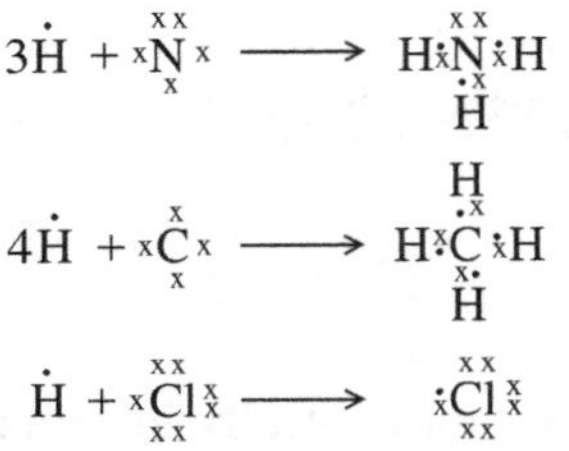

$$3\dot{H} + {}_x\ddot{N}{}_x \longrightarrow H{:}\ddot{N}{:}H$$
$$H$$

$$4\dot{H} + {}_x\ddot{C}{}_x \longrightarrow H{:}\ddot{C}{:}H$$
$$H$$

$$\dot{H} + {}_x\ddot{Cl}{}_x^x \longrightarrow \ddot{:}Cl{}_x^x$$

The N, C, and Cl atoms now have eight electrons in their outermost energy level. The H atoms complete their outer shells with two electrons. The molecules are more stable than the atoms from which they are formed.

The Use of Dashes

Instead of using electron-dot notation to show the bonding in compounds, we sometimes use dashes. The dash (—) has the same meaning as a double dot (:), but it simplifies the diagram. The dash, then, represents a single bond, or a shared pair of electrons. The following examples will show you how the dash is used in combination with dot notation.

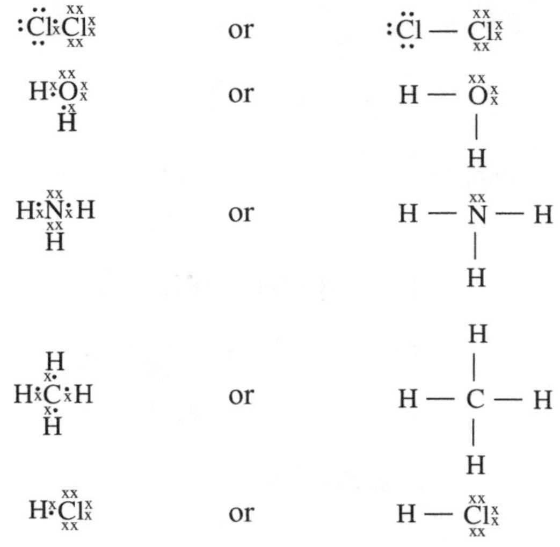

$$:\ddot{Cl}{:}\ddot{Cl}{}_x^x \quad \text{or} \quad :\ddot{Cl} — \ddot{Cl}{}_x^x$$

$$H{:}\ddot{O}{}_x^x \quad \text{or} \quad H — \ddot{O}{}_x^x$$
$$H |$$
$$ H$$

$$H{:}\ddot{N}{:}H \quad \text{or} \quad H — \ddot{N} — H$$
$$H |$$
$$ H$$

$$ H$$
$$ |$$
$$H{:}\ddot{C}{:}H \quad \text{or} \quad H — C — H$$
$$H |$$
$$ H$$

$$H{:}\ddot{Cl}{}_x^x \quad \text{or} \quad H — \ddot{Cl}{}_x^x$$

CHAPTER 7: CHEMICAL BONDING 117

This notation is especially useful when you want to represent the structures of compounds that have double and triple bonds.

The Double Bond

A **double covalent bond** is *one in which two pairs of electrons are shared between two atoms.* For example, in oxygen (O_2) the bonding is

$$\cdot\ddot{O}: + {}_x^x\!\ddot{O}{}_x^x \longrightarrow :\ddot{O}:{}_x^x\!\ddot{O}{}_x^x \quad \text{or} \quad :\ddot{O}={\ddot{O}}{}_x^x$$

By sharing two electrons, each oxygen atom obtains eight electrons in its outermost energy level. (Actually, this bonding structure for O_2 is simplified and not entirely correct, but the simplified structure communicates the general idea.)

The bonding in the compound ethylene (C_2H_4), which is used to make polyethylene plastic, can be written as

$$\begin{array}{cc} H & H \\ {}_x^{\,}C:{}_x^x\!C{}_x^x \\ H & H \end{array} \quad \text{or} \quad \begin{array}{c} H \quad H \\ | \quad | \\ C = C \\ | \quad | \\ H \quad H \end{array}$$

(In this diagram, an x is used to represent the electron in each hydrogen.) Here the hydrogen atoms are satisfied with two electrons, and each of the carbon atoms has the eight electrons it needs.

The Triple Bond

A **triple covalent bond** is *one in which three pairs of electrons are shared between two atoms.* For example, in nitrogen (N_2) the bonding can be diagrammed in this way:

$$\cdot\ddot{N}\cdot + {}_x^x\!\ddot{N}{}_x \longrightarrow \ddot{N}:{}_x^x\!\ddot{N} \quad \text{or} \quad \ddot{N}\equiv\ddot{N}$$

With the triple bond, each nitrogen atom obtains eight electrons in its outermost energy level. The original N atom has only five electrons, but when two N atoms get together by triple bonding, each has a share in eight electrons.

In the compound acetylene (C_2H_2), the bonding is

$$H{:}C{:}{}_x^x\!C{}_x^x\!H \quad \text{or} \quad H-C\equiv C-H$$

EXAMPLE 7.1

Write the electron-dot structures for each of the following covalent compounds.

(a) CO_2 (Each oxygen is bonded to the carbon, but the oxygens aren't bonded to each other.)

(b) CH_3I (The iodine and all the hydrogens are bonded to the carbon.)

(c) $COCl_2$ (This is phosgene. The oxygen and both chlorines are bonded to the carbon.)

(d) $AsCl_3$ (Each chlorine is bonded to the arsenic.)

(e) H_2S (Both hydrogens are bonded to the sulfur.)

(f) CH_2O (This is formaldehyde. The oxygen and both hydrogens are bonded to the carbon.)

(g) C_2Cl_2 (One chlorine is bonded to each carbon.)

Solution

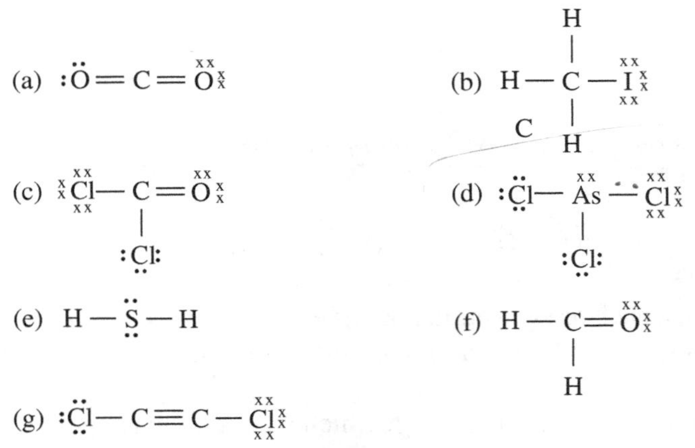

(a) $:\ddot{O}=C=\ddot{O}:$ (xx on right O)

(b)
$$H-\overset{\overset{H}{|}}{\underset{\underset{H}{|}}{C}}-\ddot{I}:$$

(c) $:\ddot{Cl}-\underset{\underset{:\ddot{Cl}:}{|}}{C}=\ddot{O}:$

(d) $:\ddot{Cl}-\underset{\underset{:\ddot{Cl}:}{|}}{\ddot{As}}-\ddot{Cl}:$

(e) $H-\ddot{S}-H$

(f) $H-\underset{\underset{H}{|}}{C}=\ddot{O}:$

(g) $:\ddot{Cl}-C\equiv C-\ddot{Cl}:$

Practice Exercise 7.1

Write the electron-dot structures for the following covalent compounds: (a) NH3 (b) HCL (c) SO2

7.3 The Coordinate Covalent Bond

In the examples of covalent bonds that we have discussed so far, both atoms furnished electrons to form the bond. This can be thought of as a "Dutch-treat" bond, in which each atom pays its share of the bill. Another type of covalent bond exists. It is called a **coordinate covalent bond**, *a covalent bond in which only one atom donates the electrons to form the bond.*

A coordinate covalent bond can be thought of as a "you-treat" bond: one atom pays the entire bill. An example of this type of bond is found in the compound phosphoric acid (H_3PO_4). A coordinate covalent bond forms between the phosphorus atom and the single oxygen atom, to which it gives two electrons.

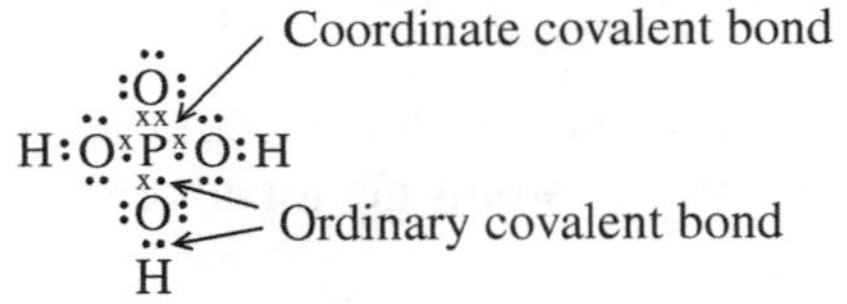

$$H:\ddot{O}:\overset{:\ddot{O}:}{\underset{\underset{H}{:\ddot{O}:}}{P}}:\ddot{O}:H$$

(Coordinate covalent bond indicated at top O; Ordinary covalent bond at bottom O)

Ordinary (not coordinate) covalent bonds form between the phosphorus atom and the other three oxygen atoms (those with hydrogens attached to them). Once formed, the coordinate covalent bond is indistinguishable from an ordinary covalent bond. However, in drawing bonding diagrams of compounds that have coordinate covalent bonds, we sometimes use an arrow to designate the coordinate covalent bond. For example, H_3PO_4 may be diagrammed as follows:

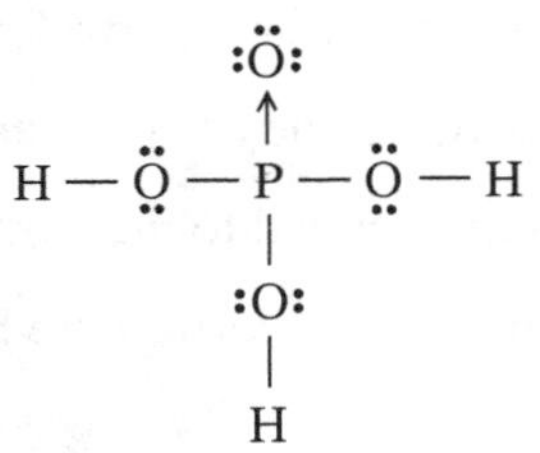

$$H-\ddot{O}-\overset{\overset{:\ddot{O}:}{\uparrow}}{\underset{\underset{H}{:O:}}{P}}-\ddot{O}-H$$

Note that the arrow does not mean that the bond is in any way different.

EXAMPLE 7.2

Draw bonding diagrams for the following covalent compounds that have coordinate covalent bonds included in their structures.

(a) SO_3 (Each oxygen is bonded to the sulfur.)
(b) H_2SO_4 (Each oxygen is bonded to the sulfur, and each of the hydrogens is bonded to a different oxygen.)

Solution

(a)
$$\ddot{\underset{..}{O}}: \\ \| \\ :\ddot{\underset{..}{O}} \leftarrow S \rightarrow \ddot{\underset{..}{O}}:$$

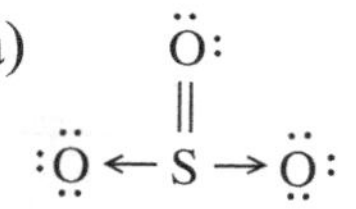

There are two coordinate covalent bonds in this compound.

(b)
$$:\ddot{O}: \\ \uparrow \\ H - \ddot{\underset{..}{O}} - S - \ddot{\underset{..}{O}} - H \\ \downarrow \\ :\ddot{O}:$$

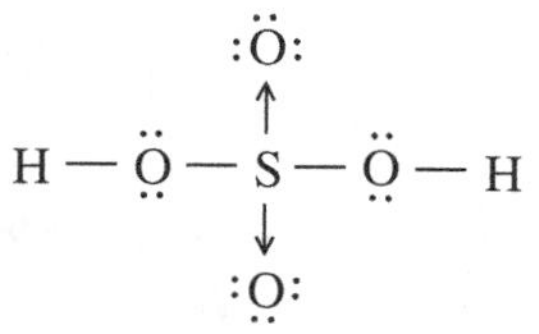

There are two coordinate covalent bonds in this compound.

Practice Exercise 7.2

Draw a bonding diagram for H_2SO_3, sulfurous acid, which has a coordinate covalent bond included in its structure. (*Hint:* Each oxygen is bonded to the sulfur, and each of the hydrogens is bonded to a different oxygen.)

7.4 Ionic Bonding

One atom can actually *give* electrons to another atom so that they both obtain an octet of electrons in their outermost energy level. This process is called *ionic bonding*. An example of ionic bonding is the formation of sodium fluoride (Figure 7.7). A bond formed by sharing electrons (a covalent bond) would help the fluorine *but not the sodium*. If a sodium atom shared a pair of electrons, it would have only two electrons in its outermost energy level:

$$Na{:}\overset{xx}{\underset{xx}{F}}{:}^{xx}$$

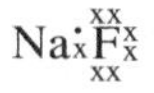

This does not satisfy the octet rule.

However, note what happens if the sodium *gives* or *transfers* its outermost electron to the fluorine. First, the fluorine obtains eight electrons in its outermost energy level. And, in the sodium atom, the second energy level (which contains eight electrons) now becomes the outermost energy level; so the sodium too has a completed octet.

But what holds the atoms together? Remember that the fluorine now has an extra electron, which used to belong to the sodium. This means that the fluorine now has a total of ten electrons and still has only nine protons. Therefore the fluorine atom *has a net electric charge* of 1− (because it has one extra electron).

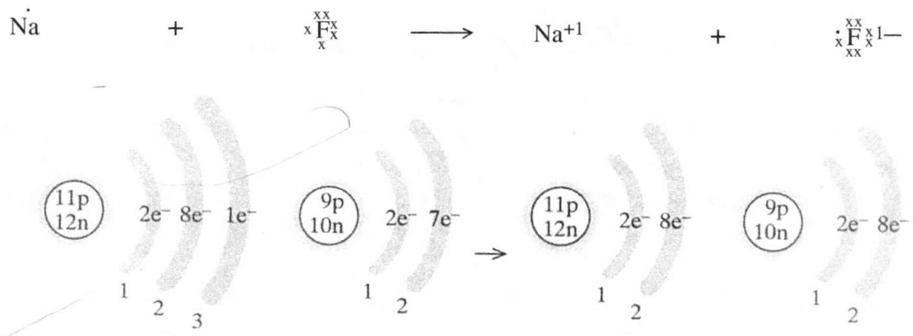

$$\dot{Na} \quad + \quad {}_x^x\overset{xx}{\underset{x}{F}}{}_x \quad \longrightarrow \quad Na^{+1} \quad + \quad {}_x^{xx}\overset{xx}{\underset{xx}{F}}{}_x{}^{1-}$$

FIGURE 7.7 Ionic bonding in sodium fluoride.

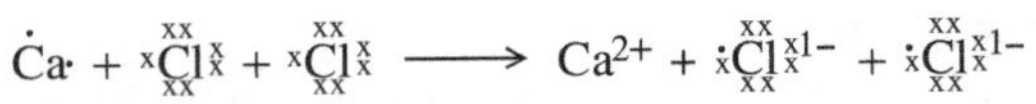

$$\dot{\text{Ca}}\cdot + {}^{\text{xx}}_{\text{xx}}\!\overset{\text{xx}}{\text{Cl}}\overset{\text{x}}{\text{x}} + {}^{\text{xx}}_{\text{xx}}\!\overset{\text{xx}}{\text{Cl}}\overset{\text{x}}{\text{x}} \longrightarrow \text{Ca}^{2+} + {}^{\text{xx}}_{\text{xx}}\!\overset{\text{xx}}{\text{Cl}}\overset{\text{x}}{\text{x}}{}^{1-} + {}^{\text{xx}}_{\text{xx}}\!\overset{\text{xx}}{\text{Cl}}\overset{\text{x}}{\text{x}}{}^{1-}$$

FIGURE 7.8   Ionic bonding in calcium chloride.

The sodium atom has given up an electron, so it now has eleven protons and only ten electrons. Therefore the sodium atom *has a net electric charge* of 1+ (because it has one less electron). The two atoms have become ions. An *ion* is *an atom or group of atoms that has gained or lost electrons and therefore has a net negative or positive charge*. A *monatomic ion* is *an ion composed of a single atom*. The positively charged sodium atom is called a *sodium ion,* written Na^{1+}. The negatively charged fluorine atom is called a fluoride ion, written F^{1-}. Opposite charges attract, so the Na^{1+} and F^{1-} ions are attracted to each other. This attraction makes them form the neutral salt sodium fluoride (NaF). The positive-negative attraction is what holds these ions together. An *ionic bond* is *a chemical bond in which one atom transfers one or more electrons to another atom and the resulting ions are held together by the attraction of opposite charges*.

Here's another example: Calcium and chlorine combine by forming an ionic bond to make calcium chloride ($CaCl_2$), but it takes two chlorines to do the job (Figure 7.8). In this compound, the calcium atom gives up its two outermost electrons, transferring one electron to each chlorine atom. The calcium atom becomes the positively charged Ca^{2+} ion, because it still has 20 protons and now only 18 electrons. Each chlorine atom now has an extra electron, for a total of 18, but it still has only 17 protons. As a result, each chlorine atom becomes a negatively charged chloride ion. The calcium ion and the chloride ions are attracted to each other, forming $CaCl_2$.

7.5 Exceptions to the Octet Rule

There are exceptions to every rule, and not all elements obey the octet rule. For example, in the compound boron trifluoride (BF_3), the boron atom has only six electrons in its outermost energy level (Figure 7.9).

Arsenic pentachloride is another exception to the rule. (This name means that one arsenic is bonded to five chlorines; *penta* is from the Greek word meaning "five.") In this compound, arsenic has *ten* outer electrons (Figure 7.10). In fact, many of the transition metals fail to follow the octet rule. Still, in the vast majority of cases, the octet rule will give you the correct bonding structure.

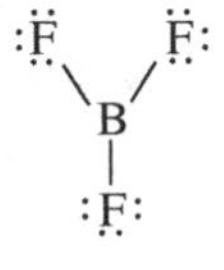

FIGURE 7.9   Covalent bonding in boron trifluoride.

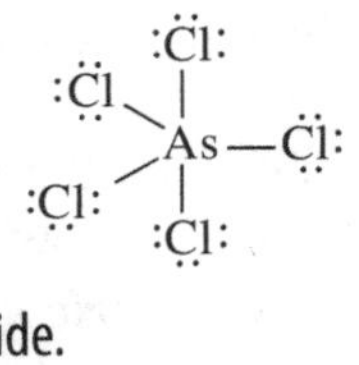

FIGURE 7.10   Covalent bonding in arsenic pentachloride.

7.6 Covalent or Ionic Bonds: The Concept of Electronegativity

At this point you may be asking, "How can I tell whether two elements bond ionically or covalently?" To answer this question, we have to introduce the concept of electronegativity. **Electronegativity** is *the attraction that an atom has for the electrons it is sharing with another atom*. The relative electronegativities for atoms of the different elements are listed in Table 7.1. (Note that no electronegativity values are listed for the Group VIIIA elements, because they tend not to form chemical compounds.)

The electronegativity scale was devised by Linus Pauling, winner of the 1954 Nobel Prize in chemistry and the 1962 Nobel Peace Prize. The scale is based on fluorine having an assigned value of 4.0. Fluorine is the most electronegative element. It tends to attract electrons more strongly than any other element does.

The metal elements generally have low electronegativities, whereas the nonmetals have high electronegativities. This means that the atoms of metals tend to lose electrons more readily than the atoms of nonmetals.

CHAPTER 7: CHEMICAL BONDING 121

TABLE 7.1 Periodic Table of Electronegativities

Period	Group IA	IIA	IIIB	IVB	VB	VIB	VIIB	VIIIB			IB	IIB	IIIA	IVA	VA	VIA	VIIA	VIIIA
1	1 H 2.1																	2 He
2	3 Li 1.0	4 Be 1.5											5 B 2.2	6 C 2.5	7 N 3.0	8 O 3.5	9 F 4.0	10 Ne
3	11 Na 0.9	12 Mg 1.2											13 Al 1.5	14 Si 1.8	15 P 2.1	16 S 2.5	17 Cl 3.0	18 Ar
4	19 K 0.8	20 Ca 1.0	21 Sc 1.3	22 Ti 1.5	23 V 1.6	24 Cr 1.6	25 Mn 1.5	26 Fe 1.8	27 Co 1.8	28 Ni 1.8	29 Cu 1.9	30 Zn 1.6	31 Ga 1.6	32 Ge 1.8	33 As 2.0	34 Se 2.4	35 Br 2.8	36 Kr
5	37 Rb 0.8	38 Sr 1.0	39 Y 1.2	40 Zr 1.4	41 Nb 1.6	42 Mo 1.8	43 Tc 1.9	44 Ru 2.2	45 Rh 2.2	46 Pd 2.2	47 Ag 1.9	48 Cd 1.7	49 In 1.7	50 Sn 1.8	51 Sb 1.9	52 Te 2.1	53 I 2.5	54 Xe
6	55 Cs 0.7	56 Ba 0.9	57–71 La–Lu	72 Hf 1.3	73 Ta 1.5	74 W 1.7	75 Re 1.9	76 Os 2.2	77 Ir 2.2	78 Pt 2.2	79 Au 2.4	80 Hg 1.9	81 Tl 1.8	82 Pb 1.8	83 Bi 1.9	84 Po 2.0	85 At 2.2	86 Rn
7	87 Fr 0.7	88 Ra 0.9	89–103 Ac–Lr	104 Rf	105 Db	106 Sg	107 Bh	108 Hs	109 Mt									

57 La 1.1	58 Ce 1.1	59 Pr 1.1	60 Nd 1.1	61 Pm 1.1	62 Sm 1.1	63 Eu 1.1	64 Gd 1.1	65 Tb 1.1	66 Dy 1.1	67 Ho 1.1	68 Er 1.1	69 Tm 1.1	70 Yb 1.1	71 Lu 1.2
89 Ac 1.1	90 Th 1.3	91 Pa 1.5	92 U 1.7	93 Np 1.3	94 Pu 1.3	95 Am 1.3	96 Cm 1.3	97 Bk 1.3	98 Cf 1.3	99 Es 1.3	100 Fm 1.3	101 Md 1.3	102 No 1.3	103 Lr

6 C 2.5 ← Atomic number / Electronegativity

In fact, the atoms of nonmetals have a strong tendency to pick up electrons. By looking at Table 7.1, you can see the trend for electronegativity, which generally increases as one moves from left to right across a given period of elements and decreases as one moves from the top to the bottom of a group.

To decide whether a bond is ionic or covalent, we find the difference between the electronegativities of the elements forming the bond. Chemists generally agree that if the difference between the electronegativities of the two elements is 2.0 or greater, the bond between the two elements is primarily ionic. If the difference is less than 2.0, the bond is primarily covalent. (Note that 2.0 is simply a convenient number to use as the cutoff point between ionic and covalent bonding. There is nothing special about this number. In fact, we will discuss a more sophisticated way of looking at ionic and covalent bonds shortly.)

EXAMPLE 7.3

Determine whether the bonds in the following compounds are ionic or covalent (use table 7.1):

(a) H_2S (b) KCl   (c) MgO   (d) H_2

Solution

To solve these problems, we use the specific strategy just described.

(a) There are two H—S bonds in hydrogen sulfide (H_2S). We want to determine whether these are ionic or covalent. From Table 7.1, we find that the electronegativity of H is 2.1, and that of S is 2.5. The difference in electronegativities is

$$2.5 - 2.1 = 0.4$$

Because the difference is only 0.4, the bond between hydrogen and sulfur is covalent. (If the difference is less than 2.0, the bond is considered covalent.)

(b) From Table 7.1, we find that the electronegativity of K is 0.8, and that of Cl is 3.0. The difference in electronegativities is

$$3.0 - 0.8 = 2.2$$

Because the difference is 2.2, the bond between potassium and chlorine is ionic.

(c) From Table 7.1, we find that the electronegativity of Mg is 1.2, and that of O is 3.5. The difference in electronegativities is

$$3.5 - 1.2 = 2.3$$

Because the difference is 2.3, the bond between magnesium and oxygen is ionic.

(d) From Table 7.1, we find that the electronegativity of H is 2.1. The difference between the electronegativities of the two H atoms is

$$2.1 - 2.1 = 0.0$$

The bond between the two H atoms is very definitely a covalent bond. (It's as covalent as it can get!)

Practice Exercise 7.3

Determine whether the bonds in each of the following compounds are ionic or covalent (use Table 7.1):

(a) H_2O (b) KF   (c) MgS   (d) Cl_2

7.7 Ionic Percentage and Covalent Percentage of a Bond

So far we have thought of compounds as either ionically bonded or covalently bonded. We decided which *type* of bond by calculating the difference in electronegativities of the elements forming the bond. A difference of less than 2.0 between the electronegativities of the two elements meant that the bond was covalent, whereas a difference of 2.0 or more meant that the bond was ionic. But this is an oversimplified approach.

Chemists find it better to express chemical bonds as a certain percentage ionic and a certain percentage covalent. Table 7.2 relates difference in electronegativities to the ionic and covalent percentages of any bond. With this table, we can state that the hydrogen-oxygen bond in water (with an electronegativity difference of 1.4) is 39% ionic and 61% covalent. (The percentages listed in Table 7.2 come directly from Pauling's work.)

EXAMPLE 7.4

Determine the ionic and covalent percentages of the bonds in each of the following compounds:

(a) NH_3 (b) SO_2 (c) HCl   (d) NaF

Solution

(a) Look up the electronegativities of nitrogen and hydrogen, determine the difference, and use Table 7.2 to find the ionic and covalent percentages of the bond. The electronegativity of N is 3.0, and that of H is 2.1. The difference between these electronegativities is

$$3.0 - 2.1 = 0.9$$

This corresponds to a bond that is 19% ionic and 81% covalent.

(b) The electronegativity of S is 2.5, and that of O is 3.5. The difference between these electronegativities is

$$3.5 - 2.5 = 1.0$$

This corresponds to a bond that is 22% ionic and 78% covalent.

(c) The electronegativity of H is 2.1, and that of Cl is 3.0. The difference between these electronegativities is
$$3.0 - 2.1 = 0.9$$
This corresponds to a bond that is 19% ionic and 81% covalent.

(d) The electronegativity of Na is 0.9, and that of F is 4.0. The difference between these electronegativities is
$$4.0 - 0.9 = 3.1$$
This corresponds to a bond that is 91% ionic and 9% covalent.

TABLE 7.2 The Relationship Between Electronegativity Difference and the Ionic Percentage and Covalent Percentage of a Chemical Bond

DIFFERENCE IN ELECTRONEGATIVITY	IONIC PERCENTAGE	COVALENT PERCENTAGE
0.0	0.0	100
0.1	0.5	99.5
0.2	1.0	99.0
0.3	2.0	98.0
0.4	4.0	96.0
0.5	6.0	94.0
0.6	9.0	91.0
0.7	12.0	88.0
0.8	15.0	85.0
0.9	19.0	81.0
1.0	22.0	78.0
1.1	26.0	74.0
1.2	30.0	70.0
1.3	34.0	66.0
1.4	39.0	61.0
1.5	43.0	57.0
1.6	47.0	53.0
1.7	51.0	49.0
1.8	55.0	45.0
1.9	59.0	41.0
2.0	63.0	37.0
2.1	67.0	33.0
2.2	70.0	30.0
2.3	74.0	26.0
2.4	76.0	24.0
2.5	79.0	21.0
2.6	82.0	18.0
2.7	84.0	16.0
2.8	86.0	14.0
2.9	88.0	12.0
3.0	89.0	11.0
3.1	91.0	9.0
3.2	92.0	8.0

124   ESSENTIAL CONCEPTS OF CHEMISTRY

Practice Exercise 7.4

Determine the ionic and covalent percentages of the bonds in each of the following compounds: (a) K_2O (b) $CaCl_2$ (c) Mg_3N_2 (d) HF

EXAMPLE 7.5

List the following compounds in order of decreasing covalence of their bonds: H_2S, H_2O, CO_2, and N_2.

Solution

Using Tables 7.1 and 7.2, determine the percentage covalence of each compound. In the compound H_2S, the H—S bond has an electronegativity difference of 0.4, which means it is 96% covalent.

In the compound H_2O, the O—H bond has an electronegativity difference of 1.4, which means it is 61% covalent.

In the compound CO_2, the C=O bond has an electronegativity difference of 1.0, which means it is 78% covalent.

In the compound N_2, the N≡N bond has an electronegativity difference of 0.0, which means it is 100% covalent.

Therefore the order of covalence is

N_2	H_2S	CO_2	H_2O
Most covalent			Least covalent

Practice Exercise 7.5

List the following compounds in order of decreasing covalence of their bonds: CO_2, $AsCL_3$, CS_2, and H_2S.

7.8 Shapes and Polarities of Molecules

Up to now, we have used only two-dimensional diagrams to show how atoms are bonded together in molecules. But our world is *three*-dimensional, and atoms and molecules, like other object, have three dimensions. In fact, the three-dimensional shapes of some molecules lead to additional properties of covalently bonded compounds.

Three-Dimensional Characteristics of Molecules

If we could enlarge a single molecule of any diatomic element (for instance, H_2, O_2, or N_2), it would appear to be linear (in a line) (Figure 7.11). This would also be true for covalently bonded compounds consisting of two *un*like atoms, such as HCl (Figure 7.12).

But if we could enlarge a single molecule of water, we would see a bent molecule (Figure 7.12). Experiments have shown that the angle between the hydrogen atoms is about 105 degrees.

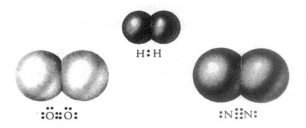

FIGURE 7.11   Three-dimensional models of some diatomic elements.

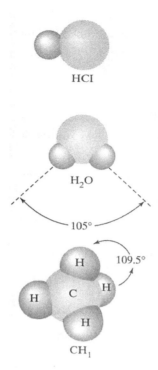

FIGURE 7.12 Three-dimensional models of hydrogen chloride (HCl), water (H_2O), and methane (CH_4) molecules.

If we looked at a molecule of methane (CH_4), we would see a pyramid-shaped symmetrical molecule (Figure 7.12). The angle between neighboring hydrogen atoms in methane has been shown to be 109.5 degrees.

Molecules of other compounds exhibit many other shapes. The important thing about these shapes is that they can affect bonding and the properties of the molecules.

The Not-So-Covalent Bond

Recall that when two atoms form a covalent bond, they share electrons. But sometimes they don't share the electrons equally. For example, in the compound hydrogen bromide (HBr), the bromine atom has a greater attraction for the electrons than the hydrogen atom does. In other words, bromine has a greater electronegativity than hydrogen. Because of the bromine's greater attraction for electrons, the two electrons being shared spend more time with the bromine than with the hydrogen. This unequal sharing makes the hydrogen seem to have a partial positive charge, and the bromine a partial negative charge (Figure 7.13). This kind of covalent bond, in which there is unequal sharing of electrons, is called a **polar covalent bond**. This is what we meant when we said a bond was a certain percentage ionic.

Most covalent bonds are polar. However, bonds formed by the diatomic elements (H—H, O=O, and so on) are nonpolar. This is because there is no difference in electronegativity between the two atoms forming the bond. In a **nonpolar covalent bond**, the electrons are shared equally by the atoms.

EXAMPLE 7.6

Determine whether each of the following covalent bonds is polar or nonpolar. (*Hint*: Use Table 7.1 to determine the electronegativity of each element.) (a) N—O bond (b) S—O bond (c) Cl—Cl bond (d) P—O bond (e) C—S bond

Solution

We look up the electronegativity value of each element in Table 7.1. If a difference in electronegativity exists between the two atoms forming the bond, then the bond is polar. If no difference in electronegativity exists, then the bond is nonpolar.

FIGURE 7.13 A three-dimensional drawing of HBr. The δ^+ indicates that the hydrogen side of the molecule has a partial positive charge. The δ^- shows that the bromine side of the molecule has a partial negative charge. (δ is the lower-case Greek letter delta.)

(a) N————O An electronegativity difference exists, so the bond is polar.
    3.0   3.5
    0.5 difference

(b) S————O An electronegativity difference exists, so the bond is *polar*.
    2.5   3.5
    1.0 difference

(c) Cl————Cl No electronegativity difference exists, so the bond is *nonpolar*.
    3.0   3.0
    0.0 difference

(d) P————O An electronegativity difference exists, so the bond is *polar*.
    2.1   3.5
    1.4 difference

(e) C————S No electronegativity difference exists, so the bond is *nonpolar*.
    2.5   2.5
    0.0 difference

Practice Exercise 7.6

Determine whether each of the following covalent bonds is polar or nonpolar. (*Hint:* Use Table 7.1.) (a) F—F bond (b) C—O bond (c) B—O bond

Polar and Nonpolar Molecules

Just as bonds can have polarity, molecules can also have polarity. Even if a molecule contains polar bonds, the molecule itself must not necessarily be a polar molecule. To determine whether or not a molecule is polar we must look at the geometry of the molecule and the polarity of its bonds.

As we mentioned earlier, molecular geometry tells us how the atoms of a molecule are arranged in space in relation to one another. You are familiar with the linear arrangement of the diatomic elements, the bent shape characteristic of a water molecule, and the pyramidal shape of the methane molecule. Such information is necessary when determining molecular polarity. We will look at the polarities of three molecules—HBr, H_2O, and CCl_4—to understand the combined effects of molecular geometry and bond polarity.

The linear HBr molecule (Figure 7.13) contains a polar covalent bond. There is a shift in electronic charge such that the hydrogen side of the molecule is positive and the bromine side of the molecule is negative. A **polar molecule** (or **dipole**) is a molecule that is positive on one side and negative on the other side. Thus HBr is a polar molecule.

H₂O   CCl₄

FIGURE 7.14 Three-dimensional models of water and carbon tetrachloride molecules. The water molecule is polar; the carbon tetrachloride is nonpolar.

And then there is water, a compound with two polar bonds. A water molecule is also polar, because there is a center of positive charge between the two hydrogen atoms and a center of negative charge on the oxygen atom (Figure 7.14).

A molecule of carbon tetrachloride (CCl_4) has four carbon-chlorine bonds. Each bond is *a polar* covalent bond. But the molecule itself is *nonpolar* (Figure 7.14). Here's why: The carbon tetrachloride molecule is shaped like a pyramid. The positive charge is on the carbon atom, but so is the center of the negative charges of the four Cl atoms. The two charges coincide and cancel each other, so the molecule is not polar.

EXAMPLE 7.7

Determine whether the bonds in each of the following molecules are polar or nonpolar. Then determine whether the molecule itself is polar or nonpolar.

(a) F_2 (linear in shape)

(b) $CHCl_3$, chloroform (shaped like a pyramid)

(c) *cis*-dichloroethene (The *cis* means that the two chlorines are on the same side of the double bond.)

$$\begin{array}{c} H \quad\quad\quad H \\ \diagdown\diagup \\ C=C \\ \diagup\diagdown \\ Cl \quad\quad\quad Cl \end{array}$$

(d) *trans*-dichloroethene (The *trans* means that the two chlorines are on opposite sides of the double bond.)

$$\begin{array}{c} H \quad\quad\quad Cl \\ \diagdown\diagup \\ C=C \\ \diagup\diagdown \\ Cl \quad\quad\quad H \end{array}$$

Solution

(a) Fluorine is a diatomic element; therefore the F—F bond is nonpolar and the F_2 molecule is nonpolar.

(b) The bonding in chloroform is similar to the bonding in carbon tetrachloride. Each bond in chloroform is a polar covalent bond. But the molecule of chloroform, unlike that of carbon tetrachloride, is polar. This is because the center of positive charge and the center of negative charge do not coincide. The molecule is not symmetrical:

$$\begin{array}{c} \delta^+H \quad\quad\quad Cl \\ \diagdown\diagup \\ C \\ \diagup\diagdown \\ Cl \quad\quad \delta^-Cl \end{array}$$

(c) In *cis*-dichloroethene, each C—H and C—Cl bond is polar. The C=C bond is nonpolar. The molecule is polar because it has a center of positive charge between the two hydrogens and a center of negative charge between the two chlorines. Because the centers of positive charge and negative charge do not coincide, the molecule is polar.

$$\begin{array}{c} H \\ \diagdown \\ Cl \end{array} C \overset{\delta+}{\underset{\delta-}{=}} C \begin{array}{c} H \\ \diagup \\ \diagdown Cl \end{array}$$

(d) In *trans*-dichloroethene, the C—H and C—Cl bonds are polar and the C=C bond is nonpolar. The molecule is also nonpolar because it is shaped in such a way that the center of positive charge coincides with the center of negative charge. The charges coincide as shown in the double bond between the two carbon atoms.

$$\begin{array}{c} H \\ \diagdown \\ Cl \end{array} C \overset{\delta\pm}{=} C \begin{array}{c} Cl \\ \diagup \\ H \end{array}$$

Practice Exercise 7.7

Determine whether the bonds in each of the following molecules are polar or nonpolar. Then determine whether the molecule itself is polar nonpolar.

Molecule	Shape
BF_3	F—B(—F)—F (trigonal)
Cl_2O	Cl—O—Cl (bent)
CO_2	O=C=O

SUMMARY

In a Lewis (electron-dot) diagram of an element, the symbol for an atom of the element is surrounded by dots. The dots correspond to the electrons in the outermost energy level of an atom of the element. Electron-dot diagrams are used to show how atoms combine to form molecules.

It has long been known that Group VIIIA elements tend to be very stable. Scientists believe this stability is due to the fact that each element in this group (except helium) has eight electrons in its outermost energy level. The atoms of other elements can achieve this same sort of stability by obtaining eight electrons in their outermost energy level. To do so, two atoms may share electrons in a covalent bond. A single covalent bond is formed when each atom donates one electron to the bond, a double covalent bond is formed when each atom donates two electrons, and a triple covalent bond is formed when each atom donates three electrons.

In a coordinate covalent bond, only one atom donates the two electrons required to form the bond. However, both atoms obtain an octet of electrons in their outermost shell.

Elements that are found in molecules with two atoms each (such as H_2) are called diatomic elements; these elements are hydrogen, oxygen, nitrogen, and the Group VIIA elements. The bonds between their atoms are covalent bonds.

An ion is an atom or group of atoms that has gained or lost electrons and therefore has a negative or positive electric charge. In ionic bonding, atoms of one element donate electrons to the atoms of a second element. Both become ions, but they are ions of opposite charges. These opposite charges attract, and this attraction holds the atoms together in a molecule.

Electronegativity is the attraction that an atom has for the electrons it is sharing with another atom. If the difference between the electronegativities of two elements is 2.0 or greater, the bond between the two elements is said to be ionic. If the difference is less than 2.0, the bond is covalent. The concept of electronegativity is also used in determining the ionic and covalent percentages of bonds.

Because of their three-dimensional character, some molecules exhibit unequal sharing of electrons, resulting in polar covalent bonds. Such molecules may have a positive side and a negative side; if so, they are called polar molecules, or dipoles.

KEY TERMS

chemical bond (**Introduction**)
coordinate covalent bond (**7.3**)
covalent bond (**7.2**)
diatomic element (**7.2**)
double covalent bond (**7.2**)
electron-dot structure (**7.1**)
electronegativity (**7.6**)
Ion (**7.4**)

ionic bond (**7.4**)
monatomic ion (**7.4**)
nonpolar covalent bond (**7.8**)
octet rule (**7.2**)
polar covalent bond (**7.8**)
polar molecule (dipole) (**7.8**)
single covalent bond (**7.2**)
triple covalent bond (**7.2**)

SELF-TEST EXERCISES

LEARNING GOAL 1

Lewis Structures

◂ *1*. Write the electron-dot structure for each of the following atoms:

(a) As   (b) Cl   (c) Ra   (d) Cs   (e) Rn

◂ 2. Write the electron-dot structure for each of the following elements:

(a) F   (b) Ca   (c) K   (d) C   (e) B

3. Write the electron-dot structure for (a) an oxygen atom, (b) an oxide ion.
4. Write the electron-dot structure for (a) a bromine atom, (b) a fluoride ion.
5. Write the electron-dot structure for (a) a magnesium atom, (b) a magnesium ion.
6. Write the electron-dot structure for (a) a barium atom, (b) a barium ion.

◂ 7. Atoms are said to be *isoelectronic* when they have the same electron configuration. Write the electron-dot structures of two ions that are isoelectronic with an argon atom.
8. Write the electron-dot structure for each of the following elements:

(a) Si   (b) Te   (c) At

9. Write the electron-dot structure for each of the following ions:

(a) magnesium ion   (b) sodium ion   (c) aluminum ion

10. Write the electron-dot structure for each of the following ions:

(a) calcium ion   (b) cesium ion   (c) boron ion

11. Write the electron-dot structures of two ions that are isoelectronic (have the same electron configuration) as a calcium ion.
12. Write the electron-dot structure for each of the following ions:

(a) sulfide ion   (b) telluride ion   (c) nitride ion

LEARNING GOAL 2

Covalent Bonds

◀ 13. Explain why covalent bonds are called electron-sharing bonds.
14. Define the term *coordinate covalent bond*.
15. When each atom donates *two* electrons to form a covalent bond, the bond formed is a _____ (single, double, quadruple) covalent bond.
16. When each atom donates *three* electrons to form a covalent bond, the bond formed is a _____ (single, triple, sextuple) covalent bond.

LEARNING GOAL 3

Diatomic Elements

17. Name the diatomic elements from memory.
◀ 18. Which of the diatomic elements contain single covalent bonds? Which contain double bonds? Which contain triple bonds?

LEARNING GOAL 4

Electron-Dot Structures of Covalent Compounds

◀ *19*. Write the electron-dot structure for each of the following covalent compounds:

(a) $CHBr_3$ (b) C_2H_2 (c) PH_3 (d) HCN

20. Write the electron-dot structure for each of the following covalent compounds:

(a) HCl (b) CBr_4 (c) SO_3 (d) $SbCl_3$

21. Write the electron-dot structure for each of the following covalent compounds:

(a) CF_4 (b) C_2H_4 (c) AsH_3 (d) H_2S

◀ 22. Write the electron-dot structure for each of the following covalent compounds:

(a) CS_2 (b) CO_2 (c) Cl_2O (d) C_3H_8

◀ *23*. Not all compounds follow the octet rule. Draw the bonding diagram for the covalent compound PCl_5. How many electrons surround the P atom in this compound?
24. Not all compounds follow the octet rule. Draw the bonding diagram for the compound BF_3. How many electrons surround the B atom in this compound?
◀ *25*. Draw the bonding diagram for each of the following covalent compounds, which have coordinate covalent bonds: (a) HNO_3 (b) H_2SO_3
26. Draw the bonding diagram for the compound chloric acid, $HClO_3$, which has a coordinate covalent bond. (*Hint:* The chlorine is bonded to three oxygen atoms, and the hydrogen is bonded to one of the oxygen atoms.)

LEARNING GOAL 5

Ionic Bonds

◀ 27. Define the term *ionic bond*.
28. How is an ionic bond different from a covalent bond?

LEARNING GOAL 6

Ionic and Covalent Bonds

29. Determine whether each of the following compounds is bonded ionically or covalently:

    (a) CsCl (b) CH_4 (c) HCl (d) SO_3

30. Determine whether each of the following compounds is bonded ionically or covalently:

    (a) MgO (b) $BaCl_2$ (c) BaO (d) CaF_2 (e) SO_2

31. Find the ionic and covalent percentages of the bonds in each of the following compounds:

    (a) CH_4 (b) N_2 (c) FeO (d) Al_2O_3

32. Find the ionic and covalent percentages of the bonds in each of the following compounds:

    (a) HCl (b) NH_3 (c) KI (d) CS_2

◀ 33. List the following compounds in order of *decreasing* covalence: HI, HCl, HF, and HBr.
34. List the following compounds in order of *decreasing* covalence: Br_2, SO_2, NH_3, HCl.
35. Find the ionic and covalent percentages of the bonds in each of the following compounds:

    (a) CF_4 (b) AsH_3 (c) CsCl (d) CuO

36. Find the ionic and covalent percentages of the bonds in each of the following compounds:

    (a) CO_2 (b) F_2 (c) $MgCl_2$ (d) BaF_2

37. List the following compounds in order of *increasing* covalence: H_2Se, H_2O, H_2Te, and H_2S.
◀ 38. List the following compounds in order of *increasing* covalence: HF, NH_3, BH_3, CH_4, and H_2O.
39. Which of the following compounds is the *most* ionic? (a) LiCl (b) NaCl (c) KCl (d) RbCl (e) CsCl
40. Which of the following compounds is the *most* ionic? HF HCl HBr HI
41. Determine whether the bonds in each of the following compounds are ionic or covalent (use Table 7.1):

    (a) SO_3 (b) $BiCl_3$ (c) LiF (d) NO

42. Determine whether the bonds in each of the following compounds are ionic or covalent (use Table 7.1):

    (a) LiH (b) OF_2 (c) CaI_2 (d) MgH_2

LEARNING GOALS 7 AND 8

Polar and Nonpolar Bonds and Molecules

◀ 43. Explain why some molecules are polar and others are nonpolar.
44. Determine whether the bond in each of the following molecules is polar or nonpolar:

    (a) H_2 (b) NO (c) HCl

◀ 45. Determine whether the *bonds* in each of the following molecules are polar or nonpolar. Then determine whether the *molecule* itself is polar or nonpolar.

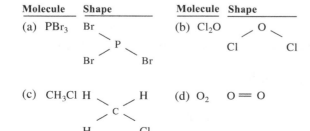

132 ESSENTIAL CONCEPTS OF CHEMISTRY

46. Determine whether the *bonds* in each of the following molecules are polar or nonpolar. Then determine whether the *molecule* is polar or nonpolar.

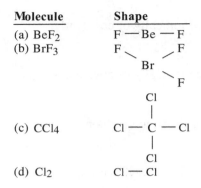

Molecule	Shape
(a) BeF_2	F — Be — F
(b) BrF_3	(see figure)
(c) CCl_4	(see figure)
(d) Cl_2	Cl — Cl

47. Determine whether the *bonds* in each of the following molecules are polar or nonpolar. Then determine whether the *molecule* itself is polar or nonpolar.

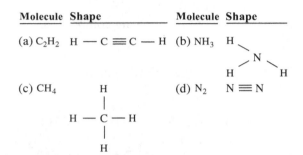

Molecule	Shape	Molecule	Shape
(a) C_2H_2	H — C ≡ C — H	(b) NH_3	(see figure)
(c) CH_4	(see figure)	(d) N_2	N ≡ N

◀ 48. Determine whether the *bonds* in each of the following molecules are polar or nonpolar. Then determine whether the *molecule* is polar or nonpolar.

Molecule	Shape
(a) H_2	H — H
(b) CO_2	O = C = O
(c) C_6H_6	(benzene ring structure)
(d) CH_2O	H — C = O with H below

EXTRA EXERCISES

49. "In both ionic and covalent bonds, *electron deficiencies are satisfied.*" Explain what is meant by this statement.

◀ 50. Draw the electron-dot diagram for each of the following:

(a) $(PH_4)^{1+}$ (b) CO_2 (c) OF_2

51. Explain why the formula for sodium chloride, NaCl, involves one sodium ion and one chloride ion but the formula for calcium chloride, $CaCl_2$, has two chloride ions with every one calcium ion.
52. Determine the order of increasing bond polarity among the following groups: O=O, N—O, C—O
53. Which of the following bonds is the most covalent? C—H, C—O, N—O, C—S
◀ 54. Draw the electron-dot diagram for each of the following: (a) $(NH_4)^{1+}$ (b) ClBr
55. Predict the compound that is *most likely* to be formed from each of the following pairs of elements:

   (a) Mg and O   (b) Al and O   (c) Cs and S

56. Use the octet rule to explain the following:

(a) The formula unit of aluminum chloride is $AlCl_3$.
(b) The formula unit of sodium oxide is Na_2O.
(c) The formula unit of magnesium nitride is Mg_3N_2.

57. Explain how a molecule that has polar covalent bonds can be a nonpolar molecule.
◀ 58. Use the octet rule to predict the formula unit of each of the following compounds:

   (a) K and S   (b) Ca and Br   (c) Li and O   (d) Al and Cl

59. Explain why a covalently bonded molecule with the formula H_3 should not exist.
60. Using the octet rule, predict the formula and draw the electron-dot diagram for each of the covalently bonded molecules that form between the following atoms:

   (a) C and F   (b) As and Cl   (c) I and Cl   (d) Te and H.

CHAPTER 8

CHEMICAL NOMENCLATURE
The Names and Formulas of Chemical Compounds

Sometimes, the names and chemical formulas of substances can be very useful to individuals in many fields outside of the chemical industry. As an example, take this story, which involved a family in the pest-control business and several local physicians.

Tom Daniels, a four-year-old boy, wandered into the family garage to play "explorer." He had been warned by his parents not to play in the garage because his dad, Ed Daniels, was in the pest-control business and kept many chemicals at home.

Tom was in the garage for only a few minutes when he came running into the house crying to his mother. "Mom," he said, "my mouth is burning." His mother detected a garlicky odor and she also noticed a white powder on his hands and face. Just then Tom vomited some material that had a bloody dark color. Working swiftly, Mrs. Daniels brushed some white powder from Tom's body and placed it in a plastic bag. She washed Tom's hands and face and had him rinse his mouth, making sure that he did not swallow anything. Then she took Tom into the garage and had him point to the container that he was playing with. Lifting the cover, she discovered a white powder with the same odor that was emanating from Tom—it was rat poison. Mrs. Daniels asked Tom if he had swallowed any of the powder and he said that he may have swallowed some.

Mrs. Daniels called Tom's pediatrician, Dr. Williams, and told him what had happened. Dr. Williams told Mrs. Daniels to take Tom and the samples of powder to the emergency room of their local hospital. Within 10 minutes Tom and his mother were at the hospital, where Dr. Bellen, the emergency room physician, had been alerted by a call from Dr. Williams, and was waiting for Tom to arrive. She took a history from Mrs. Williams and began to examine Tom. She also read the label on the container of rat poison and had a sample sent to the laboratory. (The powder that was brushed from Tom's body and placed in a plastic bag was also sent to the laboratory.) The label indicated that the active ingredient in the rat poison was white phosphorus.

Dr. Williams, Tom's pediatrician, soon arrived at the hospital. Both he and Dr. Bellen consulted with the poison control center, which advised that the best treatment for phosphorus poisoning was to give the patient frequent dosages of 0.2% copper(II) sulfate solution. This solution, if given in time, would prevent systemic poisoning by phosphorus. (This prevention occurs because the copper bonds with the phosphorus and envelops the phosphorus particles with an insoluble coating of copper(II) phosphide.) They were also advised by the poison control center that after treatment with the copper(II) sulfate, the patient should ingest a laxative solution of sodium sulfate, so that the enveloped phosphorus particles are quickly eliminated from the body. Liquid petrolatum is also administered to the patient to coat the stomach. By the time the physicians finished talking to the poison control center, the laboratory had confirmed that the powder was a phosphorus-based poison, and the physicians immediately gave Tom the prescribed treatment.

Mrs. Daniels was told that Tom would have to stay in the hospital for a few days for observation, because in some cases of phosphorus poisoning there is a quiet period from one to three days, followed by symptoms such

136 ESSENTIAL CONCEPTS OF CHEMISTRY

as nausea, vomiting, an enlarged liver, skin eruptions, jaundice, and reduced blood pressure. The physicians commended Mrs. Daniels for her quick action at home that may have prevented the poison from getting past Tom's stomach. She was told that, based on his symptoms and vital signs, Tom may not have swallowed too much of the poison, and that the copper(II) sulfate treatment should handle the small amount that might have entered his stomach or intestines.

During Tom's stay in the hospital his diet was free of all edible oils and fats, or substances containing them, such as milk, because these substances promote the absorption of phosphorus. Mrs. Daniels was advised to maintain the same diet for several days after Tom arrived home.

Fortunately for Tom, the physicians were correct. Tom had no additional adverse reactions to the poison other than some GI tract irritation, and he was released from the hospital after five days of observation. Home again, Tom found that his dad had cleaned up the garage—the Daniels family wasn't taking any more chances with poisons.

LEARNING GOALS

After you've studied this chapter, you should be able to:

1. Distinguish between the common name of a compound and its systematic name.
2. Write the formula of a binary compound containing two nonmetals when you are given the name of the compound.
3. Write the formula of a binary compound containing a metal with a fixed oxidation number and a nonmetal when you are given the name of the compound.
4. Write the formula of a binary compound containing a metal with a variable oxidation number and a nonmetal when you are given the name of the compound.
5. Write the formula of a polyatomic ion when you are given its name, and write the name of a polyatomic ion when you are given its formula.
6. Write the formula of a ternary or higher compound when you are given the name of the compound.
7. Write the name of a binary compound containing two nonmetals when you are given the formula of the compound.
8. Write the name of a binary compound containing a metal with a fixed or variable oxidation number and a nonmetal when you are given the formula of the compound.
9. Find the oxidation numbers of less familiar elements when you are given chemical formulas of compounds containing these elements.
10. Write the name of a ternary or higher compound when you are given the formula of the compound.
11. Write the formula of an inorganic acid when you are given its name, and write the name of an inorganic acid when you are given its formula.
12. Write the formula of a compound when you are given its common name, and write the common name of a compound when you are given its formula or systematic name.

INTRODUCTION

One of the most important topics to master in your study of chemistry is **chemical nomenclature**, which refers to the system for naming chemical compounds and writing their formulas. In any introductory college course, learning the basic vocabulary of the discipline is extremely important. This allows you to discuss the subject matter using the language of the discipline. Learning to write the names and formulas of chemical compounds will build your chemistry vocabulary. In fact, chemical nomenclature is a major part of the language of chemistry. In this chapter, you will learn how to write the formula of a compound when you are given its name. You will also learn how to name a compound when you are given its formula.

CHAPTER 8: CHEMICAL NOMENCLATURE   137

As we begin our study of chemical nomenclature, you should be aware that there are two ways of designating chemical compounds: the **systematic chemical name** and the **common name**. The rules governing the systematic names have been developed by the *International Union of Pure and Applied Chemistry (IUPAC)*. The IUPAC is made up of chemists from all over the world. Members of the IUPAC Commission on the Nomenclature of Inorganic Compounds first met in 1921 to develop a system for naming compounds. This commission meets periodically to refine this system as new types of compounds are synthesized or discovered.

There are no rules that govern the common names of compounds. Most common names, such as table salt (for sodium chloride, NaCl) and water (for H_2O), have been derived from common usage or simply handed down through chemical history. We'll have more to say about common names of compounds later in this chapter. Let's begin our study of nomenclature by learning how to write the formulas of compounds from their systematic chemical names.

8.1 Writing the Formulas of Compounds from Their Systematic Names

In developing the rules for naming inorganic compounds, the IUPAC wanted to make sure that each chemical formula had a unique chemical name and vice versa. The commission also decided that the more *positive portion* of the compound should be written first, and the more *negative portion* last. Therefore, in the formula of a compound composed of two ions, the positive ion precedes the negative ion. That is why the formula for sodium chloride is always written NaCl, not ClNa. In the formula for a covalent compound, the element with the lower electronegativity (the more positive portion) generally precedes the element with the higher electronegativity (the more negative portion). Thus, the formula for water is written as H_2O, not OH_2.

These simple rules can be used to write the formulas of compounds from their systematic names. We'll begin with **binary compounds** (compounds made up of two different elements) and then move on to **ternary** and higher compounds (those made up of three or more elements).

8.2 Writing the Formulas of Binary Compounds Containing Two Nonmetals

The formula of a compound containing two nonmetals can easily be deduced from the compound's name, because the name of the compound will contain the name of both elements. For binary compounds, the ending of the second element is always *-ide*. Therefore chlor*ine* becomes chlor*ide*, sul*fur* becomes sulf*ide*, and ox*ygen* becomes ox*ide*. In addition, the number of atoms of each element can be deduced from Greek prefixes that are part of the compound's name. Table 8.1 lists the Greek prefixes used in naming compounds. Note that the Greek prefix *mono-* (meaning "one") is rarely used for the first-named element, though it is used for the second-named element. *The absence of a Greek prefix for the first-named element means that there is only one atom in the formula unit of that compound.* The following example will clarify this point.

TABLE 8.1 Greek Prefixes and Their Meanings

Mono-	=	1
Di-	=	2
Tri-	=	3
Tetra-	=	4
Penta-	=	5
Hexa-	=	6
Hepta-	=	7
Octa-	=	8
Nona-	=	9
Deca-	=	10

EXAMPLE 8.1

Write the formulas of the following binary compounds composed of two nonmetals:

(a) carbon monoxide   CO
(b) carbon dioxide CO_2
(c) phosphorus trichloride PCl_3
(d) dinitrogen monoxide N_2O

Solution

(a) Carbon monoxide is a compound containing carbon and oxygen. Ox*ygen* becomes ox*ide* because it is the second-named element. Notice that there is no prefix with carbon, and the prefix *mono-* appears with oxide. There is one atom of each element in the formula unit: CO.

(b) Carbon dioxide is a compound containing carbon and oxygen. Ox*ygen* becomes ox*ide* because it is the second-named element. Notice that there is no prefix with carbon, and the prefix *di-* appears with oxide. There is one atom of carbon and two atoms of oxygen in the formula unit: CO_2.

(c) Phosphorus trichloride is a compound containing phosphorus and chlorine. Chlor*ine* becomes chlor*ide* because it is the second-named element. Notice that there is no prefix with phosphorus, and the prefix *tri-* appears with chloride. There is one atom of phosphorus and three atoms of chlorine in the formula unit: PCl_3.

(d) Dinitrogen monoxide is a compound containing nitrogen and oxygen. Ox*ygen* becomes ox*ide* because it is the second-named element. Notice the prefix *di-* on nitrogen and the prefix *mono-* on oxide. There are two atoms of nitrogen and one atom of oxygen in the formula unit: N_2O.

Practice Exercise 8.1

Write the formulas of the following binary compounds composed of two nonmetals:

(a) Sulfur dioxide SO_2
(b) Dinitrogen pentoxide N_2O_5
(c) Dinitrogen tetroxide N_2O_4
(d) Phosphorus pentabromide PBr_5

8.3 Writing the Formulas of Binary Compounds Containing a Metal and a Nonmetal

Before we learn how to write the formulas of binary compounds containing a metal and a nonmetal, we must learn about *oxidation numbers*. Chemists who have studied how elements combine to form compounds have discovered certain trends. It seems that elements tend to form ions with specific charges or they tend to form only a certain number of covalent bonds. To describe this phenomenon, chemists have devised a system to indicate how elements combine to form compounds. This system involves the assignment of oxidation numbers to the substances involved in forming the compound. An **oxidation number** is *the positive or negative number that expresses the combining capacity of an element in a particular compound*. Oxidation numbers are not always real, but are "tools" created by chemists for bookkeeping purposes.

The oxidation number can be positive or negative, depending on whether the element tends to attract electrons strongly or give them up. In carbon tetrachloride (Figure 8.1), the carbon has an oxidation number of 4+ and the chlorine has an oxidation number of 1−. Elements with high electronegativity values usually have negative oxidation numbers, and elements with low electronegativity values usually have positive oxidation numbers.

$$\begin{array}{c} \text{Cl} \\ | \\ \text{Cl} - \text{C} - \text{Cl} \\ | \\ \text{Cl} \end{array}$$

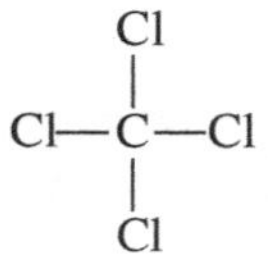

FIGURE 8.1   Covalent bonding in carbon tetrachloride.

Tables 8.2 and 8.3 list the oxidation numbers of some important ions. Note that for many monatomic ions, the oxidation number is the same as the charge on the ion. In other words, because the sodium atom tends to give up one electron when forming a chemical compound, it has a charge of 1+ as a sodium ion. Therefore its oxidation number is also 1+. Elements in Group IA, which tend to become ions with 1+ charges, all have oxidation numbers of 1+. Elements in Group IIA, which tend to become ions with 2+ charges, all have oxidation numbers of 2+. Elements in Group IIIA, which tend to become ions with 3+ charges, all have oxidation numbers of 3+.

Group IVA elements do not usually form ions. However, they still have oxidation numbers, because they do form chemical compounds. The most common oxidation numbers of the Group IVA elements are 4+ and 4−.

The Group VA elements are a bit more difficult to explain. When they form simple ions, they usually do so by gaining three electrons. Therefore they have a charge of 3−. In this case, their oxidation number is also 3−. However, in other compounds in which they bond covalently, their oxidation numbers can be anywhere from 5+ to 5−.

Elements in Group VIA, which tend to become ions with 2− charges, all have oxidation numbers of 2−. (Remember, Group VIA elements tend to gain two electrons to complete their octet.) Elements in Group VIIA, which tend to become ions with 1− charges, all have oxidation numbers of 1−.

Once you know the oxidation numbers of the various ions, you can predict the chemical formulas of any compounds they form. This is because all chemical compounds must be electrically neutral; in other words, they must have no net (overall) charge. If you want to become proficient in writing and naming compounds, you should memorize the names and symbols of the ions in Tables 8.2 and 8.3. A good idea is to use index cards to prepare a set of flash cards. Review these cards every day until you have memorized the names and symbols of the ions.

TABLE 8.2 Oxidation Numbers of Positive Ions Frequently Used in Chemistry

1+		2+	
Hydrogen	H^{1+}	Calcium	Ca^{2+}
Lithium	Li^{1+}	Magnesium	Mg^{2+}
Sodium	Na^{1+}	Barium	Ba^{2+}
Potassium	K^{1+}	Zinc	Zn^{2+}
Mercury(I)*	Hg^{1+} (also called mercurous)	Mercury(II)	Hg^{2+} (also called mercuric)
Copper(I)	Cu^{1+} (also called cuprous)	Tin(II)	Sn^{2+} (also called stannous)
Silver	Ag^{1+}	Iron(II)	Fe^{2+} (also called ferrous)
Ammonium	$(NH_4)^{1+}$	Lead(II)	Pb^{2+} (also called plumbous)
Rubidium	Rb^{1+}	Copper(II)	Cu^{2+} (also called cupric)
Cesium	Cs^{1+}	Strontium	Sr^{2+}
		Nickel(II)	Ni^{2+}
		Chromium(II)	Cr^{2+} (also called chromous)
		Cobalt(II)	Co^{2+} (also called cobaltous)
		Manganese(II)	Mn^{2+} (also called manganous)
3+		4+	
Aluminum	Al^{3+}	Tin(IV)	Sn^{4+} (also called stannic)
Iron(III)	Fe^{3+} (also called ferric)	Lead(IV)	Pb^{4+} (also called plumbic)
Bismuth(III)	Bi^{3+}	Manganese(IV)	Mn^{4+}
Chromium(III)	Cr^{3+} (also called chromic)		
Cobalt(III)	Co^{3+} (also called cobaltic)		

*Note that the mercury(I) ion is a diatomic ion. In other words, you never find the Hg^{1+} ion alone, but always as Hg^{1+}–Hg^{1+}. The two Hg^{1+} ions are bonded to each other.

TABLE 8.3 Oxidation Numbers of Negative Ions Frequently Used in Chemistry

1−		2−		3−	
Fluoride	F^{1-}	Oxide	O^{2-}	Nitride	N^{3-}
Chloride	Cl^{1-}	Sulfide	S^{2-}	Phosphide	P^{3-}
Hydroxide	$(OH)^{1-}$	Sulfite	$(SO_3)^{2-}$	Phosphate	$(PO_4)^{3-}$
Nitrite	$(NO_2)^{1-}$	Sulfate	$(SO_4)^{2-}$	Arsenate	$(AsO_4)^{3-}$
Nitrate	$(NO_3)^{1-}$	Carbonate	$(CO_3)^{2-}$	Borate	$(BO_3)^{3-}$
Acetate	$(C_2H_3O_2)^{1-}$	Chromate	$(CrO_4)^{2-}$		
Bromide	Br^{1-}	Dichromate	$(Cr_2O_7)^{2-}$		
Iodide	I^{1-}	Oxalate	$(C_2O_4)^{2-}$		
Hypochlorite	$(ClO)^{1-}$				
Chlorite	$(ClO_2)^{1-}$				
Chlorate	$(ClO_3)^{1-}$				
Perchlorate	$(ClO_4)^{1-}$				
Permanganate	$(MnO_4)^{1-}$				
Cyanide	$(CN)^{1-}$				
Hydrogen sulfite	$(HSO_3)^{1-}$				
Hydrogen sulfate	$(HSO_4)^{1-}$				
Hydrogen carbonate	$(HCO_3)^{1-}$				

Binary Compounds Containing a Metal with a Fixed Oxidation Number and a Nonmetal

In the examples that follow, we will consider only metals that have fixed oxidation numbers in combination with nonmetals. These metals are the A-group elements of Groups IA, IIA, and IIIA. The nonmetals we will consider are the A-group elements of Groups VA, VIA, and VIIA. Although you can find the oxidation number of many of these elements in Tables 8.2 and 8.3, it is easier to remember that Group IA elements take on a charge of 1+, Group IIA elements take on a charge of 2+, and Group IIIA elements take on a charge of 3+. (*Note:* Although hydrogen is a nonmetal, it is a Group IA element and takes on a charge of 1+ when it enters into most chemical combinations.) For the nonmetals, Group VA elements take on a charge of 3−, Group VIA elements take on a charge of 2−, and Group VIIA elements take on a charge of 1−.

Greek prefixes are not used in the names of binary compounds of this type. Therefore, to write the correct formula, you must choose the correct subscripts for the formula unit by noting the charges of the metal and non-metal portions. The subscripts you choose must cancel out the charges. The next example will show you how to do this.

EXAMPLE 8.2

Write the chemical formula for sodium chloride.

Solution

Sodium is a Group IA element, and therefore it takes on a charge of 1+ when it enters into chemical combination. Chlorine is a Group VIIA element, and therefore it takes on a charge of 1− when it enters into chemical combination. Thus the formula unit for sodium chloride is

$$Na_1^{1+} \ Cl_1^{1-}$$

CHAPTER 8: CHEMICAL NOMENCLATURE

The subscripts show that we need one ion of each element to obtain a neutral compound. The subscript 1 is usually not written. Likewise, the charges 1+ and 1− are left out, giving us the chemical formula NaCl.

Practice Exercise 8.2

Write the chemical formula for potassium fluoride.

Let's look more closely at how to choose the proper subscripts for a chemical formula. We know that a formula unit has a positive portion and a negative portion. The oxidation number tells us *how* positive or negative each portion is. Our job is to choose the correct subscripts to balance the charges, so that the formula unit is electrically neutral. Here are some guidelines to help you choose the correct subscripts.

1. Write the symbol of each element of the compound, along with its oxidation number. The element with the positive oxidation number is written first. For example, for the compound calcium chloride, write

$$Ca^{2+}Cl^{1-}$$

2. You want the positive side of the compound to balance the negative side. You can bring this about by crisscrossing the numbers.

$$Ca^{2+}\!\!\!\diagdown\!\!\!Cl^{1-} \quad \text{or} \quad Ca_1^{2+}Cl_2^{1-} \quad \text{or} \quad CaCl_2$$

3. Note that the subscript numbers are written without charge (that is, without a plus or minus sign). For example, the formula for aluminum oxide is

$$Al^{3+}\!\!\!\diagdown\!\!\!O^{2-} \quad \text{or} \quad Al_2^{3+}O_3^{2-} \quad \text{or} \quad Al_2O_3$$

4. Also note that subscript numbers should be in least-common-denominator form. For example, the formula for barium oxide is

$$Ba^{2+}\!\!\!\diagdown\!\!\!O^{2-} \quad \text{or} \quad Ba_2^{2+}O_2^{2-} \quad \text{or} \quad Ba_1^{2+}O_1^{2-} \quad \text{or} \quad BaO$$

EXAMPLE 8.3

Write the formula for barium chloride.

Solution

The periodic table helps us to obtain the charges of barium and chlorine when they enter into chemical combination. We begin by writing $Ba^{2+}Cl^{1-}$. Then we choose the right subscripts to balance the charges:

$$Ba_1^{2+}Cl_2^{1-} \quad \text{or} \quad BaCl_2$$

Practice Exercise 8.3

Write the formula for magnesium oxide.

EXAMPLE 8.4

Write the formula for aluminum sulfide.

Solution

First we write $Al^{3+}S^{2-}$. Then we choose the right subscripts to balance the charges:

$$Al_2^{3+}S_3^{2-} \quad \text{or} \quad Al_2S_3$$

Practice Exercise 8.4

Write the formula for strontium sulfide.

142   ESSENTIAL CONCEPTS OF CHEMISTRY

EXAMPLE 8.5

Write the chemical formula for sodium phosphide.

Solution

First we write $Na^{1+}P^{3-}$. Then we choose the right subscripts to balance the charges:

$$Na_3^{1+}P_1^{3-} \quad \text{or} \quad Na_3P$$

Practice Exercise 8.5

Write the formula for potassium oxide.

EXAMPLE 8.6

Write the chemical formula for aluminum nitride.

Solution

First we write $Al^{3+}N^{3-}$. Then we choose the right subscripts to balance the charges:

$$Al_1^{3+}N_1^{3-} \quad \text{or} \quad AlN$$

Practice Exercise 8.6

Write the formula for aluminum bromide.

Binary Compounds Containing a Metal with a Variable Oxidation Number and a Nonmetal

Table 8.2 shows that some atoms have variable oxidation numbers; that is, they can form more than one kind of ion when they enter into chemical combination. This is especially true of the transition metals (the ones in the middle of the periodic table—the B-group elements), because of their unique electron configurations. An element such as copper can combine with other elements in two different ways. It can combine as Cu^{1+} or Cu^{2+}.

There are two methods to name these ions. In the IUPAC *Stock system*, the oxidation number of the metal ion, in the form of a Roman numeral, is used as part of the name. The Roman numeral is placed in parentheses and immediately follows the name of the metal. Therefore Cu^{1+} is called copper(I) (read as "copper-one"), and Cu^{2+} is called copper(II) (read as "copper-two"). Under the Stock system, the oxidation number of a metal that has a variable oxidation number will *always* be part of the compound's name.

The Stock system is now the system of choice, but the older *-ous* and *-ic* system for naming these ions is still used. In the older system, the Latin name of the metal is used along with the suffix *-ous* or *-ic*. The ion with the lower oxidation number is given the suffix *-ous*, and the ion with the higher oxidation number is given the suffix *-ic*. For example, Fe^{2+} is called the ferr*ous* ion, and Fe^{3+} is called the ferr*ic* ion. (Note the suffixes and the use of *ferr-*, not *iron*, as the main stem of the element.) Another example is copper: Cu^{1+} is called the cupr*ous* ion, and Cu^{2+} is called the cupr*ic* ion. The disadvantage of this system is that you have to memorize which ion of a given element has the higher or lower oxidation number to apply the correct suffix. Table 8.2 lists some additional examples of positive ions and their names, using the rules for both systems.

Although you may find the Stock system easier to use, you must also become familiar with the older system, because it is still in use. In the examples that follow, you will practice writing the formulas of compounds when you are given the Stock name or the older *-ous* or *-ic* name.

EXAMPLE 8.7

Write the formula of each of the following compounds whose metal ions have variable oxidation numbers:

(a) copper(I) chloride

(b) copper(II) chloride

(c) iron(III) sulfide

(d) vanadium(V) oxide

(e) mercuric bromide

(f) ferrous nitride

Solution

(a) The (I) tells us that copper is Cu^{1+}. We already know that chloride is Cl^{1-}, because chlorine is a Group VIIA element. Therefore we have

$$Cu^{1+}Cu^{1-} \quad \text{or} \quad CuCl$$

(b) The (II) tells us that copper is Cu^{2+}. We already know that chloride is Cl^{1-}. Therefore we have

$$Cu_1^{2+}Cu_2^{1-} \quad \text{or} \quad CuCl_2$$

(c) The (III) tells us that iron is Fe^{3+}. We already know that sulfide is S^{2-}, because sulfur is a Group VIA element. Therefore we have

$$Fe_2^{3+}S_3^{2-} \quad \text{or} \quad Fe_2S_3$$

(d) The (V) tells us that vanadium is V^{5+}. We already know that oxide is O^{2-}, because oxygen is a Group VIA element. Therefore we have

$$V_2^{5+}O_5^{2-} \quad \text{or} \quad V_2O_5$$

(e) Table 8.2 tells us that the mercuric ion is Hg^{2+}. We already know that bromide is Br^{1-}, because bromine is a Group VIIA element. Therefore we have

$$Hg_1^{2+}Br_2^{1-} \quad \text{or} \quad HgBr_2$$

(f) Table 8.2 tells us that the ferrous ion is Fe^{2+}. We already know that nitride is N^{3-}, because nitrogen is a Group VA element. Therefore we have

$$Fe_3^{2+}N_2^{3-} \quad \text{or} \quad Fe_3N_2$$

Practice Exercise 8.7

Write the formula of each of the following compounds whose metal ions have variable oxidation numbers:

(a) mercury(II) oxide

(b) iron(III) bromide

(c) cobalt(II) iodide

(d) manganese (IV) oxide

(e) cuprous sulfide

(f) ferric sulfide

Before we leave this topic, we should point out that some transition metals do not have variable oxidation numbers. Therefore neither the name of the compound nor the group that it is in will tell you the charge of the ion in that compound. Two important examples are zinc and silver. You should *memorize* that zinc has an oxidation number of 2+ (Zn^{2+}) and silver has an oxidation number of 1+ (Ag^{1+}).

EXAMPLE 8.8

Write the formulas of the following compounds: (a) silver sulfide (b) zinc phosphide

144 ESSENTIAL CONCEPTS OF CHEMISTRY

Solution

(a) We know that silver is Ag^{1+}. We already know that sulfide is S^{2-}, because sulfur is a Group VIA element. Therefore we have

$$Ag_2^{1+}S_1^{2-} \quad \text{or} \quad Ag_2S$$

(b) We know that zinc is Zn^{2+}. We already know that phosphide is P^{3-}, because phosphorus is a Group VA element. Therefore we have

$$Zn_3^{2+}P_2^{3-} \quad \text{or} \quad Zn_3P_2$$

Practice Exercise 8.8

Write the formulas of the following compounds: (a) silver chloride (b) zinc oxide

8.4 Polyatomic Ions

Over the years, in studying the composition of many compounds, chemists have found that certain groups of covalently bonded atoms appear over and over again. Because these groups are electrically charged, they are not molecules but ions. More specifically, they are *polyatomic ions* (ions that are combinations of many atoms); the prefix *poly-* is from the Greek word for "many." A **polyatomic ion** is a *charged group of covalently bonded atoms*.

Examples of polyatomic ions are nitrate, $(NO_3)^{1-}$; sulfite, $(SO_3)^{2-}$; sulfate, $(SO_4)^{2-}$; carbonate, $(CO_3)^{2-}$; and phosphate, $(PO_4)^{3-}$. Polyatomic ions are common in minerals, plants, animals, and human beings. *They never exist in an uncombined state* but are always part of a chemical compound. For example, one can't isolate nitrate ions and put them in a bottle, but there are hundreds of chemical compounds made up, in part, of nitrate ions. Some examples of nitrate compounds are potassium nitrate, KNO_3 (used in chemical fertilizers); calcium nitrate, $Ca(NO_3)_2$ (used in chemical fertilizers and also in matches); and silver nitrate, $AgNO_3$ (used in photography, mirror manufacturing, hair dyeing, and silver plating, and as an external medicine).

Tables 8.2 and 8.3 list the more common polyatomic ions. You should memorize the name, symbol, and charge of each ion. That information will be useful as you learn to write the formulas of compounds containing these ions in the next section.

8.5 Writing the Formulas of Ternary and Higher Compounds

In writing the formulas of compounds containing three or more elements (usually in the form of polyatomic ions), we follow much the same procedure we used to write formulas of binary compounds. The only difference is that we must use the formula and charge of the polyatomic ion in deriving the formulas of ternary and higher compounds. Most of the polyatomic ions that we'll encounter have negative charges and are listed in Table 8.3. The only common polyatomic ion with a positive charge is the ammonium ion, $(NH_4)^{1+}$, which is listed in Table 8.2.

The names of most polyatomic ions end in *-ite* or *-ate*. For example, $(SO_3)^{2-}$ is called the sulf*ite* ion, and $(SO_4)^{2-}$ is called the sulf*ate* ion. Notice that the *-ate* ion has one more oxygen than the corresponding *-ite* ion. There are two polyatomic ions in Table 8.3 that do not end in *-ite* or *-ate*. They are the hydrox*ide* ion, $(OH)^{1-}$, and the cyan*ide* ion, $(CN)^{1-}$.

In the next example, you will practice writing the formulas of compounds containing polyatomic ions. If you've already memorized the polyatomic ions, try writing the formulas using only the periodic table as a reference. Otherwise, look up the formulas and charges for the polyatomic ions in Tables 8.2 and 8.3.

EXAMPLE 8.9

Write the formula of each of the following ternary compounds:

(a) aluminum sulfate

(b) calcium arsenate

(c) copper(II) phosphate

(d) sodium nitrate

Solution

(a) Aluminum is a Group IIIA element, and therefore it takes on a charge of 3+ when it enters into chemical combination. Sulfate ion has the formula $(SO_4)^{2-}$. Thus the formula for aluminum sulfate is

$$Al_2^{3+}(SO_4)_2^{2-} \quad \text{or} \quad Al_2(SO_4)_3$$

(b) Calcium is a Group IIA element, and therefore it takes on a charge of 2+ when it enters into chemical combination. Arsenate ion has the formula $(AsO_4)^{3-}$. Thus the formula for calcium arsenate is

$$Ca_3^{2+}(AsO_4)_2^{3-} \quad \text{or} \quad Ca_3(AsO_4)_2$$

(c) Copper(II) has a charge of 2+. Phosphate ion has the formula $(PO_4)^{3-}$. Therefore the formula for copper(II) phosphate is

$$Cu_3^{2+}(PO_4)_2^{3-} \quad \text{or} \quad Cu_3(PO_4)_2$$

(d) Sodium is a Group IA element, and therefore it takes on a charge of 1+ when it enters into chemical combination. Nitrate ion has the formula $(NO_3)^{1-}$. Thus the formula for sodium nitrate is

$$Na_1^{1+}(NO_3)_1^{1-} \quad \text{or} \quad NaNO_3$$

Notice that the parentheses are dropped when the polyatomic ion has a subscript of 1.

Practice Exercise 8.9

Write the formula of each of the following ternary compounds:

(a) ferric nitrite

(b) barium phosphate

(c) copper(II) sulfate

(d) potassium chromate

8.6 Writing the Names of Binary Compounds Containing Two Nonmetals

Now let's turn our attention to writing the names of compounds when we are given their formulas. We will begin by learning how to name binary compounds composed of two nonmetals.

In Section 8.2, you learned how to write the formulas of these compounds from their names. You need only to reverse the process to write the name of the compound from the formula. Remember that Greek prefixes are used to designate the number of atoms in a formula unit of the compound.

EXAMPLE 8.10

Write the names of the following binary compounds composed of two nonmetals: (a) SO_3 (b) NO (c) N_2O (d) CO_2

Solution

(a) SO_3 is called sulfur trioxide. This compound contains one sulfur atom and three oxygen atoms. There is no prefix used for the sulfur portion of the compound. The prefix *tri-* is used for the oxygen portion of the compound. (Also, remember that in naming a binary compound, you change ox*ygen* to ox*ide* because it is the second-named element.)

(b) NO is called nitrogen monoxide. This compound contains one nitrogen atom and one oxygen atom. There is no prefix used for the nitrogen portion of the compound. The prefix *monz-* is used for the oxygen portion of the compound. (Remember the rule for *mono-*: The prefix *mono-* is used for the second-named element, but rarely appears with the first-named element.)

146 ESSENTIAL CONCEPTS OF CHEMISTRY

(c) N_2O is called dinitrogen monoxide. This compound contains two nitrogen atoms and one oxygen atom. The prefix *di-* is used for the nitrogen portion of the compound. The prefix *mono-* is used for the oxygen portion of the compound.

(d) CO_2 is called carbon dioxide. This compound contains one carbon atom and two oxygen atoms. There is no prefix used for the carbon portion of the compound. The prefix *di-* is used for the oxygen portion of the compound.

Practice Exercise 8.10

Write the names of the following binary compounds composed of two nonmetals: (a) P_2O_5 (b) OF_2 (c) SO_2 (d) CO

8.7 Writing the Names of Binary Compounds Containing a Metal and a Nonmetal

We already know how to write the formulas of binary compounds composed of a metal and a nonmetal when we are given their names (Section 8.3). Once again, we can simply reverse this process to write the name of such a compound from its formula. Remember that for these compounds, *no Greek prefixes are used.* For compounds that contain metals with fixed oxidation numbers, just name the elements. (Don't forget to change the ending of the second-named element to *-ide.*) For compounds that contain metals with variable oxidation numbers, the Roman numeral, which represents the oxidation number of the metal, must be given as part of the compound's name (or the *-ous* and *-ic* suffix system may be used).

Finding Oxidation Numbers

Tables 8.2 and 8.3 list oxidation numbers only for some of the more common ions. However, once you know these, you can use them to find the oxidation numbers of less common ions from the compounds they form. You need only remember that every compound must be neutral. That is, the sum of the oxidation numbers of all the atoms in the compound must be zero.

EXAMPLE 8.11

What is the oxidation number of cobalt in $CoCl_3$?

Solution

We know that each chloride ion has a charge of 1−. The molecule contains three chloride ions, for a total charge of 3−. For the compound to be neutral, the single cobalt ion must have a charge of 3+: $Co_1^{3+}Cl_3^{1-}$.

Practice Exercise 8.11

What is the oxidation number of copper in $CuCl_2$?

EXAMPLE 8.12

Find the oxidation number of each underlined element:

(a) $\underline{Mn}O_2$ (b) $K\underline{Mn}O_4$ (c) $\underline{Cs}_2SO_3$ (d) $\underline{Ga}_2O_3$

We follow the same procedure used in the previous example.

Solution

(a) We know that the oxidation number of oxygen is 2−: $Mn_1^{?}O_2^{2-}$ For the compound to be electrically neutral, the manganese must have a charge of 4+: $Mn_1^{4+}O_2^{2-}$

(b) We know that the oxidation number of potassium is 1+, and that of oxygen is 2−: $K_1^{1+}Mn_1^{?}O_4^{2-}$

For the compound to be electrically neutral, the manganese must have a charge of 7+: $K_1^{1+}Mn_1^{7+}O_4^{2-}$

(c) We know that the oxidation number of the sulfite group is 2–: $Cs_2^?(SO_3)_1^{2-}$
For the compound to be electrically neutral, the cesium must have a charge of 1+: $Cs_2^{1+}(SO_3)_1^{2-}$

(d) We know that the oxidation number of oxygen is 2–: $Ga_2^?O_3^{2-}$ For the compound to be electrically neutral, the gallium must have a charge of 3+: $Ga_2^{3+}O_3^{2-}$

Practice Exercise 8.12

Find the oxidation number of each underlined element: (a) $\underline{Sn}F_4$ (b) $\underline{W}Cl_5$

EXAMPLE 8.13

Write the names of the following binary compounds composed of a metal and a nonmetal:

(a) $AlCl_3$ (b) Na_2O (c) Mg_3N_2 (d) K_3P (e) $CoCl_3$ (f) Fe_2O_3 (g) FeO (h) MnO_2

Solution

(a) $AlCl_3$ is called aluminum chloride. All we have to do is name the elements and remember to change the name of chlor*ine* to chlor*ide*.

(b) Na_2O is called sodium oxide. All we have to do is name the elements and remember to change the name of ox*ygen* to ox*ide*.

(c) Mg_3N_2 is called magnesium nitride. All we have to do is name the elements and remember to change the name of nitro*gen* to nitr*ide*.

(d) K_3P is called potassium phosphide. All we have to do is name the elements and remember to change the name of phosph*orus* to phosph*ide*.

(e) $CoCl_3$ is called cobalt(III) chloride (or cobaltic chloride). Cobalt is a metal with a variable oxidation number, and a Roman numeral representing that oxidation number must be part of the compound's name. To determine the oxidation number of cobalt in this compound, we go through the following reasoning: We know that each chloride ion has a charge of 1–. This molecule contains three chloride ions, for a total charge of 3–. For the compound to be neutral, the single cobalt ion must have a charge of 3+:

$$Co_1^{3+}Cl_3^{1-}$$

(f) Fe_2O_3 is called iron(III) oxide (or ferric oxide). Iron is a metal with a variable oxidation number, and a Roman numeral representing that oxidation number must be part of the compound's name. To determine the oxidation number of iron in this compound, we go through the following reasoning: We know that each oxide ion has a charge of 2–. This molecule contains three oxide ions, for a total charge of 6–. For the compound to be neutral, the two iron ions must have a total charge of 6+. Therefore each iron ion must have a charge of 3+:

$$Fe_2^{3+}O_3^{2-}$$

(g) FeO is called iron(II) oxide (or ferrous oxide). Iron is a metal with a variable oxidation number, and a Roman numeral representing that oxidation number must be part of the compound's name. To determine the oxidation number of iron in this compound, we go through the following reasoning: We know that each oxide ion has a charge of 2–. This molecule contains one oxide ion, for a total charge of 2–. For the compound to be neutral, the single iron ion must have a charge of 2+:

$$Fe_1^{2+}O_1^{2-}$$

(h) MnO_2 is called manganese(IV) oxide. Manganese is a metal with a variable oxidation number, and a Roman numeral representing that oxidation number must be part of the compound's name. To determine the oxidation number of manganese in this compound, we go through the following reasoning: We know that each oxide ion has a charge of 2–. This molecule contains two oxide ions, for a total charge of 4–. For the compound to be neutral, the single manganese ion must have a charge of 4+:

$$Mn_1^{4+}O_2^{2-}$$

148 ESSENTIAL CONCEPTS OF CHEMISTRY

Practice Exercise 8.13

Write the names of the following binary compounds composed of a metal and a nonmetal:

(a) RbCl   (b) Ga_2S_3 (c) SrO   (d) ZnI_2 (e) $NiCl_2$ (f) FeI_3 (g) HgS   (h) Cu_2O

8.8 Writing the Names of Ternary and Higher Compounds

You learned how to write the formulas of ternary and higher compounds from their names in Section 8.5. You can just reverse the process to write the name of this type of compound from the formula. Remember that for these compounds, *no Greek prefixes are used*, but you must include the name of the polyatomic ion. For compounds that contain metals with fixed oxidation numbers, simply name the element and the polyatomic ion. For compounds that contain metals with variable oxidation numbers, the Roman numeral, which represents the oxidation number of the metal, must be given as part of the compound's name (or the *-ous* and *-ic* suffix system may be used).

EXAMPLE 8.14

Write the names of the following ternary and higher compounds:

(a) $Al_2(CrO_4)_3$ (b) Li_2SO_3 (c) $Mg(NO_2)_2$ (d) $(NH_4)_3PO_4$ (e) $Cr(OH)_2$ (f) $Fe_2(Cr_2O_7)_3$

Solution

(a) $Al_2(CrO_4)_3$ is called aluminum chromate. All we have to do is name the metal, aluminum, and the polyatomic ion, chromate.

(b) Li_2SO_3 is called lithium sulfite. All we have to do is name the metal, lithium, and the polyatomic ion, sulfite.

(c) $Mg(NO_2)_2$ is called magnesium nitrite. All we have to do is name the metal, magnesium, and the polyatomic ion, nitrite.

(d) $(NH_4)_3PO_4$ is called ammonium phosphate. All we have to do is name two polyatomic ions, ammonium and phosphate.

(e) $Cr(OH)_2$ is called chromium(II) hydroxide (or chromous hydroxide). Chromium is a metal with a variable oxidation number, and a Roman numeral representing that oxidation number must be part of the compound's name. Therefore we have chromium(II). The polyatomic ion is hydroxide.

(f) $Fe_2(Cr_2O_7)_3$ is called iron(III) dichromate (or ferric dichromate). Iron is a metal with a variable oxidation number, and a Roman numeral representing that oxidation number must be part of the compound's name. Therefore we have iron(III). The polyatomic ion is dichromate.

Practice Exercise 8.14

Write the names of the following ternary and higher compounds: (a) $(NH_4)_2O$ (b) $Mg(CN)_2$ (c) $Al(NO_2)_3$ (d) $Zn_3(PO_4)_2$ (e) CaC_2O_4 (f) $Ni_3(BO_3)_2$

8.9 Writing the Names and Formulas of Inorganic Acids

Svante Arrhenius, a well-known Swedish chemist of the late 1800s and early 1900s, developed a classical definition of an acid. According to Arrhenius, an *acid* is a substance that releases hydrogen ions, H^{1+}, in aqueous solutions. Two major classes of inorganic acids exist, and the naming rules differ for the two classes. First we'll examine the rules for naming the non-oxygen-containing acids. Then we'll examine the rules for naming the oxygen-containing acids.

Non-Oxygen-Containing Acids

These acids usually consist of hydrogen plus a nonmetal ion. For example, HCl, which is hydrogen chloride gas in its pure form, becomes hydrochloric acid when dissolved in water. The covalently bonded HCl molecules ionize in water to form hydrogen ions and chloride ions. We derive the names of acids such as HCl by adding the prefix *hydro-* to the name of the nonmetal, which is given an *-ic* suffix. For example,

$$\underset{\text{Prefix}}{hydro} —— \underset{\substack{\text{Name of} \\ \text{nonmetal}}}{chlor} —— \underset{\text{Suffix}}{ic} —— acid$$

EXAMPLE 8.15

Name each of the following non-oxygen-containing acids:

(a) HF (b) HBr (c) H_2S (d) HCN

Solution

We'll give the name of each substance first as a covalent compound and then as an acid in aqueous solution.

NAME AS COVALENT COMPOUND	NAME AS ACID IN AQUEOUS SOLUTION
(a) Hydrogen fluoride	Hydrofluoric acid
(b) Hydrogen bromide	Hydrobromic acid
(c) Hydrogen sulfide	Hydrosulfuric acid
(d) Hydrogen cyanide	Hydrocyanic acid

Practice Exercise 8.15

Name each of the following non-oxygen-containing acids:

(a) HCl (b) H_2Se (c) HI

Oxygen-Containing Acids

Oxygen-containing acids formed from hydrogen and polyatomic ions (containing oxygen) are named as follows:

1. If the polyatomic ion has an *-ate* suffix, we name the acid by replacing the *-ate* suffix with an *-ic* suffix and adding the word *acid*. For example, HNO_3, composed of a hydrogen ion and nit*rate* ion, is called nit*ric* acid in aqueous solution.

EXAMPLE 8.16

Name each of the following oxygen-containing acids: a) H_2SO_4 (b) HC_2H_3O (c) $HBrO_3$

Solution

These three acids are composed of hydrogen plus a polyatomic ion that has the suffix *-ate*.

(a) H_2SO_4 is called sulfur*ic* acid. (SO_4^{2-} is the sulf*ate* ion.)

(b) $HC_2H_3O_2$ is called acet*ic* acid. ($C_2H_3O_2^{1-}$ is the acet*ate* ion.)

(c) $HBrO_3$ is called brom*ic* acid. (BrO_3^{1-} is the brom*ate* ion.)

Note that H_2SO_4 is an exception to our rule, in that the whole name of the element is used rather than the name of the sulfate group.

150 ESSENTIAL CONCEPTS OF CHEMISTRY

Practice Exercise 8.16

Name each of the following oxygen-containing acids: (a) H_3PO_4 (b) $HClO_3$

2. If the polyatomic ion has an *-ite* suffix, we name the acid by replacing the *-ite* suffix with an *-ous* suffix and adding the word *acid*. For example, HNO_2, composed of a hydrogen ion and a nit*rite* ion, is called nitr*ous* acid.

EXAMPLE 8.17

Name each of the following oxygen-containing acids:

(a) H_2SO_3 (b) H_3PO_3 (c) $HBrO_2$

Solution

These three acids are composed of hydrogen plus a polyatomic ion that has the suffix *-ite*.

(a) H_2SO_3 is called sulfur*ous* acid. (SO_3^{2-} is the sulf*ite* ion.)
(b) H_3PO_3 is called phosphor*ous* acid. (PO_3^{3-} is the phosph*ite* ion.)
(c) $HBrO_2$ is called brom*ous* acid. (BrO_2^{1-} is the brom*ite* ion.)

Note that H_2SO_3 is an exception to our rule, in that the whole name of the element is used rather than the name of the sulfite group.

Practice Exercise 8.17

Name each of the following oxygen-containing acids:

(a) $HClO_2$ (b) HNO_2

3. Some elements form more than two oxygen-containing acids. Most notable are the acids formed by the elements chlorine, bromine, and iodine. This occurs because these elements can form four polyatomic ions. For example, bromine forms the following ions:

BrO_4^{1-}, perbrom*ate* ion BrO_3^{1-}, brom*ate* ion
BrO_2^{1-}, brom*ite* ion BrO^{1-}, hypobrom*ite* ion

Note that the first two ions listed have the suffix *-ate* and that the last two ions listed have the suffix *-ite*. Therefore we use rules 1 and 2. Keep the name of the polyatomic ion, but change the *-ate* suffix to *-ic* or the *-ite* suffix to *-ous*.

EXAMPLE 8.18

Name each of the following oxygen-containing acids:

(a) $HBrO_4$ (b) $HBrO_3$ (c) $HBrO_2$ (d) $HBrO$

Solution

(a) $HBrO_4$ is called perbrom*ic* acid. (BrO_4^{1-} is the perbrom*ate* ion.)
(b) $HBrO_3$ is called brom*ic* acid. (BrO_3^{1-} is the brom*ate* ion.)
(c) $HBrO_2$ is called brom*ous* acid. (BrO_2^{1-} is the brom*ite* ion.)
(d) $HBrO$ is called hypobrom*ous* acid. (BrO^{1-} is the hypobrom*ite* ion.)

Practice Exercise 8.18

Name each of the following oxygen-containing acids:

(a) $HClO_4$ (b) $HClO_3$ (c) $HClO_2$ (d) $HClO$

8.10 Common Names of Compounds

As we stated at the beginning of this chapter, there are no rules governing the common names of compounds. Most common names have been derived from common usage or simply handed down through chemical history. Many common names are still in use, mostly because the systematic names are too complex for everyday use. For example, baking soda is a common substance that has a variety of uses around the house, but hardly anyone would call it sodium hydrogen carbonate. The same could be said of Epsom salts, whose systematic name is magnesium sulfate heptahydrate. Table 8.4 lists the common names, systematic chemical names, and formulas of several familiar substances.

SUMMARY

The system for naming chemical compounds and designating their formulas is called chemical nomenclature. Chemical compounds may be referred to by their systematic chemical names or their common names. The rules for the systematic names have been developed by the International Union of Pure and Applied Chemistry (IUPAC). There are no rules that govern the common names. These names have simply been handed down through chemical history.

According to IUPAC rules for writing the formula of a compound, the more positive portion of the compound is written first. In a compound composed of two ions, the positive ion precedes the negative ion. In a covalent compound, the element with the lower electronegativity (the more positive portion) precedes the element with the higher electronegativity (the more negative portion).

Greek prefixes are used to write the names of binary compounds composed of two nonmetals. For binary compounds, the ending of the second element is always -*ide*. Oxidation numbers are used to write the formulas of binary compounds containing a metal and a nonmetal. An oxidation number is the positive or negative number that expresses the combining capacity of an element in a particular compound. Some metals (typically the A-group metals) have fixed oxidation numbers. Other metals (typically the B-group, or transition, metals) have variable oxidation numbers and can combine with a nonmetal element in more than one way. The periodic table can be used to determine the oxidation number of an A-group metal: The group number of the A-group metal is its common oxidation number. The oxidation number of metals with variable oxidation numbers can be determined from the name of the compound, either through the Roman numeral attached to the name of the metal or through the -*ous* or -*ic* suffix attached to the Latin name of the metal.

TABLE 8.4 Common Names of Some Chemical Compounds

COMMON NAME	FORMULA	SYSTEMATIC CHEMICAL NAME
Baking soda	$NaHCO_3$	Sodium hydrogen carbonate
Borax	$Na_2B_4O_7 \cdot 10H_2O$	Sodium tetraborate decahydrate
Dry ice	CO_2	Carbon dioxide
Epsom salts	$MgSO_4 \cdot 7H_2O$	Magnesium sulfate heptahydrate
Gypsum	$CaSO_4 \cdot 2H_2O$	Calcium sulfate dihydrate
Laughing gas	N_2O	Nitrous oxide
Marble	$CaCO_3$	Calcium carbonate
Milk of magnesia	$Mg(OH)_2$	Magnesium hydroxide
Muriatic acid	HCl	Hydrochloric acid
Oil of vitriol	H_2SO_4	Sulfuric acid
Quicklime	CaO	Calcium oxide
Saltpeter	$NaNO_3$	Sodium nitrate

Ternary and higher compounds contain polyatomic ions. A polyatomic ion is a charged group of covalently bonded atoms. The names of most polyatomic ions end in *-ite* or *-ate*. We use the formula and charge of the polyatomic ion and the rules for writing the formulas of binary compounds in deriving the formulas of ternary and higher compounds.

A number of chemicals are known by their common names, either because of tradition or because the systematic names are too complex for convenient reference. Because there are no rules that govern the common names of compounds, the formulas of these compounds may be associated with their names only through memorization.

KEY TERMS

binary compound (**8.1**)
chemical nomenclature (**Introduction**)
common name (**Introduction**)
oxidation number (**8.3**)

polyatomic ion (**8.4**)
systematic chemical name (**Introduction**)
ternary compound (**8.1**)

SELF-TEST EXERCISES

LEARNING GOAL 1

Distinguishing Between Common Name and Systematic Name

◀ 1. Explain why chemists developed systematic names for compounds, when common names were already in existence.

◀ 2. Define or explain the following terms:

(a) IUPAC (b) Stock system (c) binary compound (d) ternary compound

3. The common name for H_2O is water. What is its systematic name?
4. What is the common name and the systematic name of solid CO_2?

LEARNING GOAL 2

Formulas of Binary Compounds Containing Two Nonmetals

5. Write the number that corresponds to each of the following prefixes:

(a) *hepta-* (b) *di-* (c) *tri-* (d) *octa-* (e) *mono-*

6. Write the Greek prefix for each of the following numbers:

(a) 3 (b) 7 (c) 2 (d) 9

7. Write the formulas of the following binary compounds composed of two nonmetals:

(a) diphosphorus pentasulfide
(b) chlorine dioxide
(c) dinitrogen tetroxide
(d) dichlorine heptoxide

8. Write the formulas of the following binary compounds composed of two nonmetals:

(a) carbon tetrachloride
(b) tetraphosphorus decaoxide
(c) phosphorus pentabromide
(d) selenium dioxide

LEARNING GOAL 3

Formulas of Binary Compounds Containing a Metal with a Fixed Oxidation Number and a Nonmetal

    9. Write the formulas of the following binary compounds composed of a metal with a fixed oxidation number and a nonmetal:

        (a) aluminum sulfide
        (b) lithium oxide
        (c) sodium nitride
        (d) strontium phosphide
        (e) silver bromide
        (f) zinc oxide

    10. Write the formulas of the following binary compounds composed of a metal with a fixed oxidation number and a nonmetal:

        (a) lithium iodide
        (b) calcium oxide
        (c) strontium bromide
        (d) potassium phosphide
        (e) rubidium sulfide
        (f) barium nitride

LEARNING GOAL 4

Formulas of Binary Compounds Containing a Metal with a Variable Oxidation Number and a Nonmetal

    11. Give the Roman numerals that correspond to the following Arabic numerals:

        (a) 9   (b) 8   (c) 7   (d) 6

    12. Give the Arabic numerals that correspond to the following Roman numerals:

        (a) III   (b) VIII   (c) VI   (d) IV

◀ *13*. Give a definition or an explanation of the following suffixes:

        (a) *-ic*   (b) *-ous*   (c) *-ide*

◀ 14. Explain the difference between

        (a) ferrous and ferric   (b) cuprous and cupric   (c) cobaltous and cobaltic   (d) stannous and stannic

    15. Write the formulas of the following binary compounds composed of a metal with a variable oxidation number and a nonmetal:

        (a) copper(I) sulfide
        (b) mercury(II) chloride
        (c) iron(II) oxide
        (d) tin(II) iodide
        (e) cobaltic bromide
        (f) mercuric nitride

    16. Write the formulas of the following binary compounds composed of a metal with a variable oxidation number and a nonmetal:

        (a) uranium(VI) fluoride
        (b) iron(III) nitride
        (c) manganous chloride
        (d) ferrous sulfide
        (e) cuprous oxide
        (f) mercurous chloride

LEARNING GOAL 5

Formulas and Names of Polyatomic Ions

◀ *17.* Write the formulas of the following polyatomic ions:
   (a) ammonium ion
   (b) hydroxide ion
   (c) borate ion
   (d) hydrogen carbonate ion
   (e) oxalate ion
   (f) sulfite ion

◀ 18. Write the formulas of the following polyatomic ions:
   (a) cyanide ion
   (b) chlorite ion
   (c) chlorate ion
   (d) chromate ion
   (e) dichromate ion
   (f) phosphate ion

19. Write the names of the following polyatomic ions:
   (a) $(AsO_4)^{3-}$ (b) $(MnO_4)^{1-}$ (c) $(SO_3)^{2-}$ (d) $(NH_4)^{1+}$ (e) $(HSO_4)^{1-}$ (f) $(C_2H_3O_2)^{1-}$

20. Write the names of the following polyatomic ions:
   (a) $(NO_2)^{1-}$ (b) $(NO_3)^{1-}$ (c) $(ClO_4)^{1-}$ (d) $(CN)^{1-}$ (e) $(BO_3)^{3-}$ (f) $(ClO)^{1-}$

LEARNING GOAL 6

Formulas of Ternary and Higher Compounds

21. Write the formulas of the following ternary and higher compounds:
   (a) mercury(II) phosphate
   (b) tin(II) arsenate
   (c) iron(III) acetate
   (d) lithium phosphate
   (e) aluminum sulfite
   (f) zinc nitrite

22. Write the formulas of the following ternary and higher compounds:
   (a) cesium hydroxide
   (b) copper(I) arsenate
   (c) ammonium sulfate
   (d) potassium carbonate
   (e) ferric cyanide
   (f) cuprous sulfate

◀ *23.* Write the formulas of the following ternary and higher compounds:
   (a) lead(II) sulfate
   (b) cobalt(II) phosphate
   (c) ammonium dichromate
   (d) calcium oxalate
   (e) stannous nitrate
   (f) magnesium hydrogen sulfate

◀ 24. Write the formulas of the following ternary and higher compounds:

   (a) potassium chlorate
   (b) zinc hydroxide
   (c) zinc phosphate
   (d) silver nitrite
   (e) mercuric nitrate
   (f) cuprous sulfite

LEARNING GOAL 7

Names of Binary Compounds Containing Two Nonmetals

25. Write the names of the following binary compounds composed of two nonmetals:

   (a) P_2S_5 (b) CO (c) SiO_2 (d) ClO_2

26. Write the names of the following binary compounds composed of two nonmetals:

   (a) N_2O (b) NO_2 (c) SO_3 (d) N_2O_5

LEARNING GOAL 8

Names of Binary Compounds Containing a Metal and a Nonmetal

◀ *27.* Write the names of the following binary compounds containing metals with fixed oxidation numbers:

   (a) Al_2O_3 (b) NaI (c) $ZnCl_2$ (d) Mg_3N_2 (e) Ag_2S (f) LiI

◀ 28. Write the names of the following binary compounds containing metals with fixed oxidation numbers:

   (a) Cs_2O (b) Al_2S_3 (c) BaI_2 (d) $GaCl_3$ (e) K_2O (f) MgS

LEARNING GOAL 9

Determining Oxidation Numbers

29. Find the oxidation number of the underlined element or polyatomic ion:

   (a) Na$\underline{ClO_4}$ (b) $\underline{Pb}_3O_4$ (c) $\underline{Ge}S_2$ (d) $\underline{V}OCl_3$ (e) Ca$(\underline{HCO_3})_2$

◀ 30. Find the oxidation number of the underlined element:

   (a) H$\underline{Cl}$O   (b) H$\underline{Cl}$O$_2$ (c) H$\underline{Cl}$O$_3$ (d) H$\underline{Cl}$O$_4$

31. Find the oxidation number of the underlined element or polyatomic ion:

   (a) $\underline{V}_2O_5$ (b) $\underline{In}_2O$ (c) $\underline{N}_2O$ (d) $\underline{Pb}P_5$ (e) $Mg_3(\underline{BO_3})_2$ (f) $Al(\underline{ClO_3})_3$

32. Find the oxidation number of the underlined element or polyatomic ion:

   (a) $H_3\underline{B}O_3$ (b) $H\underline{Br}O_3$ (c) $H\underline{I}O_3$ (d) $\underline{Cr}PO_3$ (e) $Al(\underline{HCO_3})_3$ (f) Li$\underline{Cl}$O

◀ *33.* Write the names of the following binary compounds containing metals with variable oxidation numbers:

   (a) OsO_4 (b) Hg_6P_2 (c) FeS (d) $CoCl_2$ (e) Cu_3N (f) Cu_2O

◀ 34. Write the names of the following binary compounds containing metals with variable oxidation numbers:

   (a) CuS (b) $AuBr_3$ (c) FeO (d) Cu_3P_2 (e) V_2O_5 (f) MnO_2

LEARNING GOAL 10

Names of Ternary and Higher Compounds

35. Write the names of the following ternary and higher compounds:

(a) Ag_2CO_3 (b) $Hg_3(PO_4)_2$ (c) $Fe_2(SO_4)_3$ (d) $NaNO_3$ (e) $CuCrO_4$ (f) $Zn(OH)_2$

36. Write the names of the following ternary and higher compounds:

(a) $CaCO_3$ (b) $NaNO_2$ (c) $NaOH$ (d) $Mg(OH)_2$ (e) $K_2Cr_2O_7$ (f) NH_4I

◄ *37.* Write the names of the following ternary and higher compounds:

(a) $CaSO_4$ (b) KCN (c) $AlPO_4$ (d) $Cu_2C_2O_4$ (e) $Fe_2(CrO_4)_3$ (f) $Cu(NO_2)_2$

◄ 38. Write the names of the following ternary and higher compounds:

(a) Rb_2SO_4 (b) $Fe(C_2H_3O_2)_2$ (c) $Mg_3(BO_3)_2$ (d) $KMnO_4$ (e) $Bi_2(SO_4)_3$ (f) $(NH_4)_2C_2O_4$

LEARNING GOAL 11

Formulas and Names of Inorganic Acids

39. Write the formulas of the following inorganic acids:

(a) hypochlorous acid
(b) hydrobromic acid
(c) nitric acid
(d) chlorous acid
(e) perbromic acid
(f) sulfuric acid

40. Write the formulas of the following inorganic acids:

(a) acetic acid
(b) sulfurous acid
(c) chloric acid
(d) hydrofluoric acid
(e) phosphoric acid
(f) nitrous acid

◄ *41.* Name the following compounds as inorganic acids:

(a) HCN (b) H_2S (c) $HBrO_3$ (d) H_2SO_4 (e) $HClO$ (f) $HBrO_2$

◄ 42. Name the following compounds as inorganic acids:

(a) HIO_4 (b) $HClO_3$ (c) HBr (d) HF (e) $HClO_2$ (f) HIO

LEARNING GOAL 12

Common Names and Formulas of Compounds

43. Write the formula of each of the following compounds designated by their common names:

(a) milk of magnesia
(b) oil of vitriol
(c) saltpeter
(d) laughing gas

44. Write the formula of each of the following compounds designated by their common names:

    (a) Epsom salts
    (b) muriatic acid
    (c) marble
    (d) table salt

45. Write the common name of the compound designated by each formula:

   (a) CaO   (b) $CaCO_3$ (c) $CaSO_4 \cdot 2H_2O$ (d) H_2SO_4

46. Write the systematic chemical name of each of the following substances:

   (a) dry ice   (b) saltpeter   (c) baking soda   (d) borax

EXTRA EXERCISES

◀ *47.* Predict the compound that is *most likely* to be formed from each of the following pairs of elements:

   (a) Sr and O   (b) Al and N   (c) Rb and S

48. Write the formula of the compound formed from each of the following pairs of ions:

   (a) Al^{3+} and N^{3-}
   (b) V^{5+} and O^{2-}
   (c) Fe^{3+} and $(OH)^{1-}$
   (d) $(NH_4)^{1+}$ and S^{2-}

◀ *49.* Name each of the compounds in Exercise 48.

50. Write the formulas of the following compounds:

    (a) tin(II) ion plus sulfide ion
    (b) copper(I) ion plus phosphate ion
    (c) iron(III) ion plus nitrite ion
    (d) magnesium ion plus acetate ion

51. Complete the table following Exercise 53 by writing the correct formula of the compound formed by each metal ion in combination with each of the given nonmetal ions.

52. Write the correct name of each of the following compounds:

    (a) $KClO_2$ (b) $Fe(CN)_3$ (c) P_2S_5 (d) CCl_4 (e) $HC_2H_3O_2$ (in water)   (f) SnO_2

53. Write the correct formula of each of the following compounds:

    (a) gallium fluoride
    (b) palladium(II) nitrate
    (c) gold(III) phosphide
    (d) lanthanum(III) acetate
    (e) plutonium(IV) oxide
    (f) ruthenium(VIII) oxide

	NONMETAL ION		
Metal ion	*Bromide*	*Sulfate*	*Phosphate*
Sodium	_____	_____	_____
Calcium	_____	_____	_____
Aluminum	_____	_____	_____

158 ESSENTIAL CONCEPTS OF CHEMISTRY

CUMULATIVE REVIEW Chapter 7–8

1. Write the electron-dot structure for (a) a calcium atom, (b) a calcium ion.
2. Write the electron-dot structure for (a) a chlorine atom, (b) a chloride ion.
3. Distinguish among an ionic bond, a covalent bond, and a coordinate covalent bond.
4. Draw the electron-dot diagram for C_3H_8.
◀ 5. List the following compounds in order of decreasing covalence: HCl, $BaCl_2$, CaF_2, and $CaCl_2$.
6. Which of the following compounds is most ionic? LiF, NaF, RbF, or CsF
7. Write the chemical formula of each of the following compounds:

   (a) copper(I) sulfite
   (b) osmium tetroxide
   (c) ammonium carbonate
   (d) dinitrogen monoxide

◀ 8. Find the oxidation number of each underlined element:

   (a) $\underline{W}Cl_5$ (b) $\underline{Sn}F_4$ (c) $\underline{V}_2O_5$ (d) $\underline{In}I_3$

9. Write the name of each of the following compounds:

   (a) NH_4I (b) CaC_2 (c) Cu_2CrO_4

10. Determine the formula of a hypothetical compound that would be formed between elements X and Y if X has three electrons in the outer shell and Y has six electrons in the outer shell.
11. The electron–dot structure for carbon dioxide is True or false?

$$:\ddot{O}\mathbin{{\ast}\mkern-4mu{\ast}}C\mathbin{{\ast}\mkern-4mu{\ast}}\ddot{O}:$$

12. An atom having the electron configuration $1s^2 2s^2 2p^6 3s^2$ would form an ion with a 2+ charge. True or false?
13. A Br^{1-} ion, an atom of Kr, and a Rb^{1+} ion have the same electron structure. True or false?
14. The bonds in a molecule of are nonpolar. True or false?

$$\begin{array}{c} H \\ \diagdown \\ N \\ \diagup \diagdown \\ H H \end{array}$$

15. A molecule of $CHCl_3$ is polar. True or false?
16. Write the chemical formula for the combination of atoms that have the following electron configuration:

$$1s^2 2s^2 2p^4 \quad \text{and} \quad 1s^2 2s^2 2p^6 3s^2$$

◀ 17. Which of the following bonds is most covalent?

C—H, C—Cl, or C—O

18. Write the formula of the compound formed from Na^{1+} and SO_4^{2-}.
◀ 19. Predict the oxidation number that the element with atomic number 16 is most likely to have.
20. Draw the electron-dot diagram for each of the following:

   (a) NO_3^{1-} (b) NH_4^{1+}

◀ 21. Calculate the oxidation number for the element indicated in the following compounds or ions:

   (a) S in $(SO_4)^{2-}$ (b) Mn in MnO_2 (c) Mn in $(MnO_4)^{-1}$ (d) N in HNO_2

22. Use the symbol δ^+ to show a partial positive charge and the symbol δ^- to show a partial negative charge for each of the atoms in the following compounds

   (a) H_2O (b) HI   (c) Cl_2O (d) PCl_5

23. Write the electron-dot structure for each of the following compounds:

   (a) F_2 (b) C_2Cl_2 (c) PCl_5 (d) C_2Cl_4

◀ 24. State the common oxidation number for each of the following elements when they enter into chemical combination. Base your answer on the element's position in the periodic table.

   (a) cesium   (b) astatine   (c) barium   (d) gallium

25. Write the formulas of the compounds formed by the following ions:

   (a) ferrous, Fe^{2+}, and phosphate, $(PO_4)^{3-}$
   (b) mercuric, Hg^{2+}, and cyanide, $(CN)^{1-}$
   (c) zinc, Zn^{2+}, and bicarbonate, $(HCO_3)^{1-}$
   (d) ferric, Fe^{3+}, and dichromate, $(Cr_2O_7)^{2-}$

26. Write the formulas of the following compounds:

   (a) indium sulfide
   (b) magnesium arsenate
   (c) gallium oxide
   (d) rubidium selenide
   (e) iron(II) oxide
   (f) cobalt(II) chloride
   (g) copper(I) nitride
   (h) copper(I) oxide

27. Write the formulas of the following compounds:

   (a) periodic acid
   (b) chloric acid
   (c) hydrobromic acid
   (d) chlorous acid

◀ 28. What is a coordinate covalent bond? Explain.

◀ 29. Is it possible for nonpolar molecules to contain polar bonds? Explain.

◀ 30. Element number 116 has not yet been discovered. But using what you know about periodic trends, answer the following questions:

   (a) In what family (group) would this element be placed?
   (b) How many electrons would be in its outermost energy level?
   (c) What would be its most common oxidation number?

CHAPTER 9

CALCULATIONS INVOLVING CHEMICAL FORMULAS

Let's journey to the beginning of the nineteenth century. John Dalton had just published his atomic theory. A belief that matter was made up of discrete particles called atoms was beginning to gain favor among many chemists, believers of Dalton's theory, who were called atomists. However, there were also many respected chemists at the time who did not believe in the existence of atoms. Both the atomists and non-atomists continued to debate this issue as new chemical research was published.

It was about this time (1808) that the French chemist Joseph Gay-Lussac performed experiments that indicated an amazing regularity in the way that (elemental) gases combined to form gaseous products. In one of these experiments, Gay-Lussac discovered that the volume ratio of hydrogen gas to oxygen gas that was required to form water (in the form of steam) was 1.9989 to 1.0000. Because this result was very close to a simple 2-to-1 ratio, Gay-Lussac attributed the slight difference in his experimental value to experimental error. Gay-Lussac discovered that simple numerical ratios held between reactants and products, as long as both gases were measured under identical conditions of temperature and pressure. For example,

2 volumes of hydrogen gas + 1 volume of oxygen gas → 2 volumes of steam
or 1 volume of nitrogen gas + 3 volumes of hydrogen gas → 2 volumes of ammonia gas

These experiments gave rise to the *Law of Combining Volumes*. This law stated that the volumes of gases in chemical reactions are related by simple whole numbers. To the modern-day chemist these experiments would be viewed as adding evidence to the existence of atoms, but to Dalton these observations appeared to be in conflict with his atomic theory. Dalton held that particles in gaseous elements must be atoms. Therefore, according to Dalton 1 atom of hydrogen should unite with 1 atom of oxygen to form 1 molecule of steam.

$$H + O \rightarrow HO$$

The Law of Combining Volumes appeared to be in conflict with atomic theory, yet there was much evidence to support both. Chemists of the day were asking how both Dalton's atomic theory and Gay-Lussac's Law of Combining Volumes could be valid. It was the Italian chemist Amedeo Avogadro in 1811 who resolved the apparent conflict when he presented his two-part hypothesis:

1. Equal volumes of gases under the same conditions of temperature and pressure contain an equal number of particles.

2. The ultimate *physical* units of elemental substances may be different from their ultimate *chemical* units.

Avogadro's first assumption, although important to resolving the conflict between Dalton and Gay-Lussac, had been stated by others. His second assumption, however, was original and showed miraculous insight into

162 ESSENTIAL CONCEPTS OF CHEMISTRY

what was happening on a molecular level. Avogadro proposed that the ultimate physical units of both hydrogen and oxygen were not *atoms* but *molecules*, each composed of two atoms. According to Avogadro's hypothesis:

$$2H_2 + 1O_2 \rightarrow 2H_2O$$

where H_2 represents a molecule of hydrogen gas, and O_2 represents a molecule of oxygen gas. Therefore, 2 molecules of hydrogen gas plus 1 molecule of oxygen gas produces 2 molecules of water vapor (steam).

Avogadro's insight into the formulas of gases coupled with what was occurring on the molecular level made a significant contribution both to the understanding of atomic theory and to the advancement of modern-day chemistry.

LEARNING GOALS

After you've studied this chapter, you should be able to:

1. Calculate the number of moles in a sample of an element when you are given the mass of the sample.
2. Calculate the mass, in grams, of a sample of an element when you are given the number of moles.
3. Calculate the number of atoms of an element in a sample when you are given the mass of the sample.
4. Calculate the empirical formula of a compound when you are given its percentage composition.
5. Calculate the number of moles in a sample of a compound when you are given the mass of the sample.
6. Calculate the mass, in grams, of a sample of a compound when you are given the number of moles.
7. Calculate the number of formula units of a compound in a sample when you are given the mass of the sample.
8. Determine the molecular formula of a compound from its molecular mass and its empirical formula or percentage composition data.
9. Calculate the percentage composition by mass of a compound when you are given its chemical formula.

INTRODUCTION

We began our study of chemistry by reviewing the basic structure of matter. We learned about elements, compounds, atoms, and molecules. We discussed the basic structure of the atom and the concept of chemical bonding. We also learned how to name and write the formulas of chemical compounds. We did all of this in a qualitative fashion and didn't get very involved with quantitative calculations. In this chapter, we will revisit the topic of elements and compounds. However, this time we will discuss some quantitative aspects involving these substances.

We will begin our study of calculations involving elements and compounds by learning about the *mole*, a useful concept for expressing the amount of a chemical substance. Before proceeding with this chapter, you may find it helpful to review the discussion of atomic mass and formula mass in Chapter 3. Also, because we will use the factor-unit method in our calculations, you may find it helpful to review the discussion of the factor-unit method in Chapter 2.

9.1 Gram-Atomic Mass and the Mole

The **gram-atomic mass** of an element is *its atomic mass* expressed *in grams*. For example, the atomic mass of gold is 197.0 amu, so gold's gram-atomic mass is 197.0 g. And the gram-atomic mass of carbon is 12.0 g.

One gram-atomic mass of any element contains the same number of atoms as one gram-atomic mass of any other element. For example, 197.0 g of gold contain the same number of atoms as 12.0 g of carbon and the same number of atoms as 1.0 g of hydrogen. In about 1870, scientists discovered that this number is 6.02×10^{23}. It is called **Avogadro's number**, in honor of the scientist whose thinking led to its discovery.

CHAPTER 9: CALCULATIONS INVOLVING CHEMICAL FORMULAS

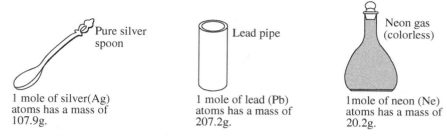

FIGURE 9.1 1 Mole of atoms of some common substances.

Later scientists took this idea one step further and defined 6.02×10^{23} atoms as 1 **mole** of atoms. Thus 1 mole is equal to 6.02×10^{23}; it is simply a particular number of items. We could just as easily talk about 1 mole of people, 1 mole of apples, or 1 mole of dollars. (By the way, if we had 1 mole of dollar bills, we would have $602,000,000,000,000,000,000,000. If we distributed our money equally among a lucky 5 billion people on the earth (there are currently 7.2 billion people on the planet), each person would get more than 120 trillion dollars:

$$\frac{\$602,000,000,000,000,000,000,000}{5,000,000,000} = 1.20 \times 10^{14}$$

A person could spend a million dollars every day, each day of the year, and never run out of money for 330,000 years.)

A mole of hydrogen atoms weighs 1.0 g; a mole of oxygen atoms weighs 16.0 g; a mole of carbon atoms weighs 12.0 g; and a mole of uranium atoms weighs 238.0 g. (To simplify the mathematics, we are rounding off the atomic masses found in the periodic table to one decimal place. We will do this throughout the text.) Figure 9.1 shows the mass of a mole of some other kinds of atoms.

The mole is very important in chemistry. It gives the chemist a convenient way to describe a large number of atoms or molecules and to relate this number of atoms or molecules to a mass (usually in grams). In fact, the mole has replaced the gram-atomic mass, which is seldom used today. But you do have to be able to convert from moles of atoms of an element to mass in grams, and vice versa. This is like converting inches to feet and feet to inches.

EXAMPLE 9.1

Do the following conversions: (a) $6\overline{0}$ in = ? ft (b) 8.0 ft = ? in

Solution

(a) There are 12 in in 1 ft, so

$$? \text{ ft} = 6\overline{0} \text{ in} \times \frac{1 \text{ ft}}{12 \text{ in}} = 5.0 \text{ ft}$$

(b) In the same way,

$$? \text{ in} = 8.0 \text{ ft} \times \frac{12 \text{ in}}{1 \text{ ft}} = 96 \text{ in}$$

Practice Exercise 9.1

Do the following conversions:

(a) 72.0 in = ? ft (b) 10.0 ft = ? in

164 ESSENTIAL CONCEPTS OF CHEMISTRY

EXAMPLE 9.2

(a) How many moles of oxygen atoms are there in $8\bar{0}$ g of oxygen?

(b) What is the mass in grams of 0.50 mole of oxygen atoms?

Solution

UNDERSTAND THE PROBLEM

(a) In this example, a specific mass of oxygen atoms is given. This mass must be converted into moles.

DEVISE A PLAN

We use the factor-unit method as described in Chapter 2. (For convenience, we will round all atomic masses to one decimal place.)

CARRY OUT THE PLAN

$$\text{Moles of O atoms} = \cancel{g} \times \frac{\text{moles of O atoms}}{\cancel{g}}$$

Then, because the atomic mass of oxygen is 16.0 amu,

$$\text{? moles of O atoms} = 8\bar{0}\,\cancel{g} \times \frac{1 \text{ moles of O atoms}}{16.0\,\cancel{g}}$$

$$= 5.0 \text{ moles of O atoms}$$

We find that $8\bar{0}$ g of oxygen atoms equals 5.0 moles.

LOOK BACK

We reason that if 1 mole of oxygen atoms contains 16.0 g, then 5 moles should have five times as many grams. Since 16.0×5.0 equals $8\bar{0}$, the answer does make sense.

UNDERSTAND THE PROBLEM

(b) In this case, the number of moles of oxygen is given and it must be converted into a mass (in grams).

DEVISE A PLAN

We use the factor-unit method.

CARRY OUT THE PLAN

Using the atomic mass of oxygen, we write

$$g = \cancel{\text{moles of O atoms}} \times \frac{g}{\cancel{\text{moles of O atoms}}}$$

$$?g = 0.50 \cancel{\text{ mole of O atoms}} \times \frac{16.0 \text{ g}}{1 \cancel{\text{ mole of O atoms}}}$$

$$= 8.0 \text{ g}$$

LOOK BACK

We know that 1 mole of oxygen has a mass of 16.0 g; therefore 0.50 mole of oxygen should have half that mass, or 8.0 g. We conclude that our answer is sensible.

Practice Exercise 9.2

(a) How many moles of oxygen atoms are there in $16\overline{0}$ g of oxygen? (b) What is the mass in grams of 2.00 moles of oxygen atoms?

$160 g \times \frac{1 m}{16.g} = \frac{160}{16} = 10$ moles of oxygen

$2 mO \times \frac{1}{16.g} = 16 \times 2 = 32 g$ of O

EXAMPLE 9.3

Find the number of moles of atoms in each of the following samples:

(a) 46 g of Na (b) 5.4 g of Al (c) 0.12 g of C (d) 23.8 g of U

(a) $46 g \times \frac{1m}{23g} = \frac{46}{23} = 2m$ (b) $5.4 g \times \frac{1m}{27g}$ $\frac{5.4}{27} = 0.2$ (c) $0.12 g \times \frac{1}{12}$ $\frac{.12}{12} = .01$ (d) $23.8 g \times \frac{1m}{238} = \frac{23.8}{238} = .10 m$ of U

Solution

We need to look up the atomic mass of each element in the periodic table inside the front cover.

(a) The atomic mass of Na is 23.0 amu. Therefore

$$\text{? moles of Na atoms} = 46 \text{ g} \times \frac{1 \text{ mole of Na atoms}}{23.0 \text{ g}}$$

$$= 2.0 \text{ moles of Na atoms}$$

(b) The atomic mass of Al is 27.0 amu. Therefore

$$\text{? moles of Al atoms} = 5.4 \text{ g} \times \frac{1 \text{ mole of Al atoms}}{27.0 \text{ g}}$$

$$= 0.20 \text{ moles of Al atoms}$$

(c) The atomic mass of C is 12.0 amu. Therefore

$$\text{? moles of C atoms} = 0.12 \text{ g} \times \frac{1 \text{ mole of C atoms}}{0.12 \text{ g}}$$

$$= 0.010 \text{ mole of C atoms}$$

(d) The atomic mass of U is 238.0 amu. Therefore

$$\text{? moles of U atoms} = 23.8 \text{ g} \times \frac{1 \text{ mole of U atoms}}{238.0 \text{ g}}$$

$$= 0.100 \text{ mole of U atoms}$$

Practice Exercise 9.3

Find the number of moles of atoms in each of the following samples:

(a) 11.50 g of Na (b) 2.70 g of Al (c) $28\overline{0}$ g of N (d) 0.0238 g of U

$11.50 g \times \frac{1}{23g} = \frac{11.50}{23} = .5 m$ of Na

$2.70 g \times \frac{1}{27} = \frac{2.70}{27} = .1 m$ of Al

$280 g \times \frac{1}{14}$ $280/14 = 20 m$

$0.0238 \times \frac{1}{238} = 0.0001 m$

EXAMPLE 9.4

Find the mass in grams of each of the following:

(a) 0.20 mole of Zn atoms (b) 4.0 moles of Br atoms (c) 1.50 moles of Ca atoms

$0.20 \times \frac{1m}{65} = 13 g$ of Zn

$4.0 m \times \frac{1}{80}$ $4.0 \times 80 = 320 g$ of Br

$1.50 m \times \frac{1}{40}$ $1.50 \times 40 = 60 g$ of Ca

166 ESSENTIAL CONCEPTS OF CHEMISTRY

Solution

We need to look up the atomic mass of each element in the periodic table.

(a) The atomic mass of Zn is 65.4 amu. Therefore

$$? \text{ g} = 0.20 \text{ mole of Zn atoms} \times \frac{65.4 \text{ g}}{1 \text{ mole of Zn atoms}} = 13 \text{g}$$

(b) The atomic mass of Br is 79.9 amu. Therefore

$$? \text{ g} = 4.0 \text{ mole of Br atoms} \times \frac{79.9 \text{ g}}{1 \text{ mole of Br atoms}} = 320 \text{ g}$$

(c) The atomic mass of Ca is 40.1 amu. Therefore

$$? \text{ g} = 1.50 \text{ mole of Ca atoms} \times \frac{40.1 \text{ g}}{1 \text{ mole of Ca atoms}} = 60.2 \text{ g}$$

Practice Exercise 9.4

Find the mass in grams of each of the following:

(a) 15.0 moles of Zn atoms (b) 0.200 mole of Br atoms (c) 3.00 moles of Ca atoms

EXAMPLE 9.5

How many *atoms* of each element are there in the samples given in Example 9.4?

Solution

We know that 1 mole of atoms of any element is 6.02×10^{23} atoms of that element; in other words, there are

$$\frac{6.02 \times 10^{23} \text{ atoms}}{1 \text{ mole of atoms}}$$

for any element. But for this example, we will round off Avogadro's number to 6.0×10^{23}.

(a) For the zinc,

$$? \text{ Zn atoms} = 0.20 \text{ mole of Zn atoms} \times \frac{6.0 \times 10^{23} \text{ Zn atoms}}{1 \text{ mole of Zn atoms}}$$

$$= 1.2 \times 10^{23} \text{ Zn atoms}$$

(b) For the bromine,

$$? \text{ Br atoms} = 4.0 \text{ moles of Br atoms} \times \frac{6.0 \times 10^{23} \text{ Br atoms}}{1 \text{ mole of Br atoms}}$$

$$= 24 \times 10^{23} \text{ Br atoms (or } 2.4 \times 10^{24})$$

(c) For the calcium,

$$? \text{ Ca atoms} = 1.50 \text{ moles of Ca atoms} \times \frac{6.0 \times 10^{23} \text{ Ca atoms}}{1 \text{ mole of Ca atoms}}$$

$$= 9.0 \times 10^{23} \text{ Ca atoms}$$

Practice Exercise 9.5

How many *atoms* of each element are there in the samples given in Practice Exercise 9.4?

9.2 Empirical Formulas

Chemists prepare hundreds of new compounds in their search for substances that may be beneficial in medicine, agriculture, industry, and the home. A first step in determining the nature of such a new or unknown compound is to obtain its empirical formula. The **empirical formula** of a compound is *the simplest whole-number ratio of the atoms that make up a formula unit of the compound*. For example, the empirical formula of water is H_2O. The subscripts indicate that the ratio of hydrogen atoms to oxygen atoms in this molecule is 2 to 1, often written 2:1. (The lack of a subscript on the O is taken to mean a subscript of 1.)

To find the empirical formula of a compound, the chemist measures the percentage by mass of each element in the compound. From the percentage of each element, the chemist then determines (1) the number of moles of atoms of each element in 100 g of the compound and (2) the ratio of the moles of atoms. (We take 100 g of compound merely as a convenience.)

EXAMPLE 9.6

Determine the empirical formula of a compound whose composition is 50.05% S and 49.95% O by mass.

Solution

If we had 100 g of the compound, 50.05 g would be sulfur and 49.95 g would be oxygen. All we have to do now is convert these masses to moles of atoms and then find their whole-number ratio.

$$? \text{ moles of S atoms} = 50.05 \text{ g} \times \frac{1 \text{ mole of S atoms}}{32.1 \text{ g}}$$

$$= 1.56 \text{ moles of S atoms}$$

$$? \text{ moles of O atoms} = 49.95 \text{ g} \times \frac{1 \text{ mole of O atoms}}{16.0 \text{ g}}$$

$$= 3.12 \text{ moles of O atoms}$$

Therefore the formula may be written as $S_{1.56}O_{3.12}$, but this formula does not have whole-number subscripts. One way to obtain a formula with whole-number subscripts is to *divide* all the subscripts by the *smallest* subscript. This gives us

$$S_{1.56/1.56}O_{3.12/1.56} \text{ or } SO_2$$

This is an empirical formula, because it has the lowest possible ratio of whole-number subscripts. (The compound is called sulfur dioxide.)

Practice Exercise 9.6

Determine the empirical formula of a compound whose composition is 88.9% oxygen and 11.1% hydrogen by mass.

EXAMPLE 9.7

Find the empirical formula of a compound whose composition is 3.1% H, 31.5% P, and 65.4% O by mass.

168 ESSENTIAL CONCEPTS OF CHEMISTRY

Solution

In 100 g of this compound, there are 3.1 g of H, 31.5 g of P, and 65.4 g of O. Therefore there are

$$? \text{ moles of H atoms} = 3.1 \text{ g} \times \frac{1 \text{ mole of H atoms}}{1.0 \text{ g}}$$

$$= 3.1 \text{ moles of H atoms}$$

$$? \text{ moles of P atoms} = 31.5 \text{ g} \times \frac{1 \text{ mole of P atoms}}{31.0 \text{ g}}$$

$$= 1.02 \text{ moles of P atoms}$$

$$? \text{ moles of O atoms} = 65.4 \text{ g} \times \frac{1 \text{ mole of O atoms}}{16.0 \text{ g}}$$

$$= 4.09 \text{ moles of O atoms}$$

This formula may be written $H_{3.1}P_{1.02}O_{4.09}$, but this formula does not have whole-number subscripts. Therefore we divide each subscript by the smallest subscript to obtain

$$H_{3.1/1.02}P_{1.02/1.02}O_{4.09/1.02} \quad \text{or} \quad H_3PO_4$$

Practice Exercise 9.7

Find the empirical formula of a compound whose composition is 51.9% Cr and 48.1% S by mass.

EXAMPLE 9.8

Determine the empirical formula of a compound whose composition is 23.8% C, 5.9% H, and 70.3% Cl by mass.

Solution

In 100 g of this compound, there are 23.8 g of C, 5.9 g of H, and 70.3 g of Cl. Therefore there are

$$? \text{ moles of C atoms} = 23.8 \text{ g} \times \frac{1 \text{ mole of C atoms}}{12.0 \text{ g}}$$

$$= 1.98 \text{ moles of C atoms}$$

$$? \text{ moles of H atoms} = 5.9 \text{ g} \times \frac{1 \text{ mole of H atoms}}{1.0 \text{ g}}$$

$$= 5.9 \text{ moles of H atoms}$$

$$? \text{ moles of Cl atoms} = 70.3 \text{ g} \times \frac{1 \text{ mole of Cl atoms}}{35.5 \text{ g}}$$

$$= 1.98 \text{ moles of Cl atoms}$$

One possible formula is $C_{1.98}H_{5.9}Cl_{1.98}$, but this does not have whole-number subscripts. Again we divide by the smallest subscript to obtain

$$C_{1.98/1.98}H_{5.9/1.98}Cl_{1.98/1.98} \quad \text{or} \quad CH_3Cl$$

Practice Exercise 9.8

Determine the empirical formula of a compound whose composition is 41.5% Zn, 17.8% N, and 40.7% O by mass.

After you divide the subscripts in determining an empirical formula, you may round subscripts to the nearest whole number if they are within 0.1 of the whole number. Otherwise you must search for a factor that will give you whole numbers. For example, suppose a compound contains 68.4% Cr and 31.6% O by mass. This means that 100 g of the compound contains 68.4 g (or 1.32 moles) of Cr and 31.6 g (or 1.98 moles) of O. Therefore the formula is

$$Cr_{1.32}O_{1.98}$$

Following our usual procedure, we divide by the smallest number of moles (the smallest subscript) and obtain

$$Cr_{1.32/1.32}O_{1.98/1.32} \quad \text{or} \quad CrO_{1.5}$$

The formula $CrO_{1.5}$ does not have *whole*-number subscripts. What do we do now? Should we simply round off the 1.5 to the number 2 and make the empirical formula CrO_2? The answer is NO! We may round off only when the number is within 0.1 of the whole number. Therefore we must search for a factor that will give us a whole-number ratio. When we look carefully at $Cr_1O_{1.5}$, we see that if we multiply each subscript by 2, we can get whole-number subscripts. $Cr_1O_{1.5}$ becomes Cr_2O_3, and the empirical formula is Cr_2O_3.

9.3 Gram-Formula Mass and the Mole

We're now going to learn how to apply the concept of the *mole* to compounds. Because some compounds are composed of molecules and other compounds are composed of ions, we'll use the term *formula mass* to help us with this application. Recall from Chapter 3 that the formula mass of a compound is *the sum of the atomic masses of all the atoms or ions that make up a formula unit of the compound*. The formula mass of a compound in grams is called the **gram-formula mass**.* This is the mass of a collection of 6.02×10^{23} formula units of the compound. Therefore *1 gram-formula mass of a compound is the mass in grams of 1 mole of molecules for a molecular compound or 1 mole of formula units for an ionic compound.*

Consider some examples. Water (H_2O) is a compound composed of molecules, each containing two hydrogen atoms and one oxygen atom. The formula mass of water is 18.0 amu—two times the atomic mass of hydrogen (2×1.0) plus the atomic mass of oxygen (16.0). So the mass of 1 mole of water molecules is 18.0 g (Figure 9.2). Calcium chloride ($CaCl_2$) is a compound composed of ions; the formula unit of the compound consists of one calcium ion and two chloride ions. The formula mass of calcium chloride is 111.1 amu—the

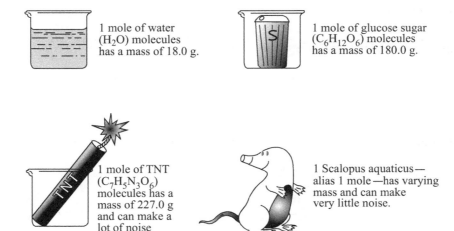

FIGURE 9.2 One mole of molecules of some common substances.

---

*The term **molar mass** is used in some texts as a general term to describe gram-formula mass and gram-atomic mass of a substance. In this text, we will continue to use the more traditional terms, *gram-formula mass* and *gram-atomic mass*.

170 ESSENTIAL CONCEPTS OF CHEMISTRY

atomic mass of calcium (40.1) plus two times the atomic mass of chlorine (2 × 35.5). So the mass of 1 mole of calcium chloride formula units is 111.1 g.

It is important to be able to convert the number of grams of a compound into the number of moles, and vice versa. You will need to perform these calculations when we study the topic of chemical stoichiometry in Chapter 11. The following examples will give you some practice.

EXAMPLE 9.9

Find the number of moles in each of the following:

(a) 32 g of CH_4 (b) 0.32 g of CH_4 (c) 81 g of H_2O (d) 37 g of $Ca(OH)_2$

Solution

We first find the formula mass of each compound and then use the factor-units method to determine the number of moles.

(a) The formula mass of CH_4 is 16.0 amu. Therefore we have

$$? \text{ moles of } CH_4 \text{ molecules} = 32 \text{ g} \times \frac{1 \text{ mole of } CH_4 \text{ molecules}}{16.0 \text{ g}}$$

$$= 2.0 \text{ moles of } CH_4 \text{ molecules}$$

(b) The formula mass of CH_4 is 16.0 amu. Therefore we have

$$? \text{ moles of } CH_4 \text{ molecules} = 0.32 \text{ g} \times \frac{1 \text{ mole of } CH_4 \text{ molecules}}{16.0 \text{ g}}$$

$$= 0.020 \text{ mole of } CH_4 \text{ molecules}$$

(c) The formula mass of H_2O is 18.0 amu. Therefore

$$? \text{ moles of } H_2O \text{ molecules} = 81 \text{ g} \times \frac{1 \text{ mole of } H_2O \text{ molecules}}{18.0 \text{ g}}$$

$$= 4.5 \text{ moles of } H_2O \text{ molecules}$$

(d) The formula mass of $Ca(OH)_2$ is 74.1 amu. Therefore

$$? \text{ moles of } Ca(OH)_2 \text{ formula units} = 37 \text{ g} \times \frac{1 \text{ mole of } Ca(OH)_2 \text{ formula units}}{74.1 \text{ g}}$$

$$= 0.50 \text{ mole of } Ca(OH)_2 \text{ formula units}$$

Practice Exercise 9.9

Find the number of moles in each of the following: (a) $9\bar{0}$ g of H2O (b) 0.016 g CH_4

EXAMPLE 9.10

Determine the number of grams in each of the following:
(a) 6.00 moles of butane (C_4H_{10}) (b) 0.025 mole of CO_2 (c) 7.00 moles of Al_2O_3 (d) 0.400 mole of Cu_3N

Solution

We first find the formula mass of each compound and then use the factor-unit method to determine the number of grams.

CHAPTER 9: CALCULATIONS INVOLVING CHEMICAL FORMULAS

(a) The formula mass of C_4H_{10} is 58.0 amu. Therefore we have

$$? \text{ g } C_4H_{10} = 6.00 \text{ moles} \times \frac{58.0 \text{ g}}{1 \text{ mole}} = 348 \text{ g}$$

(b) The formula mass of CO_2 is 44.0 amu. Therefore we have

$$? \text{ g } CO_2 = 0.025 \text{ mole} \times \frac{44.0 \text{ g}}{1 \text{ mole}} = 1.1 \text{ g}$$

(c) The formula mass of Al_2O_3 is 102.0 amu. Therefore we have

$$? \text{ g } Al_2O_3 = 7.00 \text{ moles} \times \frac{102.0 \text{ g}}{1 \text{ mole}} = 714 \text{ g}$$

(d) The formula mass of Cu_3N is 204.5 amu. Therefore we have

$$? \text{ g } Cu_3N = 0.400 \text{ mole} \times \frac{204.5 \text{ g}}{1 \text{ mole}} = 81.8 \text{ g}$$

Practice Exercise 9.10

Determine the number of grams in each of the following:

(a) 4.00 moles of NO_2 (b) 0.050 mole of $CaCO_3$ (c) 2.50 moles of $Ca(C_2H_3O_2)_2$ (d) 0.060 mole of $(NH_4)_2SO_4$

EXAMPLE 9.11

In Example 9.9, how many formula units are present in each sample?

Solution

To obtain our factor unit, we recall that there are 6.0×10^{23} formula units in 1 mole of a compound.

(a) For the first sample of CH_4, composed of molecules,

$$? \text{ molecules of } CH_4 = 2.0 \text{ moles} \times \frac{6.0 \times 10^{23} \text{ molecules of } CH_4}{1 \text{ mole}}$$
$$= 12 \times 10^{23} \text{ molecules of } CH_4$$

(b) For the second sample of CH_4, composed of molecules,

$$? \text{ molecules of } CH_4 = 0.020 \text{ mole} \times \frac{6.0 \times 10^{23} \text{ molecules of } CH_4}{1 \text{ mole}}$$
$$= 1.20 \times 10^{23} \text{ molecules of } CH_4$$

(c) For the H_2O, composed of molecules,

$$? \text{ molecules of } H_2O = 4.5 \text{ moles} \times \frac{6.0 \times 10^{23} \text{ molecules of } H_2O}{1 \text{ mole}}$$
$$= 27 \times 10^{23} \text{ molecules of } H_2O$$

(d) For the Ca(OH)$_2$, composed of ions,

$$? \text{ formula units of Ca(OH)}_2 = 0.50 \text{ mole} \times \frac{6.0 \times 10^{23} \text{ formula units of Ca(OH)}_2}{1 \text{ mole}}$$

$$= 3.0 \times 10^{23} \text{ formula units of Ca(OH)}_2$$

Practice Exercise 9.11

In Practice Exercise 9.10, how many *formula units* are present in each sample?

9.4 Molecular Formulas

The **molecular formula** of a compound is *a formula that shows the actual number of atoms of each element that are in one molecule of that compound*. The molecular formula is found from the percentages by mass of the elements *and* the molecular mass of the compound.

EXAMPLE 9.12

A compound is composed of $4\bar{0}\%$ C, 6.6% H, and 53.4% O by mass. The molecular mass of the compound is 180.0. Determine its empirical and molecular formulas.

Solution

Find the empirical formula first, by the usual method.

$$? \text{ molecules of C atoms} = 4\bar{0} \text{ g} \times \frac{1 \text{ mole of C atoms}}{12.0 \text{ g}}$$

$$= 3.3 \text{ moles of C atoms}$$

$$? \text{ molecules of H atoms} = 6.6 \text{ g} \times \frac{1 \text{ mole of H atoms}}{1.0 \text{ g}}$$

$$= 6.6 \text{ moles of H atoms}$$

$$? \text{ molecules of O atoms} = 53.4 \text{ g} \times \frac{1 \text{ mole of O atoms}}{16.0 \text{ g}}$$

$$= 3.3 \text{ moles of O atoms}$$

Hence one possible formula is $C_{3.3}H_{6.6}O_{3.3}$, but this does not have whole-number subscripts. Therefore we divide by the smallest number of moles to get

$$C_{3.3/3.3}H_{6.6/3.3}O_{3.3/3.3} \quad \text{or} \quad CH_2O$$

This is the empirical formula. But it certainly is not the molecular formula of the compound, because CH$_2$O does not have a molecular mass of 180.0. What is the molecular formula of the compound? It could be any formula wherein the ratio of C:H:O is 1:2:1. Is it $C_2H_4O_2$, $C_3H_6O_3$, $C_4H_8O_4$, $C_5H_{10}O_5$, $C_6H_{12}O_6$, or what? Remember, the correct formula is the one that has a molecular mass of 180.0. Let's determine the molecular mass of each proposed formula.

CHAPTER 9: CALCULATIONS INVOLVING CHEMICAL FORMULAS

MOLECULAR FORMULA	MOLECULAR MASS (AMU)
CH_2O	30.0
$C_2H_4O_2$	60.0
$C_3H_6O_3$	90.0
$C_4H_8O_4$	120.0
$C_5H_{10}O_5$	150.0
$C_6H_{12}O_6$	180.0

It is clear from our trial-and-error method that the molecular formula is $C_6H_{12}O_6$. It's the only one that has a molecular mass of 180.0. But is there an easier way to do this? Yes. We simply ask ourselves, "How many times must we take the molecular mass of the empirical formula (30.0 in this problem) to get the true molecular mass (180.0 in this problem)?" The answer is obviously *six* times. It takes six times the mass of the empirical formula, CH_2O, to give the true molecular mass, so the true molecular formula must be $C_6H_{12}O_6$.

Practice Exercise 9.12

A compound has a molecular mass of 180.0 and is composed of 60.0% C, 4.48% H, and 35.5% O by mass. Calculate the empirical and molecular formulas of this common compound, which is called aspirin.

9.5 Percentage Composition by Mass

You have seen how to determine the empirical formula of a compound from its percentage composition. There will also be times when you will have to do the opposite—find the percentage by mass of each element in a compound. To do so, you need to know the empirical or molecular formula of the compound and the atomic mass of its constituent elements.

EXAMPLE 9.13

Find the percentage by mass of H and O in water.

Solution

In each mole of water (H_2O), there are 2 moles of H, having a mass of 1.0 g/mole, and 1 mole of O, having a mass of 16.0 g/mole. Therefore

$$\text{Percent H} = \frac{\text{mass of hydrogen in 1 mole of water}}{\text{mass of 1 mole of water}} \times 100$$

$$= \frac{2.0 \text{ g of H}}{18.0 \text{ g of } H_2O} \times 100 = 11\%$$

$$\text{Percent O} = \frac{\text{mass of oxygen in 1 mole of water}}{\text{mass of 1 mole of water}} \times 100$$

$$= \frac{16.0 \text{ g of H}}{18.0 \text{ g of } H_2O} \times 100 = 89\%$$

Note that the sum of the percentages must be 100.

Practice Exercise 9.13

Find the percentage by mass of K and Br in potassium bromide (KBr).

EXAMPLE 9.14

Find the percentage by mass of each element in the following compounds: (a) C_8H_{18} (b) NH_3 (c) FeO (d) Fe_2O_3

Solution

We must find the formula mass of each compound and the percent contribution of each element to that mass.

(a) C_8H_{18} contains 96.0 g of C + 18.0 g of H = 114.0 g in each mole. Therefore

$$\text{Percent C} = \frac{96.0 \text{ g of C}}{114.0 \text{ g of } C_8H_{18}} \times 100 = 84.2\%$$

$$\text{Percent H} = \frac{18.0 \text{ g of H}}{114.0 \text{ g of } C_8H_{18}} \times 100 = 15.8\%$$

(b) NH_3 contains 14.0 g of N + 3.0 g of H = 17.0 g in each mole. Therefore

$$\text{Percent N} = \frac{14.0 \text{ g of N}}{17.0 \text{ g of } NH_3} \times 100 = 82.4\%$$

$$\text{Percent H} = \frac{3.0 \text{ g of H}}{17.0 \text{ g of } NH_3} \times 100 = 17.6\% \text{ (round of 18\%)}$$

(c) FeO contains 55.8 g of Fe + 16.0 g of O = 71.8 g in each mole. Therefore

$$\text{Percent Fe} = \frac{55.8 \text{ g of Fe}}{71.8 \text{ g of FeO}} \times 100 = 77.7\%$$

$$\text{Percent O} = \frac{16.0 \text{ g of O}}{71.8 \text{ g of FeO}} \times 100 = 22.3\%$$

(d) Fe_2O_3 contains 111.6 g of Fe + 48.0 g of O = 159.6 g in each mole. Therefore

$$\text{Percent Fe} = \frac{111.6 \text{ g of Fe}}{159.6 \text{ g of } Fe_2O_3} \times 100 = 69.2\%$$

$$\text{Percent O} = \frac{48.0 \text{ g of O}}{159.6 \text{ g of } Fe_2O_3} \times 100 = 30.1\%$$

Practice Exercise 9.14

Find the percentage by mass of each element in the following compounds:

(a) $C_{22}H_{44}$ (b) NaCl (c) C_6H_6

EXAMPLE 9.15

How many grams of sulfur are there in 256 g of SO_3?

Solution

Determine the percentage of sulfur in SO_3 and multiply this by the amount of the sample (256 g). SO_3 contains 32.1 g of S + 48.0 g of O = 80.1 g/mole. Therefore

$$\text{Percent S} = \frac{32.1 \text{ g of S}}{80.1 \text{ g of SO}_3} \times 100 = 40.1\%$$

So 40.1% of the 256 g of SO_3 is sulfur (S); there are 0.401 × 256 g = 102.7 g of sulfur (or 103 g of sulfur to three significant figures).

Practice Exercise 9.15

How many grams of carbon are there in $1\overline{00}$ g of CH_4?

SUMMARY

Each element has its own symbol, and every compound has its own chemical formula. Each element has a unique atomic mass. One mole of anything is equal to 6.02×10^{23} things (Avogadro's number of things): 1 mole of an element contains 6.02×10^{23} atoms of that element, and 1 mole of a compound contains 6.02×10^{23} formula units of that compound.

One mole of atoms of an element has a mass equal to the atomic mass of that element in grams. One mole of formula units of a compound has a mass equal to the formula mass of that compound in grams.

The empirical formula of a compound is the simplest whole-number ratio of the atoms that make up a formula unit of the compound. The molecular formula of a compound is the actual number of atoms of each element that are in one molecule (or formula unit) of the compound. The empirical formula of a compound can be found from percentage composition data. The molecular formula of a compound can be found from its empirical formula and its molecular mass. The percentage composition of a compound may be determined from its empirical or molecular formula.

KEY TERMS

Avogadro's number (**9.1**)
empirical formula (**9.2**)
gram-atomic mass (**9.1**)
gram-formula mass (**9.3**)

molar mass (**9.3**)
mole (**9.1**)
molecular formula (**9.4**)

SELF-TEST EXERCISES

LEARNING GOAL 1

Moles from Mass of an Element

1. Explain the importance of the mole concept in chemistry.
2. (a) Using scientific notation, write the number that represents 1 mole.
   (b) Write this number without the use of scientific notation.
◀ *3*. A sample of gold (Au) has a mass of 19.7 g. How many moles of gold atoms is this?
4. How many moles are there in 5.62 g of silicon?
◀ *5*. The atomic department store is selling uranium atoms the reduced price of 100,000,000 atoms for 1 cent. What would be the cost in dollars of 1 mg of uranium?
◀ 6. A sample of sulfur has a mass of 0.963 g. How many moles of sulfur atoms are there in this sample?
7. Determine the number of *moles* of atoms in each of the following samples:

   (a) 6.2 g of Al   (b) 239.5 g of Ti   (c) 0.06075 g of Mg   (d) 3,570.0 g of U

◀ 8. Datermine the number of *moles* of atoms in each of the following samples:

    (a) 2397.0 g of Cu   (b) 13.79 g of W   (c) 0.00399 g of Ar   (d) 16.19 g of Ag

LEARNING GOAL 2

Mass, in Grams, from Moles of an Element

◀ 9. Mercury (Hg), element number 80 in the periodic table, has an atomic mass of 200.6. What mass (in grams) of Hg would you need to obtain 0.250 mole?

10. What is the mass in grams of 1.50 moles of rubidium (Rb) aoms?

11. What is the mass in grams of 0.750 mole of cobalt (Co) atoms?

◀ 12. How many grams of Cr are contained in 0.0250 mole of Cr?

13. Determine the number of grams in each of the following:

    (a) 4.600 moles of Ni   (b) 0.00300 mole of Br_2 (c) 200.0 moles of Ca   (d) 0.04000 mole of S

14. Determine the number of grams in each of the following:

    (a) 0.400 mole of Rn   (b) 19.0 moles of F_2 (c) 0.0350 mole of Hg   (d) 7.200 moles of Yb

LEARNING GOAL 3

Number of Atoms of an Element from Mass of a Sample

◀ 15. After you've finished eating your lunch, you recycle the aluminum (Al) foil in which it was wrapped. If the aluminum foil weighed 2.00 g, how many atoms of aluminum did you recycle?

◀ 16. At the end of an experiment, you find that there are 13.8 g of sodium left in the reaction vessel. How many atoms of sodium are left?

*17. Diamond is made up of carbon atoms. A 2-carat diamond costs $1,800. What is the cost of 1 carbon atom in the diamond? (Assume that 1 carat weighs 0.2 g.)

*18. How many atoms are present in a sample of water molecules that has a mass of $1,8\overline{00}$ g?

19. The radius of a silver atom is 1.34 angstroms, or 1.34 Å (1 Å = 10^{-8} cm). If the atoms of 0.1 mole of silver atoms were placed in a straight line, touching each other, what distance (in miles) would they cover?

*20. How many *atoms* of Cl_2 are present in a sample of Cl_2 that has a mass of $35\overline{0}$ g?

LEARNING GOAL 4

Empirical Formula from Percentage Composition

21. Sodium, the eleventh element in the periodic table (whose name means "headache" in medieval Latin), is a very dangerous chemical when uncombined. But when 39.6% sodium is combined with 60.4% chlorine by mass, a common compound is formed. What is the empirical formula of this compound, and what is its common name?

22. When 15.8% Al is combined with 28.2% S and 56.1% O by mass, a compound is formed. Determine the empirical formula of this compound.

◀ 23. Cholesterol, a compound suspected of causing hardening of the arteries, has a molecular mass of 386 and the following percentage composition by mass: 84.0% C, 11.9% H, and 4.1% O. What is the molecular formula of cholesterol?

24. When 24.2% Ca is combined with 17.1% N and 58.5% O by mass, a common compound is formed. What are its empirical formula and name?

CHAPTER 9: CALCULATIONS INVOLVING CHEMICAL FORMULAS

25. Determine the empirical formula of a compound that is 80.0% C and 20.0% H by mass. The molecular mass of this compound is 30.0. What is its molecular formula?
◄ 26. When 60.56% C is combined with 11.18% H and 28.26% N by mass, the compound spherophysine, which is used for treating high blood pressure, is formed. What is the empirical formula of this compound?
27. Determine the empirical formula of a compound that is 54.55% C, 9.09% H, and 36.36% O by mass. The molecular mass of this compound is 88. What is its molecular formula?
28. When 64.81% C is combined with 13.60% H and 21.59% O by mass, a compound that is used as a surgical anesthetic is formed. The name of this compound is diethyl ether. What is its empirical formula?

LEARNING GOAL 5

Moles from Mass of a Compound

29. How many moles of $Ba_3(PO_4)_2$ are there in 451.4 g of $Ba_3(PO_4)_2$?
30. A sample of $CaSO_4$ has a mass of 0.1705 g. How many moles of $CaSO_4$ are there in this sample?
31. How many moles of HCl are there in 100.0 g of this compound?
32. How many moles of O_2 molecules are there in 6.40 g of oxygen gas?
◄ 33. A sample of laughing gas (N_2O) has a mass of 2.20 g. How many moles of N_2O is this?
◄ 34. Milk of magnesia, $Mg(OH)_2$, is used as an antacid in small doses and as a laxative in large doses. The typical laxative dose for an adult is about 2.00 g. How many moles of $Mg(OH)_2$ is this?

LEARNING GOAL 6

Mass, in Grams, from Moles of a Compound

35. Determine the number of grams in each of the following:

(a) 2.50 moles of CH_4 (b) 0.0400 mole of SO_3 (c) 0.00600 mole of $Al_2(SO_4)_3$ (d) 0.020 mole of $(NH_4)_3PO_4$

36. What is the mass in grams of 10.0 moles of H_2O?
37. How many grams of SO_2 are contained in 0.20 mole of SO_2?
38. How many grams of H_2SO_4 are there in 4.00 moles of this compound?
◄ 39. Portable gas grills use propane (C_3H_8) as a fuel. A cylinder of propane contains 51.6 moles of C_3H_8. How many grams of propane are in the cylinder? How many pounds of propane are in the cylinder?
◄ 40. Baking soda ($NaHCO_3$) has many uses around the house. A box of baking soda contains 5.40 moles of $NaHCO_3$. How many grams is this? How many pounds is this?

LEARNING GOAL 7

Number of Formula Units of a Compound from Mass of a Sample

41. How many *molecules* of CO_2 are there in 198.0 g of CO_2?
42. How many *molecules* of SO_2 are there in 6.40 g of SO_2?
43. How many formula units of NaCl are there in 14.6 g of NaCl?
◄ 44. How many molecules of C_3H_8 are contained in the tank of propane described in Exercise 39?
◄ 45. How many formula units of $NaHCO_3$ are contained in the box of baking soda described in Exercise 40?
◄ 46. How many formula units of magnesium nitride are contained in 20.2 g of Mg_3N_2?

LEARNING GOAL 8

Molecular Formula of a Compound

47. According to Nobel Prize winner Linus Pauling, vitamin C, also known as ascorbic acid, is a cure and preventative for colds. The empirical formula of vitamin C is $C_3H_4O_3$, and its molecular mass is 176.0. What is its molecular formula?
48. A compound is found to have the empirical formula HF and a molecular mass of 40.0. What is its molecular formula?
49. The empirical formula of a very important compound, called benzene, is CH. Its molecular mass is 78.0. What is its molecular formula?
50. The compound fumaric acid has the empirical formula CHO. Its molecular mass is 116.0. What is its molecular formula?
51. The plastic melamine, used to make Melmac® dishes, has the empirical formula CH_2N_2. Its molecular mass is 126.0. What is its molecular formula?
52. The compound benzidine, used in the manufacture of dyes, has the empirical formula C_6H_6N. Its molecular mass is 184.0. What is its molecular formula?
◄ 53. Dextrose, a type of sugar, has the empirical formula CH_2O. Acetic acid, which is found in vinegar, shares the same empirical formula as dextrose. The molecular mass of dextrose is 180.0, whereas the molecular mass of acetic acid is 60.00. Write the molecular formula of each compound.
54. A compound of iron and sulfur with a formula mass of 208.0 has an empirical formula of Fe_2S_3. What is its molecular formula?

LEARNING GOAL 9

Percentage Composition from Chemical Formula

◄ 55. Find the percentages of K, Cr, and O by mass in potassium chromate (K_2CrO_4).
56. Find the percentages by mass of the elements in each of the following compounds:

    (a) C_2H_6 (b) NO_2 (c) CO_2

57. Determine the percentages by mass of the elements in the following compounds:

    (a) BF_3 (b) UF_6 (c) $C_3H_8O_3$

58. Determine the percentage by mass of oxygen in each of the following compounds:

    (a) H_2O (b) CO (c) H_2SO_4

*59. Some sodium chloride (NaCl) and some sugar ($C_{12}H_{22}O_{11}$) are mixed together. The mixture is analyzed together and found to contain $2\overline{0}\%$ chlorine by mass. The mixture has a mass of 50 g. How much sugar is present?
60. How many moles of nitrogen atoms can be obtained from $1,7\overline{00}$ grams of NH_3?
61. Determine the percentages by mass of the elements in the following compounds:

    (a) SO_2 (b) CH_4 (c) $C_{12}H_{22}O_{11}$ (d) $CaCO_3$

◄ 62. How many grams of hydrogen can be obtained from $18\overline{0}$ g of water?
63. How many grams of copper can be obtained from $1,6\overline{00}$ g of CuO?
64. How many grams of oxygen can be obtained from 320.0 g of SO_2?
65. How many grams of iron can be obtained from 350.0 g of Fe_2O_3?
66. Determine the percentage by mass of sulfur in H_2S.
67. How many grams of oxygen combine with $2\overline{00}$ g of sulfur in the compound SO_2? (*Hint:* Do Exercise 61 (a) to obtain the percentages by mass of S and O in SO_2.)
68. Determine the percentage by mass of hydrogen in CH_3OH.

*69. An unknown mixture contains sodium sulfide (Na_2S) and iron(III) oxide (Fe_2O_3). The mixture is analyzed and found to contain 25.0% sulfur by mass. The mixture has a mass of $5\overline{00}$ g. How many grams of each compound are present?

70. Determine the percentage by mass of N in HNO_3.

EXTRA EXERCISES

71. If you had Avogadro's number of dollars and spent 1 dollar every second of every day, how long would it take you to run out of money?

72. What is the simplest formula for the compound that has the following percentage composition by mass: 24.3% C, 4.1% H, and 71.6% Cl?

73. If the molecular mass of the compound in Exercise 72 is determined to be 99.0, what is its molecular formula?

74. What is the mass of an Fe_3O_4 sample that contains $1\overline{00}$ g of oxygen?

◀ 75. How many carbon atoms are there in a 1.75-carat diamond? (*Hint:* 1.00 carat = $2\overline{00}$ mg.)

76. Two ores of iron are hematite (Fe_2O_3) and magnetite (Fe_3O_4). Calculate the percentage by mass of iron in each ore to determine the richer source of iron.

◀ 77. Two ores of copper have the following mass percentage analyses. What is the empirical formula of each ore?

ORE A	ORE B
Cu, 63.3%	Cu, 34.6%
Fe, 11.1%	Fe, 30.4%
S, 25.6%	S, 34.9%

78. The percentages by mass of oxygen and sulfur are almost the same in the compound SO_2. Calculate the percentage by mass of each.

79. What is the mass in grams of 10 atoms of silver?

80. How many moles of phosphorus are there in 392 g of H_3PO_4?

81. In Exercise 80, how many grams of phosphorus are there?

82. How much is 1 atomic mass unit in grams?

83. Ethyl alcohol, which is found is some popular beverages, has the molecular formula C_2H_6O. You have 9.2 g of ethyl alcohol.

   (a) How many moles of ethyl alcohol do you have?
   (b) How many molecules of ethyl alcohol?
   (c) How many moles of carbon atoms?
   (d) How many moles of hydrogen atoms?
   (e) How many moles of oxygen atoms?
   (f) How many atoms of carbon?
   (g) How many atoms of hydrogen?
   (h) How many atoms of oxygen?

84. The formula for methanol, also known as wood alcohol, is CH_3OH. You have 3.2 grams of methanol.

   (a) How many moles of methanol do you have?
   (b) How many molecules of methanol?
   (c) How many moles of carbon atoms?
   (d) How many moles of hydrogen atoms?
   (e) How many moles of oxygen atoms?
   (f) How many atoms of carbon?
   (g) How many atoms of hydrogen?
   (h) How many atoms of oxygen?

180 ESSENTIAL CONCEPTS OF CHEMISTRY

<u>85</u>. Determine the number of *moles* of molecules or formula units in each of the following samples:

  (a) 7.20 g of H_2O
  (b) 260.7 g of MnO_2
  (c) 0.070 g of N_2
  (d) 1,980.0 g of $(NH_4)_2SO_4$

86. Determine the number of moles of molecules in each of the following: (a) 0.34 g of NH_3 (b) 1,335.0 g of MnO_2

◀ <u>87</u>. Glycine is the simplest amino acid. It has the molecular formula $C_2H_5O_2N$. You have 300.0 g of glycine.

  (a) How many moles of glycine do you have?
  (b) How many molecules of glycine?
  (c) How many moles of carbon atoms?
  (d) How many moles of hydrogen atoms?
  (e) How many moles of oxygen atoms?
  (f) How many moles of nitrogen atoms?
  (g) How many grams of carbon?
  (h) How many grams of hydrogen?
  (i) How many grams of oxygen?
  (j) How many grams of nitrogen?
  (k) What should be true about the sum of the numbers of grams of each element in (g), (h), (i), and (j) and the number of grams of glycine we started with?

◀ 88. Determine the number of moles of molecules in each of the following: (a) $72\overline{0}$g of H_2O (b) 13.35 g of MnO_2

CHAPTER 10

THE CHEMICAL EQUATION
Recipe for a Reaction

This is the story of Anne Williams, wife, mother, and teacher, and how she saved her husband, Tony, from serious injury when he was charging his car battery. Although Tony was the mechanic in the family, it was something that Anne remembered from her college chemistry course that saved the day.

Our story begins on a Saturday afternoon. Tony was having trouble starting his car, which was in the garage, so he decided to charge the battery with a battery charger that operated off the house current. After plugging in the unit, he attached the charger to his car battery and the battery began to charge.

Tony decided to wait in the garage as the battery charged, have a cigarette, and read the newspaper. After reading for about fifteen minutes, Tony thought he would check the battery. He was smoking his third cigarette as he approached the car and was about to lean over the battery when Anne came walking into the garage. Seeing what he was about to do she screamed, "Tony! No!" Startled by this outburst, Tony's mouth dropped open as he jerked his head up, turning away from the car toward Anne. A second later, as Tony gazed at Anne, still surprised by her warning, the battery exploded.

The lit cigarette that fell from Tony's mouth when Anne startled him had landed on the battery casing—hydrogen gas that was venting from the charging battery had ignited and exploded. Fortunately, Tony had turned away from the battery in time and was not hurt. Anne explained to Tony that flammable hydrogen gas is produced when a car battery is being charged and that the battery could have exploded in his face had he leaned over it with a lit cigarette. That's why she had screamed at him. Thanks to Anne's quick action, only the car battery had to be replaced, and Tony was saved from serious injury.

LEARNING GOALS

After you've studied this chapter, you should be able to:

1. Write formula equations from word equations, and vice versa.
2. Balance formula equations.
3. Recognize and give examples of some general types of chemical reactions (combination, decomposition, single-replacement, and double replacement).
4. Predict the products for the various types of reactions.
5. Define the terms *acid, base,* and *salt* and recognize each from its formula.
6. Use the activity series and solubility table to predict reactions.
7. Define the terms *oxidation, reduction, oxidizing agent*, and *reducing agent*.

182 ESSENTIAL CONCEPTS OF CHEMISTRY

8. Determine which substance has been oxidized and which substance has been reduced in a redox reaction.

9. Determine which substance is the reducing agent and which substance is the oxidizing agent in a redox reaction.

INTRODUCTION

Combine 4 cups of sifted flour and 1 cake of yeast soaked in $1\frac{1}{3}$ cups of 35°C water. Add 2 tablespoons of olive oil and 1 teaspoon of salt. Roll out the dough, add tomato sauce and mozzarella cheese, and bake. When these ingredients are baked, has a chemical reaction occurred? Yes. What's more, this recipe is a set of directions that tells you how to mix certain *reactants* to yield a delectable *product*, a pizza.

In a sense, we can think of a chemical reaction as a set of directions. When we follow these directions and mix the proper chemicals, we get the products that we want. In this chapter, you will study some important classes of chemical reactions. You will also learn how to write chemical equations and how to interpret the information they provide.

10.1 Word Equations

One way to represent a chemical reaction is with words. For example, we can say

$$\text{Sulfur trioxide} + \text{water} \rightarrow \text{sulfuric acid}$$

In a chemical equation such as this one, the reacting substances, or **reactants**, are shown on the left-hand side. The **products** are shown on the right-hand side. The plus sign means *and,* and the arrow means *yields* or *gives*. Our word equation, then, means that "sulfur trioxide and water react to yield sulfuric acid." (By the way, this equation represents part of the process by which "acid rain" develops from one byproduct of fuel combustion.)

Sulfur burns in oxygen with a pale blue flame to form gaseous SO_2 and SO_3, which are used to make sulfuric acid. An older name for sulfur is *brimstone*, which means "burning stone." (Rich Treptow/Photo Researchers, Inc.)

Our word equation tells us what happens when sulfur trioxide combines with water. But it tells us nothing about the chemical formulas of the reactants and products. Nor does it tell us anything about *how* the reactants combine to form the products. That's why chemists prefer to use formula notation to write chemical reactions. And, like other people, chemists prefer to use the simpler shorthand notation whenever possible.

10.2 The Formula Equation

Let us substitute the formulas for the reactants and product in our word equation:

$$\text{Sulfur trioxide} + \text{water} \rightarrow \text{sulfuric acid}$$
$$SO_3 + H_2O \rightarrow H_2SO_4$$

Now let's do some "atomic bookeping" to see whether we have complied with the Law of Conservation of Mass. On the left-hand side of this equation are one sulfur atom, four oxygen atoms, and two hydrogen atoms. And there are exactly the same numbers of these atoms on the right-hand side. So we have not "created" or "destroyed" atoms in writing the formula equation (see Figure 10.1). The equation is *balanced*. A chemical equation is **balanced** when *the total number of atoms of each element on the left-hand side is exactly equal to the total number of atoms of each element on the right-hand side*.

Let's now look at the chemical reaction in which hydrogen and oxygen form water. Because hydrogen and oxygen are both diatomic elements, we must write the equation for this chemical reaction as follows:

$$H_2 + O_2 \xrightarrow[\text{Spark}]{\text{Electric}} H_2O$$

(Many reactions occur only under special conditions. We specify these conditions by writing them above or around the arrow.) Now for the bookkeeping: There are two hydrogens on the left side and two hydrogens on the right side of the equation. However, there are two oxygens on the left side but only *one* oxygen on the

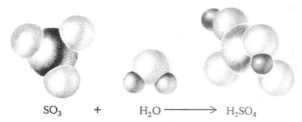

SO_3 + H_2O $\longrightarrow$ H_2SO_4

FIGURE 10.1 Models show how sulfur trioxide and water produce sulfuric acid.

right side. This equation is *not balanced*, because the number of atoms of each element on the left *does not* equal the number of atoms of each element on the right.

We can't let the equation stand this way. Every chemical equation must be balanced, because chemical reactions satisfy the Law of Conservation of Mass. As it stands, our reaction falsely implies that an atom of oxygen was lost or destroyed.

10.3 Balancing a Chemical Equation

To balance a chemical equation, we determine how many molecules of each substance are needed to satisfy the Law of Conservation of Mass. We indicate this by placing **coefficients**, or small whole numbers, before the formulas in the equation. These coefficients are usually obtained by trial and error. For example, in our equation for the formation of water, we have

$$H_2 + O_2 \xrightarrow{\text{Electric spark}} H_2O \quad \text{(not balanced)}$$

Because this unbalanced equation lacks one oxygen atom on the right, we place the coefficient 2 before the formula for water on the right:

$$H_2 + O_2 \xrightarrow{\text{Electric spark}} 2H_2O \quad \text{(not balanced)}$$

Now we have two oxygen atoms on the left and two on the right. But the coefficient 2 has unbalanced the hydrogen atoms. There are now four on the right but only two on the left. To remedy this, we use two molecules of H_2:

$$2H_2 + O_2 \xrightarrow{\text{Electric spark}} 2H_2O \quad \text{(balanced)}$$

Now the equation is balanced: there are four hydrogen atoms and two oxygen atoms on the left and the same numbers on the right. The balanced equation states that two hydrogen molecules and one oxygen molecule react to form two water molecules (Figure 10.2).

There are two rules to keep in mind when you balance a chemical equation:

The equation must be accurate. Correct formulas for all reactants and products must be shown.

The Law of Conservation of Mass must be observed. The same number of atoms of each element must appear on both sides of the balanced equation.

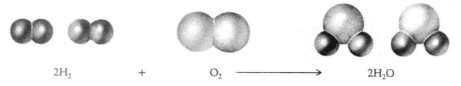

$2H_2$ + O_2 $\longrightarrow$ $2H_2O$

FIGURE 10.2 The formation of water from hydrogen and oxygen.

EXAMPLE 10.1

Write a balanced chemical equation for the following reaction:

Mercury (II) oxide $\xrightarrow{\text{Heat}}$ mercury metal + oxygen gas

184 ESSENTIAL CONCEPTS OF CHEMISTRY

Solution

We begin by writing the formulas for the reactants and products. Remember that oxygen gas is a diatomic element, so we must write O_2.

$$HgO \xrightarrow{Heat} Hg + O_2 \quad \text{(unbalanced)}$$

The mercury is balanced, but there is only one oxygen atom on the left and two oxygens on the right. We balance the oxygens by placing the coefficient 2 in front of HgO:

$$2HgO \xrightarrow{Heat} Hg + O_2 \quad \text{(unbalanced)}$$

Now the oxygens are balanced, but there are two mercury atoms on the left and only one on the right. We balance the mercury atoms by placing a 2 in front of the Hg:

$$2HgO \xrightarrow{Heat} 2Hg + O_2 \quad \text{(balanced)}$$

The balanced chemical equation shows two mercury atoms and two oxygen atoms on each side of the equation.

Practice Exercise 10.1

Write a balanced chemical equation for the following reaction:

$$\text{Copper (II) oxide} \rightarrow \text{copper metal} + \text{oxygen gas}$$

Here are two important points to remember.

1. You *cannot* balance an equation using *subscripts,* because this would change the formulas of the compounds. For example,

$$HgO \xrightarrow{Heat} Hg + O_2$$

   cannot be balanced by doing this:

$$HgO_2 \xrightarrow{Heat} Hg + O_2$$

   because HgO_2 is not the formula for mercury(II) oxide.

2. You *cannot* insert a coefficient between elements in a chemical formula. For example,

$$HgO \xrightarrow{Heat} Hg + O_2$$

   cannot be balanced by doing this:

$$Hg2O \xrightarrow{Heat} Hg + O_2$$

   because such an expression has no meaning.

After you have balanced a number of equations, you will find it easy to choose the proper coefficients. Here are some suggestions that may help:

1. Balance metal and nonmetal atoms first.
2. Balance hydrogen and oxygen atoms last.
3. Balance polyatomic ions that contain more than one atom *as a group* if they appear unchanged on both sides of the equation.

EXAMPLE 10.2

Write a balanced chemical equation for the following reaction:

$$\text{Aluminum} + \text{sulfuric acid} \rightarrow \text{aluminum sulfate} + \text{hydrogen gas}$$

Solution

We first write the correct formula for each compound:

$$Al + H_2SO_4 \rightarrow Al_2(SO_4)_3 + H_2 \quad \text{(unbalanced)}$$

There is one aluminum atom on the left-hand side but two aluminum atoms on the right-hand side. We balance the aluminums by placing the coefficient 2 in front of the Al on the left-hand side:

$$2Al + H_2SO_4 \rightarrow Al_2(SO_4)_3 + H_2 \quad \text{(unbalanced)}$$

There is one SO_4 group on the left-hand side and three SO_4 groups on the right-hand side. We balance the SO_4 groups by placing a 3 in front of the H_2SO_4:

$$2Al + 3H_2SO_4 \rightarrow Al_2(SO_4)_3 + H_2 \quad \text{(unbalanced)}$$

Now there are six hydrogens on the left-hand side and two hydrogens on the right-hand side. We balance the hydrogens by placing a 3 in front of the H_2:

$$2Al + 3H_2SO_4 \rightarrow Al_2(SO_4)_3 + 3H_2 \quad \text{(balanced)}$$

There are now two aluminum atoms, three sulfate groups, and six hydrogen atoms on each side of the equation. And so we have a balanced chemical equation.

Practice Exercise 10.2

Write a balanced chemical equation for the following reaction:

$$\text{Zinc} + \text{sulfuric acid} \rightarrow \text{zinc sulfate} + \text{hydrogen gas}$$

EXAMPLE 10.3

Write a balanced equation for each of the following reactions:

(a) Hydrogen gas + nitrogen gas $\xrightarrow[\text{Pressure}]{\text{Heat}}$ ammonia (NH_3)

(b) Sodium oxide + water $\rightarrow$ sodium hydroxide

(c) Calcium bromide + chlorine $\rightarrow$ calcium chloride + bromine

(d) Sodium hydroxide + phosphoric acid (H_3PO_4) $\rightarrow$ sodium phosphate + water

(e) Iron(III) oxide + carbon monoxide $\xrightarrow{\text{Heat}}$ iron + carbon dioxide

Solution

We use the same procedure as in the previous examples: Write the formula for each substance and then balance the equation.

(a) $H_2 + N_2 \xrightarrow[\text{Pressure}]{\text{Heat}} NH_3$ (unbalanced)

     $3H_2 + N_2 \xrightarrow[\text{Pressure}]{\text{Heat}} 2NH_3$ (balanced)

(b) $Na_2O + H_2O \rightarrow NaOH$ (unbalanced)

     $Na_2O + H_2O \rightarrow 2NaOH$ (balanced)

(c) $CaBr_2 + Cl_2 \rightarrow CaCl_2 + Br_2$ (balanced)

(d) $NaOH + H_3PO_4 \rightarrow Na_3PO_4 + H_2O$ (unbalanced)

     $3NaOH + H_3PO_4 \rightarrow Na_3PO_4 + 3H_2O$ (balanced)

186 ESSENTIAL CONCEPTS OF CHEMISTRY

(e) $Fe_2O_3 + CO \xrightarrow{Heat} Fe + CO_2$ (unbalanced)

$Fe_2O_3 + 3CO \xrightarrow{Heat} 2Fe + 3CO_2$ (balanced)

Practice Exercise 10.3

Write a balanced equation for each of the following word equations:

(a) Aluminum + chlorine gas → aluminum chloride

(b) Sodium + water → sodium hydroxide + hydrogen gas

(c) Potassium nitrate → potassium nitrite + oxygen gas

(d) Nitric acid + barium hydroxide → barium nitrate + water

10.4 Types of Chemical Reactions

The millions of known chemical reactions may be classified in many ways. We will use a classification scheme that is helpful in predicting the products of many chemical processes. In this scheme, there are four basic classes:

1. Combination reactions
2. Decomposition reactions
3. Single-replacement reactions
4. Double-replacement (or double-exchange) reactions

All of these chemical reactions, with the exception of double-replacement reactions, may also be classified as redox reactions. (*Redox* is short for "oxidation-reduction.") We have more to say about these reactions later in the chapter.

Combination Reactions

Combination reactions are *reactions in which two or more substances combine to form a more complex substance* (Figure 10.3). The general formula is

$$A + B \rightarrow AB$$

In some combination reactions, two *elements* join to form a compound. For example,

$$\underset{\text{Iron}}{4Fe} + \underset{\substack{\text{Oxygen} \\ \text{gas}}}{3O_2} \rightarrow \underset{\substack{\text{Iron(III)} \\ \text{oxide}}}{2Fe_2O_3}$$

Iron(III) oxide is iron rust. When a product made of iron contacts moist air, the iron rusts, forming a crust of iron(III) oxide. The crust flakes off, a new surface of iron becomes visible, and it too begins to rust. The iron can keep rusting until the entire piece of metal is eaten away.

A vivid example of a combination reaction! The space shuttles are launched by liquid-fuel rockets that contain liquid hydrogen and liquid oxygen, which combine to form water. The shuttle shown here is Endeavour. (NASA)

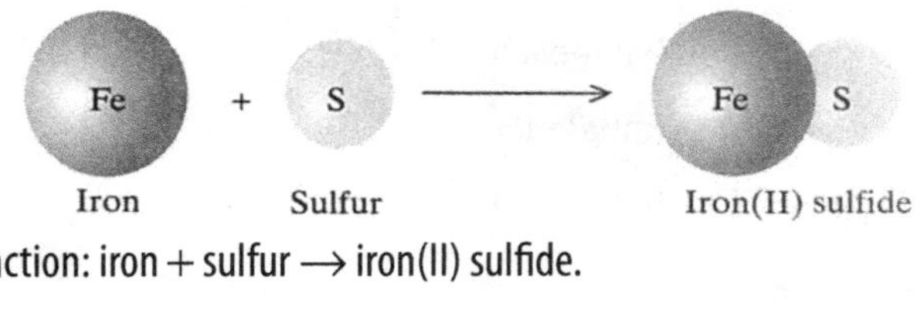

FIGURE 10.3 A combination reaction: iron + sulfur → iron(II) sulfide.

A vivid example of a combination reaction! The space shuttles are launched by liquid-fuel rockets that contain liquid hydrogen and liquid oxygen, which combine to form water. The shuttle shown here is Endeavour. (NASA)

Another example of a combination reaction that joins two elements is

$$\underset{\text{Carbon}}{C} + \underset{\text{Sulfur}}{2S} \rightarrow \underset{\text{Carbon disulfide}}{CS_2}$$

Carbon disulfide is an important chemical used in manufacturing rayon and in electronic vacuum tubes. However, it is a highly poisonous substance and should be used only in well-ventilated areas.

In some combination reactions, two *compounds* join to form a more complex compound. For example,

$$\underset{\substack{\text{Sulfur}\\\text{trioxide}}}{SO_3} + \underset{\text{Water}}{H_2O} \rightarrow \underset{\substack{\text{Sulfuric}\\\text{acid}}}{H_2SO_4}$$

$$\underset{\substack{\text{Sodium}\\\text{oxide}}}{Na_2O} + \underset{\text{Water}}{H_2O} \rightarrow \underset{\substack{\text{Sodium}\\\text{hydroxide}}}{2NaOH}$$

EXAMPLE 10.4

Complete and balance the following equations for combination reactions:

(a) $H_2 + Cl_2 \rightarrow$?

(b) $K + Br_2 \rightarrow$?

(c) $Ba + O_2 \rightarrow$?

Solution

We are told that these are combination reactions, so we know that the products are formed by combination of the elements involved. We first write the correct formula for the product. Then we balance the equation.

(a) $H_2 + Cl_2 \rightarrow HCl$ (unbalanced)

$H_2 + Cl_2 \rightarrow 2HCl$ (balanced)

(b) K + Br$_2$ → KBr   (unbalanced)

   2K + Br$_2$ → 2KBr   (balanced)

(c) Ba + O$_2$ → BaO   (unbalanced)

   2Ba + O$_2$ → 2BaO   (balanced)

Practice Exercise 10.4

Complete and balance the following equations for combination reactions:

(a) K + O$_2$ → ?

(b) Ca + O$_2$ → ?

(c) H$_2$ + Br$_2$ → ?

Decomposition Reactions

Decomposition reactions are *reactions in which a complex substance is broken down into simpler substances.* They are thus the reverse of combination reactions. The general formula for a decomposition reaction is

$$AB \rightarrow A + B$$

Some examples of common decomposition reactions follow.

1. The decomposition of a metal oxide (Figure 10.4):

$$2HgO \xrightarrow{\text{Heat}} 2Hg + O_2$$
Mercury(II) oxide → Mercury + Oxygen gas

2. The decomposition of a nonmetal oxide:

$$2H_2O \xrightarrow{\text{Electricity}} 2H_2 + O_2$$

3. The decomposition of a metal chlorate:

$$2KClO_3 \xrightarrow{\text{Heat}} 2KCl + 3O_2$$
Potassium chlorate → Potassium chloride

4. The decomposition of a metal carbonate:

$$MgCO_3 \xrightarrow{\text{Heat}} MgO + CO_2$$
Magnesium carbonate → Magnesium oxide

5. The decomposition of a metal hydroxide:

$$Ca(OH)_2 \xrightarrow{\text{Heat}} CaO + H_2O$$
Calcium hydroxide → Calcium oxide

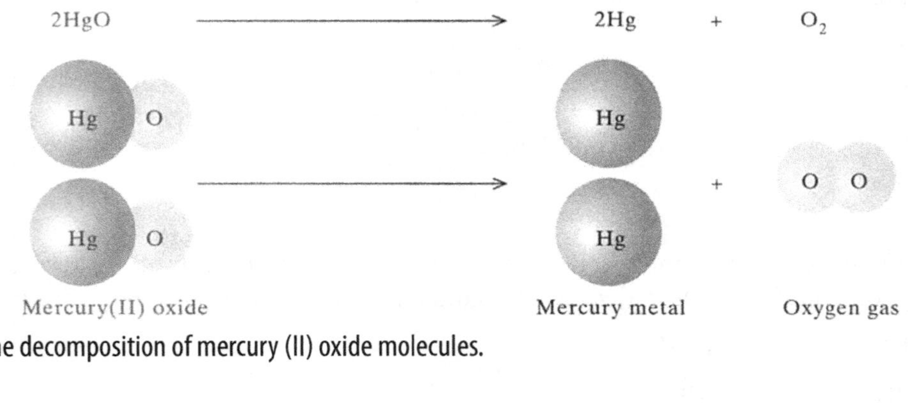

FIGURE 10.4 The decomposition of mercury (II) oxide molecules.

Decomposition reactions are a very important class of reactions. They are the ones that produce pure samples of various elements. For example, very pure hydrogen and oxygen can be obtained from the decomposition of water. Pure copper can be produced by the decomposition of copper(II) oxide.

EXAMPLE 10.5

Complete and balance the following equations for decomposition reactions:

(a) $Mg(OH)_2 \xrightarrow{\text{Heat}} ? + ?$

(b) $CaCO_3 \xrightarrow{\text{Heat}} ? + ?$

(c) $NaClO_3 \xrightarrow{\text{Heat}} ? + ?$

(d) $HCl \rightarrow ? + ?$

Solution

UNDERSTAND THE PROBLEM

We must predict the products and balance the equations for these decomposition reactions.

DEVISE A PLAN

We are told that these are decomposition reactions, so we know that the products are formed by a breakdown of the elements in the original substance. We want to balance the equation according to the Law of Conservation of Mass. Look back in the chapter for similar kinds of compounds undergoing decomposition.

CARRY OUT THE PLAN

We first write the correct formulas for the products and then balance the equation.

LOOK BACK

We look back and check to be sure we have written an accurate equation.

(a) $Mg(OH)_2 \xrightarrow{\text{Heat}} MgO + H_2O$ (balanced)

(b) $CaCO_3 \xrightarrow{\text{Heat}} CaO + CO_2$ (balanced)

(c) $NaClO_3 \xrightarrow{\text{Heat}} NaCl + O_2$ (unbalanced)

$2NaClO_3 \xrightarrow{\text{Heat}} 2NaCl + 3O_2$ (balanced)

(d) $HCl \rightarrow H_2 + Cl_2$ (unbalanced)

$2HCl \rightarrow H_2 + Cl_2$ (balanced)

Practice Exercise 10.5

Complete and balance the following equations for decomposition reactions:

(a) $Sr(OH)_2 \rightarrow ?\square + ?$

(b) $SrCO_3 \rightarrow ? + ?$

(c) $KClO_3 \rightarrow ? + ?$

(d) $KCl \rightarrow ? + ?$

Single-Replacement Reactions

A **single-replacement reaction** is *one in which an uncombined element replaces another element that is in a compound* (Figure 10.5). The general formula for this type of reaction is

$$A + BC \rightarrow AC + B$$

Examples of single-replacement reactions follow.

1. The replacement in aqueous solution of one metal by another metal that is more active (the activity of metals is discussed in Section 10.5):

$$Mg + Cu(NO_3)_2 \rightarrow Mg(NO_3)_2 + Cu$$

2. The replacement of hydrogen in water by a Group IA metal:

$$2Na + 2H_2O \rightarrow 2NaOH + H_2$$

3. The replacement of hydrogen in an acid by an active metal:

$$Zn + 2HCl \rightarrow ZnCl_2 + H_2$$

EXAMPLE 10.6

Complete and balance the following equations for single-replacement reactions:

(a) $K + H_2O \rightarrow ? + ?$

(b) $Mg + H_2SO_4 \rightarrow ? + ?$

(c) $Zn + NiCl_2 \rightarrow ? + ?$

Solution

We first write the correct formulas for the products and then balance the equation.

(a) $K + H_2O \rightarrow KOH + H_2$ (unbalanced)

   $2K + 2H_2O \rightarrow 2KOH + H_2$ (balanced)

(b) $Mg + H_2SO_4 \rightarrow MgSO_4 + H_2$ (balanced)

(c) $Zn + NiCl_2 + ZnCl_2 + Ni$ (balanced)

Practice Exercise 10.6

Complete and balance the following equations for single-replacement reactions:

(a) $Li + H_2O \rightarrow ? + ?$

(b) $Al + H_2SO_4 \rightarrow ? + ?$

(c) $Mg + Al(NO_3)_3 \rightarrow ? + ?$

FIGURE 10.5   A single-replacement reaction: zinc + copper(II) sulfate → copper + zinc sulfate.

Single-replacement reactions are also a very important class of reactions. Chemists often use single-replacement reactions to obtain various metals from compounds containing these metals. For example, you can obtain pure silver metal from the compound silver nitrate by causing a solution of silver nitrate to react with copper metal.

$$Cu + 2AgNO_3 \rightarrow Cu(NO_3)_2 + 2Ag$$

Double-Replacement Reactions

A **double-replacement** (or *double-exchange*) **reaction** is *one in which two compounds exchange ions with each other* (Figure 10.6). The general formula is

$$A^+B^- + C^+D^- \rightarrow A^+D^- + C^+B^-$$

Double-replacement reactions usually take place in water. To show that they do, we use the notation (*aq*), which stands for *aqueous*, or in water.

There are two important kinds of double-replacement reactions. One kind takes place between two salts. A **salt** is a compound composed of a positive and a negative ion. The positive ion may be a monatomic ion of a metal (for example, Na^{1+} or Ca^{2+}) or a polyatomic ion such as $(NH_4)^{1+}$. The negative ion may be a monatomic ion of a nonmetal (for example, Cl^{1-}) or a polyatomic ion such as $(SO_4)^{2-}$ or $(NO_3)^{1-}$.

A double-replacement reaction between two salts can occur if one of the *products* separates from the solution—in other words, if it is insoluble in the solution (or if a gas is formed). When a product is insoluble, it separates out of the solution in solid form (called a **precipitate**). To show this, we use the notation (*s*), meaning *solid*. Table B.4 in Appendix B tells you whether a particular product is soluble in water.

When a salt dissolves in water it **dissociates**, or breaks up into its ions. For example, when we dissolve silver nitrate ($AgNO_3$) in water, there are Ag^+ and $(NO_3)^-$ ions in the solution. There are no $AgNO_3$ units as such in the water.

Now let's see what happens when we mix an aqueous solution of silver nitrate and an aqueous solution of potassium chloride. Both salts are soluble in water, so both solutions are clear. But when we mix the two solutions and allow them to react, we get a solid precipitate. This precipitate is silver chloride, which is *not* soluble in water. The reaction is

$$KCl(aq) + AgNO_3(aq) \rightarrow KNO_3(aq) + AgCl(s)$$

The other product, potassium nitrate, remains in solution [as K^+ and $(NO_3)^-$ ions] because it *is* soluble in water.

The second important kind of double-replacement reaction takes place between an acid and a base. The simplest **acids** are *compounds that dissociate in water into hydrogen ions and negatively charged ions of a nonmetal (or negatively charged polyatomic ions)*. Acids release hydrogen ions in water. The simplest **bases** are *composed of metallic ions (or positively charged polyatomic ions) combined with hydroxide ions*. Bases release hydroxide ions in water.

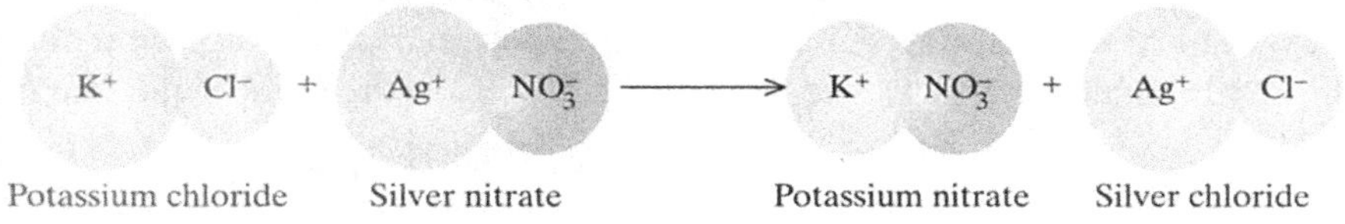

FIGURE 10.6 A double-replacement reaction: potassium chloride + silver nitrate → potassium nitrate + silver chloride.

In an acid–base reaction, the acid and base react to form a salt and water. This type of reaction is also called a **neutralization reaction**. Some examples of neutralization reactions are

$$HCl + NaOH \rightarrow NaCl + H_2O$$
Hydrochloric acid, Sodium hydroxide

$$H_2SO_4 + 2NaOH \rightarrow Na_2SO_4 + 2H_2O$$
Sulfuric acid

$$H_3PO_4 + 3LiOH \rightarrow Li_3PO_4 + 3H_2O$$
Phosphoric acid, Lithium hydroxide

We make much use of neutralization reactions in our daily lives. For example, if you have an acid stomach (caused by excess hydrochloric acid), you take a product such as milk of magnesia to neutralize the excess acidity. Milk of magnesia contains a base—magnesium hydroxide, $Mg(OH)_2$.

EXAMPLE 10.7

Complete and balance the following equations for double-replacement reactions:

(a) $BaCl_2(aq) + Na_2SO_4(aq) \rightarrow ? + ?$

(b) $H_3PO_4 + Ca(OH)_2 \rightarrow ? + ?$

Solution

As usual, we first write the correct formula for each product and then balance the equation.

(a) $BaCl_2(aq) + Na_2SO_4(aq) \rightarrow$

$\qquad NaCl(aq) + BaSO_4(s)$ (unbalanced)

$BaCl_2(aq) + Na_2SO_4(aq) \rightarrow$

$\qquad 2NaCl(aq) + BaSO_4(s)$ (balanced)

(b) $H_3PO_4 + Ca(OH)_2 \rightarrow Ca_3(PO_4)_2 + H_2O$ (unbalanced)

$2H_3PO_4 + 3Ca(OH)_2 \rightarrow Ca_3(PO_4)_2 + 6H_2O$ (balanced)

Practice Exercise 10.7

Complete and balance the following equations for double-replacement reactions:

(a) $BiCl_3 + H_2S \rightarrow ? + ?$

(b) $Fe(OH)_3 + H_2SO_4 \rightarrow ? + ?$

10.5 The Activity Series

The **activity series** (Table 10.1) is *a list of elements in decreasing order of their reactivity, or ability to react chemically in aqueous solution.* The most active elements (those that react most readily) appear at the top of the table, and reactivity decreases as we move down the list. Under ordinary conditions, each element in the table can replace any element that appears below it by taking part in a single-replacement reaction. For example, aluminum is higher in Table 10.1 than zinc, so it replaces zinc in the compound zinc nitrate:

$$2Al + 3Zn(NO_3)_2 \rightarrow 2Al(NO_3)_3 + 3Zn$$

The activity series is very helpful in determining whether a particular reaction—especially a single-replacement reaction—can take place under ordinary conditions.

EXAMPLE 10.8

Predict whether each of the following reactions can occur:

(a) $Cu + 2HCl \xrightarrow{Heat} CuCl_2 + H_2$

(b) $H_2 + CuO \xrightarrow{Heat} H_2O + Cu$

Solution

(a) Table 10.1 shows that copper is *below* hydrogen in the activity series. Therefore copper cannot replace hydrogen in the compound HCl, and

$$Cu + HCl \rightarrow \text{no reaction}$$

(b) Hydrogen is above copper in the activity series, so it can replace copper in the compound CuO. This reaction can occur:

$$H_2 + CuO \rightarrow H_2O + Cu$$

Practice Exercise 10.8

Predict whether each of the following reactions can occur:

(a) $Cu + H_2SO_4 \rightarrow CuSO_4 + H_2$

(b) $H_2 + CuCl_2 \rightarrow 2HCl + Cu$

TABLE 10.1 The Activity Series

METALS	NONMETALS
Lithium	Fluorine
Potassium	Chlorine
Calcium	Bromine
Sodium	Iodine
Magnesium	
Aluminum	
Zinc	
Chromium	
Iron	
Nickel	
Tin	
Lead	
HYDROGEN*	
Copper	
Mercury	
Silver	
Platinum	
Gold	

*Hydrogen is in capital letters because the activities of the other metals were calculated relative to the activity of hydrogen. The most active elements are at the top of the table. The activity of the elements decreases going down the table.

10.6 Oxidation–Reduction (Redox) Reactions

We've just finished classifying reactions according to one scheme—as combination, decomposition, single-replacement, or double-replacement reactions. Now we will introduce another scheme for classifying reactions—as either oxidation–reduction (redox, for short) reactions or nonredox reactions. Let's begin with some definitions.

Oxidation and Reduction

In the past, chemists defined oxidation as the combining of a substance with oxygen. (That's how the process got its name.) Now, however, the concept of oxidation is expanded to include reactions that do not involve oxygen at all. **Oxidation** is *the loss of electrons by a substance undergoing a chemical reaction.* (*Alternatively, it is an increase in the oxidation number of a substance.*) (If you don't recall the meaning of *oxidation number*, refer to Section 8.3, and note that the oxidation number of an element in the uncombined state is 0.) As an example, look at the following equation, in which the oxidation number of the iron is shown over its symbol:

$$\overset{0}{4Fe} + 3O_2 \longrightarrow \overset{3+}{2Fe_2O_3}$$

The oxidation number of the iron changes from 0 to 3+. By our definition, this means that the iron is oxidized. But what happens to the oxygen? We say that the oxygen has been *reduced*, which leads us to our definition of reduction. **Reduction** is *the gain of electrons by a substance undergoing a chemical reaction.* (*Alternatively, it is a decrease in the oxidation number of a substance.*)

In the reaction between iron and oxygen, the oxidation number of the oxygen changes from 0 to 2–, because each oxygen atom gains two electrons from the iron:

$$4Fe + \overset{0}{3O_2} \longrightarrow \overset{2-}{2Fe_2O_3}$$

Note that in this reaction, one substance (iron) is oxidized, while the other substance (oxygen) is reduced. This is not a coincidence: The processes of oxidation and reduction always occur together. In other words, oxidation cannot take place in a reaction unless there is also reduction. And reduction can't take place in a reaction unless there is also oxidation. A reaction in which oxidation and reduction take place is called an **oxidation-reduction** (or **redox**) **reaction**. All other reactions are **nonredox reactions**.

EXAMPLE 10.9

In each of the following reactions, determine which substance is oxidized and which substance is reduced:

(a) $2K + Br_2 \rightarrow 2KBr$

(b) $S + O_2 \rightarrow SO_2$

(c) $Mg + H_2SO_4 \rightarrow MgSO_4 + H_2$

(d) $2H_2O \rightarrow 2H_2 + O_2$

Solution

In each case, we use the ideas of Section 8.3 (and the fact that the oxidation number of an element in its uncombined state is 0) to write the oxidation numbers for the elements.

(a) $\overset{0}{2K} + \overset{0}{Br_2} \longrightarrow \overset{1+\;1-}{2\,K\,Br}$

The oxidation number of the potassium changes from 0 to 1+. This increase means that the potassium is oxidized. The oxidation number of the bromine changes from 0 to 1–. This decrease means that the bromine is reduced.

(b)
$$\overset{0}{S} + \overset{0}{O_2} \longrightarrow \overset{4+\ 2-}{SO_2}$$

The oxidation number of the sulfur changes from 0 to 4+, so the sulfur is oxidized. The oxidation number of the oxygen changes from 0 to 2−, so the oxygen is reduced.

(c)
$$\overset{0}{Mg} + \overset{1+\ 2-}{H_2(SO_4)} \longrightarrow \overset{2+\ 2-}{Mg(SO_4)} + \overset{0}{H_2}$$

The oxidation number of the magnesium changes from 0 to 2+, so the magnesium is oxidized. The oxidation number of the hydrogen changes from 1+ to 0. This decrease means that the hydrogen is reduced. The oxidation number of the sulfate group does not change, so it is neither oxidized nor reduced.

(d)
$$\overset{1+\ 2-}{2H_2O} \longrightarrow \overset{0}{2H_2} + \overset{0}{O_2}$$

The oxidation number of the oxygen changes from 2− to 0, so the oxygen is oxidized. The oxidation number of the hydrogen changes from 1+ to 0, so the hydrogen is reduced.

Practice Exercise 10.9

In each of the following reactions, determine which substance is oxidized and which substance is reduced:

(a) $Cu + 2AgNO_3 \rightarrow 2Ag + Cu(NO_3)_2$

(b) $2Na + Cl_2 \rightarrow 2NaCl$

(c) $Cl_2 + 2NaBr \rightarrow 2NaCl$

(d) $Zn + S \rightarrow ZnS$

Oxidizing and Reducing Agents

An oxidizing agent is *a substance that causes something else to be oxidized.* **A reducing agent** is *a substance that causes something else to be reduced.* As things work out, the oxidizing agent is the substance being reduced, and the reducing agent is the substance being oxidized. Let's take another look at the reaction in which iron and oxygen form iron(III) oxide:

$$4Fe + 3O_2 \rightarrow 2Fe_2O_3$$

The iron is the substance being oxidized, and the oxygen is the substance being reduced. Therefore the iron is the *reducing agent* and the oxygen is the *oxidizing agent*.

$$\underset{\text{Substance oxidized (reducing agent)}}{\overset{0}{4Fe}} + \underset{\text{Substance reduced (oxidizing agent)}}{\overset{0}{3O_2}} \longrightarrow \overset{3+\ 2-}{2Fe_2O_3}$$

EXAMPLE 10.10

For each reaction in Example 10.9, state which substance is the oxidizing agent and which substance is the reducing agent.

Solution

Just remember that the substance being oxidized is the reducing agent and the substance being reduced is the oxidizing agent.

(a)
$$\overset{0}{2K} + \overset{0}{Br_2} \longrightarrow \overset{1+\ 1-}{2KBr}$$

   Substance oxidized   Substance reduced
   (reducing agent)   (oxidizing agent)

(b)
$$\overset{0}{S} + \overset{0}{O_2} \longrightarrow \overset{4+\ 2-}{SO_2}$$

   Substance oxidized   Substance reduced
   (reducing agent)   (oxidizing agent)

(c)
$$\overset{0}{Mg} + \overset{1+\ 2-}{H_2(SO_4)} \longrightarrow \overset{2+\ 2-}{Mg(SO_4)} + \overset{0}{H_2}$$

   Substance oxidized   Substance reduced
   (reducing agent)   (oxidizing agent)

Actually, in this case, the whole compound, H_2SO_4, is considered the oxidizing agent.

(d)
$$\overset{1+\ 2-}{2H_2O} \longrightarrow \overset{0}{2H_2} + \overset{0}{O_2}$$

The hydrogen is reduced and the oxygen is oxidized. The compound water acts as both oxidizing agent and reducing agent.

Practice Exercise 10.10

For each reaction in Practice Exercise 10.9, state which substance is the oxidizing agent and which substance is the reducing agent.

SUMMARY

A chemical equation is a representation of a chemical reaction. A word equation indicates (in words) which reactants combine (or decompose) and which products are formed as a result of the reaction. In a formula equation, the words are replaced with chemical formulas. In addition, the formula equation must be balanced, so that the number of atoms of each element on the left side of the equation equals the number of atoms of each element on the right side. We balance an equation by placing the proper coefficients before elements or compounds in the equation. Coefficients are generally chosen by trial and error.

Chemical reactions may be classified as combination reactions, decomposition reactions, single-replacement reactions, or double-replacement reactions. Combination reactions are those in which two or more substances combine to form a more complex substance. Decomposition reactions (the reverse of combination reactions) involve the breakdown of a complex substance into simpler substances. A single-replacement reaction is one in which an uncombined element replaces another element in a compound. A double-replacement reaction is one in which two compounds exchange ions with each other.

To predict whether or not a reaction will occur under ordinary conditions, it is important to know the relative activities of the elements involved in the reaction. The activity series is a list of elements in decreasing order of activity or reactivity. It can be used to predict whether or not a single-replacement reaction will occur. That is, each element in the list can replace any element below it in a single-replacement reaction.

Another way of classifying chemical reactions is by dividing them into oxidation-reduction (redox) reactions and non-oxidation–reduction (nonredox) reactions. Oxidation is the loss of electrons by a substance undergoing a chemical reaction (in other words, an increase in the oxidation number of the substance). Reduction is the gain of electrons by a substance undergoing a chemical reaction (in other words, a decrease in the oxidation number of the substance). Oxidation and reduction always occur together. An oxidizing agent is a substance

CHAPTER 10: THE CHEMICAL EQUATION

that causes something else to be oxidized, and a reducing agent is a substance that causes something else to be reduced. A redox reaction is a reaction in which oxidation and reduction take place. In such a reaction, the oxidizing agent is reduced, and the reducing agent is oxidized.

KEY TERMS

acid (**10.4**)
activity series (**10.5**)
balanced equation (**10.2**)
base (**10.4**)
coefficient (**10.3**)
combination reaction (**10.4**)
decomposition reaction (**10.4**)
dissociate (**10.4**)
double-replacement (double-exchange) reaction (**10.4**)
neutralization reaction (**10.4**)

nonredox reaction (**10.6**)
oxidation (**10.6**)
oxidation-reduction (redox) reaction (**10.6**)
oxidizing agent (**10.6**)
precipitate (**10.4**)
product (**10.1**)
reactant (**10.1**)
reducing agent (**10.6**)
reduction (**10.6**)
salt (**10.4**)
single-replacement reaction (**10.4**)

SELF-TEST EXERCISE

LEARNING GOALS 1 AND 2

Formula and Word Equations; Balancing Equations

◀ *1*. Write balanced equations for each of the following reactions:

(a) Potassium + water → potassium hydroxide + hydrogen gas
(b) Acetic acid + calcium hydroxide → calcium acetate + water
(c) Magnesium + copper(II) nitrate → magnesium nitrate + copper
(d) Sodium oxide + water → sodium hydroxide
(e) Zinc sulfide + oxygen gas → zinc oxide + sulfur dioxide
(f) Potassium hydroxide + aluminum nitrate → aluminum hydroxide + potassium nitrate

◀ 2. Write a balanced chemical equation for each of the following reactions:

(a) Calcium hydroxide + phosphoric acid → calcium phosphate + water
(b) Strontium + oxygen gas → strontium oxide
(c) Magnesium chlorate → magnesium chloride + oxygen gas
(d) Iron(III) hydroxide + hydrochloric acid → iron(III) chloride + water
(e) Potassium + oxygen gas → potassium oxide
(f) Phosphorus + bromine → phosphorous tribromide

3. Write a balanced chemical equation for each of the following reactions:

(a) Iron + oxygen gas → iron(III) oxide
(b) Sulfuric acid + aluminum hydroxide → aluminum sulfate + water
(c) Silver nitrate + barium chloride → silver chloride + barium nitrate
(d) Copper(I) sulfide + oxygen gas → copper + sulfur dioxide

4. Write a balanced chemical equation for each of the following reactions:

(a) Hydrogen gas + oxygen gas → water
(b) Potassium hydroxide + acetic acid → potassium acetate + water
(c) Nitrogen gas + hydrogen gas → ammonia
(d) Potassium chlorate → potassium chloride + oxygen gas

LEARNING GOAL 3

Recognizing Types of Chemical Reactions

◀ 5. Identify each of the following reactions as a combination, decomposition, single-replacement, or double-replacement reaction:

(a) $4Fe + 3O_2 \rightarrow 2Fe_2O_3$
(b) $2HBr \rightarrow H_2 + Br_2$
(c) $AgNO_3 + KCl \rightarrow AgCl + KNO_3$
(d) $Zn + H_2SO_4 \rightarrow ZnSO_4 + H_2$
(e) $Mg(CN)_2 + 2HCl \rightarrow MgCl_2 + 2HCN$

6. Identify each of the following reactions as a combination, decomposition, single-replacement, or double-replacement reaction:

(a) $H_2 + Br_2 \rightarrow 2HBr$
(b) $2PbO_2 \xrightarrow{Heat} 2PbO + O_2$
(c) $2KI + Br_2 \rightarrow 2KBr + I_2$
(d) $2Fe(OH)_3 + 3H_2SO_4 \rightarrow Fe_2(SO_4)_3 + 6H_2O$
(e) $2Al + 3Cl_2 \rightarrow 2AlCl_3$
(f) $2K_3PO_4 + 3BaCl_2 \rightarrow 6KCl + Ba_3(PO_4)_2$

7. Identify the type (combination, decomposition, single-replacement, or double-replacement) of each reaction in Exercise 13.

8. Identify the type (combination, decomposition, single-replacement, or double-replacement) of each reaction in Exercise 14.

9. Identify each of the reactions in Exercise 3 as a combination, decomposition, single-replacement, or double-replacement reaction.

◀ 10. Identify each of the reactions in Exercise 2 as a combination, decomposition, single-replacement, or double-replacement reaction.

LEARNING GOAL 4

Predicting Products of Chemical Reactions

◀ 11. Complete and balance each of the following equations. Use the activity series and solubility table when necessary.

COMBINATION REACTIONS

(a) $H_2 + Br_2 \rightarrow ?$
(b) $BaO + H_2O \rightarrow ?$
(c) $Na + Cl_2 \rightarrow ?$
(d) Nitrogen gas + oxygen gas $\rightarrow$ nitrogen dioxide

DECOMPOSITION REACTIONS

(e) $CaCo_3 \xrightarrow{Heat} ? + ?$
(f) $KOH \rightarrow ? + ?$
(g) $Hg(ClO_3)_2 \xrightarrow{Heat} ? + ?$
(h) $PbCl_2 \rightarrow ? + ?$

SINGLE-REPLACEMENT REACTIONS

(i) $Li + H_2O \rightarrow ? + ?$
(j) Zinc + sulfuric acid $\rightarrow ? + ?$
(k) $Ni + Al(NO_3)_3 \rightarrow ? + ?$
(l) $Al + Hg(C_2H_3O_2)_2 \rightarrow ? + ?$

DOUBLE-REPLACEMENT REACTIONS

(m) Sulfuric acid + ammonium hydroxide $\rightarrow ? + ?$
(n) Silver nitrate + barium chloride $\rightarrow ? + ?$
(o) H_2SO_3 (sulfurous acid) $+ Al(OH)_3 \rightarrow ? + ?$
(p) $NaNO_3(aq) + KCl(aq) \rightarrow ? + ?$

12. Complete and balance each of the following equations. Use the activity series and solubility table when necessary.

COMBINATION REACTIONS

(a) $Cu + S \rightarrow ?$
(b) $CaO + H_2O \rightarrow ?$
(c) $Na + F_2 \rightarrow ?$
(d) Phosphorus + iodine $\rightarrow ?$

DECOMPOSITION REACTIONS

(e) $MgCO_3 \xrightarrow{Heat} ? + ?$
(f) $LiOH \rightarrow ? + ?$
(g) $Mg(ClO_3)_2 \rightarrow ? + ?$
(h) $Mg_3N_2 \rightarrow ? + ?$

SINGLE-REPLACEMENT REACTIONS

(i) $Al + H_3PO_4 \rightarrow ? + ?$
(j) $Zn + Pb(NO_3)_2 \rightarrow ? + ?$
(k) $Cl_2 + KBr \rightarrow ? + ?$
(l) $Br_2 + NaCl \rightarrow ? + ?$

DOUBLE-REPLACEMENT REACTIONS

(m) $AgNO_3 + H_2S \rightarrow ? + ?$
(n) Barium chloride + ammonium carbonate $\rightarrow ? + ?$
(o) $CaCO_3 + H_3PO_4 \rightarrow ? + ?$
(p) Silver nitrate + magnesium chloride $\rightarrow ? + ?$

13. Complete and balance each of the following equations. Use the activity series and solubility table when necessary.

(a) $N_2O \xrightarrow{Heat} ? + ?$
(b) $H_2CO_3 \xrightarrow{Heat} ? + ?$
(c) Sodium nitrate $\xrightarrow{Heat}$ sodium nitrite + oxygen gas
(d) $H_2 + F_2 \rightarrow ?$
(e) $N_2 + H_2 \xrightarrow[\text{High pressure}]{Heat} ?$
(f) $Ca + HCl \rightarrow ? + ?$
(g) $Cu + NiCl_2 \rightarrow ? + ?$
(h) $AgNO_3(aq) + K_3AsO_4(aq) \rightarrow ? + ?$

200 ESSENTIAL CONCEPTS OF CHEMISTRY

◀ 14. Complete and balance each of the following equations: Use the activity series and solubility table when necessary?

(a) $Na_2O \rightarrow ? + ?$
(b) $H_2 + I_2 \rightarrow ?$
(c) Phosphous + oxygen gas → disphosphorus trioxide
(d) $K_3PO_4 + BaCl_2 \rightarrow ? + ?$
(e) $KClO_3 \rightarrow ? + ?$
(f) $Ca + O_2 \rightarrow ?$
(g) $Cl_2 + NH_4I \rightarrow ? + ?$
(h) $Ag_2O \xrightarrow{Heat} ? + ?$

15. Complete and balance each of the following equation:

COMBINATION REACTIONS

(a) $K + Cl_2 \rightarrow ?$
(b) $K_2O + H_2O \rightarrow ?$
(c) $MgO + H_2O \rightarrow ?$
(d) $CaO + CO_2 \rightarrow ?$

DECOMPOSITION REACTIONS

(e) $NH_3 \rightarrow ? + ?$
(f) $SrCO_3 \rightarrow ? + ?$
(g) $NaClO_3 \rightarrow ? + ?$
(h) $Mg(OH)_2 \rightarrow ? + ?$

16. Complete and balance each of the following equations:

COMBINATION REACTIONS

(a) $K + Br_2 \rightarrow ?$
(b) $Li_2O + H_2O \rightarrow ?$
(c) $BaO + H_2O \rightarrow ?$
(d) $MgO + CO_2 \rightarrow ?$

DECOMPOSITION REACTIONS

(e) $MgCO_3 \rightarrow ? + ?$
(f) Copper(I) sulfide → ? + ?
(g) Chromium(III) carbonate → ? + ?
(h) $Al(ClO_3)_3 \xrightarrow{Heat} ? + ?$

LEARNING GOAL 5

Acids, Bases, and Salts

◀ *17*. The reaction of an acid and a base produces a salt and water. Find the acid and base needed to produce each of the following salts:

(a) NaCl (b) K_2SO_4 (c) $Al(NO_3)_3$

18. The reaction of an acid and a base produces a salt and water. Find the acid and base needed to produce each of the following salts:

(a) KCl (b) $MgSO_4$ (c) $Fe_2(SO_4)_3$

CHAPTER 10: THE CHEMICAL EQUATION

LEARNING GOAL 6

Using the Activity Series and Solubility Table

◀ 19. Complete and balance each of the following equations. Check the activity series and solubility table to be sure that the reaction can take place.

SINGLE-REPLACEMENT REACTIONS

(a) $Zn + HNO_2 \rightarrow ? + ?$
(b) $Ag + NiCl_2 \rightarrow ? + ?$
(c) $Zn + AgNO_3 \rightarrow ? + ?$
(d) $Cs + H_2O \rightarrow ? + ?$

DOUBLE-REPLACEMENT REACTIONS

(e) $HCl + Al(OH)_3 \rightarrow ? + ?$
(f) $KNO_3 + ZnCl_2 \rightarrow ? + ?$
(g) $Al(NO_3)_3 + NaOH \rightarrow ? + ?$
(h) $K_2CrO_4 + Pb(NO_3)_2 \rightarrow ? + ?$

◀ 20. Complete and balance each of the following equations. Check the activity series and solubility table to be sure that the reaction can take place.

SINGLE-REPLACEMENT REACTIONS

(a) $Cu + AgCl \rightarrow ? + ?$
(b) $Cl_2 + KF \rightarrow ? + ?$
(c) $Zn + HNO_3 \rightarrow ? + ?$
(d) $Cu + HCl \rightarrow ? + ?$

DOUBLE-REPLACEMENT REACTIONS

(e) $NH_4NO_3 + H_3PO_4 \rightarrow ? + ?$
(f) $ZnCl_2 + KOH \rightarrow ? + ?$
(g) $Ni_3(PO_4)_2 + HCl \rightarrow ? + ?$
(h) $AgNO_3 + KCl \rightarrow ? + ?$

LEARNING GOAL 7

Oxidation and Reduction; Oxidizing and Reducing Agents

◀ 21. Define the following terms and give an example of each:

(a) oxidation (b) reduction (c) oxidizing agent (d) reducing agent

22. Complete each of the following statements:

    (a) A loss of electrons is called _____.
    (b) A gain of electrons is called_____.
    (c) A substance that causes something else to be oxidized is called a(n) _____.
    (d) A substance that causes something else to be reduced is called a(n) _____.

LEARNING GOAL 8

Determining Oxidized and Reduced Substances

◄ *23.* Balance each of the following reactions and determine which substance is oxidized and which is reduced:

(a) $K + Cl_2 \rightarrow KCl$
(b) $NH_3 \rightarrow N_2 + H_2$
(c) $CuO + H_2 \rightarrow Cu + H_2O$
(d) $Sn + Cl_2 \rightarrow SnCl_4$

◄ *24.* Balance each of the following reactions and determine which substance is oxidized and which is reduced:

(a) $Na + F_2 \rightarrow NaF$
(b) $Al + C \rightarrow Al_4C_3$
(c) $Cu + S \rightarrow CuS$
(d) $Ag_2O \rightarrow Ag + O_2$

LEARNING GOAL 8

Determining Oxidizing and Reducing Agents

◄ *25.* For each of the reactions in Exercise 23, determine the oxidizing agent and the reducing agent.
◄ 26. For each of the reactions in Exercise 24, determine the oxidizing agent and the reducing agent.
27. For each single-replacement reaction in Exercise 19, determine the oxidizing agent and the reducing agent.
28. For each single-replacement reaction in Exercise 20, determine the oxidizing agent and the reducing agent.

EXTRA EXERCISES

◄ *29.* For the reaction

$$5Zn + V_2O_5 \rightarrow 5ZnO + 2V$$

list (a) the oxidizing agent, (b) the reducing agent, (c) the element oxidized, (d) the element reduced

30. Magnesium carbonate and calcium carbonate, when heated, both decompose to form carbon dioxide and the metal oxides. Write these two reactions.

31. Determine whether the following equations are balanced. If they are not, balance them.

(a) $N_2 + 3H_2 \rightarrow NH_3$
(b) $Fe + H_2O \rightarrow Fe_3O_4 + H_2$
(c) $PCl_3 + Cl_2 \rightarrow PCl_5$

◄ 32. A student is asked to balance the following:
Calcium hydroxide + hydrochloric acid → calcium chloride + water
The student works the problem and gets the following solution:

$$CaOH + HCl \rightarrow CaCl + H_2O$$

This attempt is marked incorrect. Can you explain why the answer is wrong?

33. Write a balanced equation for each of the following. If no reaction occurs, write no reaction.

(a) Carbon burning in oxygen gas to form carbon monoxide.
(b) Hydrogen gas and chlorine gas reacting to for hydrogen chloride gas
(c) Metallic sodium dropped into water to create sodium hydroxide and hydrogen gas
(d) Silver metal and hydrochloric acid reacting to produce _____

◀ __34.__ Consider an experiment in which metal A is immersed in a water solution of ions of metal B. For each of the following pairs A and B, determine whether or not a reaction occurs. If a reaction does occur, write the balanced equation, assuming that the B ions are part of nitrate salts.

	A	B
(a)	Cu	Zn
(b)	Mg	Zn
(c)	Cu	Ni
(d)	Zn	Cu

35. Explain why an oxidizing agent is reduced.
__36.__ Write balanced formula equations for the following word equations and state what type of reaction each is:

(a) Calcium hydroxide + hydrochloric acid →
(b) Zinc metal + sulfuric acid →
(c) Barium metal + oxygen gas →
(d) Cesium oxide + water →

__37.__ Balance the equation

$$NH_3(g) + O_2(g) \rightarrow N_2O_4(g) + H_2O(g)$$

__38.__ Consider the redox reaction

$$(Cr_2O_7)^{2-} + 14H^{1+} + 6Fe^{2+} \rightarrow 2Cr^3 + 7H_2O + 6Fe^{3+}$$

(a) Determine which substance is oxidized.
(b) Determine which substance is reduced.
(c) Determine the oxidizing agent.
(d) Determine the reducing agent.

__39.__ Balance the following equation and write the chemical name of each product and reactant in the balanced equation.

$$Na + P_4 \rightarrow Na_3P$$

◀ 40. Write a balanced equation for a reaction that produces magnesium chloride. (*Hint:* In other words, magnesium chloride has to be at least one of the products.)
41. Write a balanced equation for an acid-base reaction that produces lithium chloride as one of the products.
42. Are single-replacement reactions also redox reactions? Explain with examples.
43. Write a balanced equation for any reaction that produces carbon dioxide as one of the products.
44. A double-replacement reaction produces zinc phosphate. What are the reactants? Write a balanced equation for the reaction.

CHAPTER 11

STOICHIOMETRY
The Quantities in Reactions

One of the most important contributors to the advancement of modern-day chemistry is the French chemist Antoine Laurent Lavoisier (1743-1794). As the 1800s drew to a close, Lavoisier melded together some of the most important discoveries about gases, thereby establishing the quantitative aspects of chemistry.

From the very beginning Lavoisier recognized the importance of accurate measurement. In some of his most important work, around 1772, Lavoisier heated metals, such as tin and lead, in closed containers with a limited supply of air. Both metals formed a layer of *calx* on the surface and then rusted no further. A popular (but soon-to-be-proven incorrect) theory at the time, called the *phlogiston theory*, had been used by many chemists to explain these observations. According to the phlogiston theory, anything that burned contained the substance phlogiston, and as the material burned the phlogiston was released into the air. The phlogistonists explained the results of Lavoisier's experiments by stating that the air had absorbed all of the phlogiston that it could from the metal. However, experiments had shown that the calx weighed more, not less, than the metal itself. More important, when Lavoisier weighed the entire container (metal, calx, air, and all) after heating, the container weighed exactly the same as it weighed before the heating.

Based on the results of these experiments, Lavoisier conjectured that if the metal had gained weight by being partially turned into calx, then something else in the vessel must have lost an equivalent amount of weight. That something else, it seemed, would have to be air (or something that was part of air). If this were true, then a partial vacuum should exist in the container. Sure enough, when Lavoisier opened the container, air rushed in. Once that had occurred, the container and its contents weighed more. Lavoisier had shown that the conversion of a metal into its calx was not the result of the loss of the mysterious substance phlogiston, but the gain of something from the air. Now Lavoisier had to determine what in the air combined with the metal.

Fortunately, about the same time that Lavoisier was performing his experiments, another famous chemist, Joseph Priestly, had made a startling discovery. In 1774 Priestly decomposed a calx of mercury (known today as mercuric oxide) and obtained an unknown gas that he called *dephlogisticated* air (known today as oxygen gas). When he visited Lavoisier in Paris later that year, Priestly described his discovery. Lavoisier immediately saw the significance of this gas! The following year, Lavoisier published a paper in which he stated that air is not a simple substance, but a mixture of at least two substances, one of them being Priestly's dephlogisticated air (oxygen gas).

It wasn't long before Lavoisier had determined that air consisted of about 20% oxygen and 80% nitrogen. The importance of measurement had made its initial impact on the chemical sciences.

206 ESSENTIAL CONCEPTS OF CHEMISTRY

LEARNING GOALS

After you've studied this chapter, you should be able to:

1. Calculate the quantities of reactants needed or products yielded in a chemical reaction.
2. Determine which starting material is the limiting reactant in a chemical reaction.

INTRODUCTION

In Chapter 10, we wrote balanced equations showing how many atoms or molecules of each substance take part in a reaction. For example, we considered the reaction

$$N_2 + 3H_2 \xrightarrow[\text{Pressure}]{\text{Heat}} 2NH_3$$

In this reaction, one molecule of nitrogen combines with three molecules of hydrogen to form two molecules of ammonia (Figure 11.1). Industry uses this reaction to make ammonia for fertilizers, explosives, and plastics, among many other products. The industrial process based on this reaction is called the **Haber process** for the synthesis of ammonia. (To **synthesize** means to combine the parts into the whole.)

Suppose we also wanted to produce ammonia in large quantities for commercial use. To prevent waste and thereby minimize costs, we would like to combine just enough of each reactant. But how much is "just enough of each reactant"? And how much ammonia would we obtain? Or suppose we wanted to produce just 100 kilograms of ammonia. How much nitrogen and hydrogen would we need?

To answer such questions, we need to "translate" balanced chemical equations into quantities of reactants and products. In this chapter, you will learn how to do so by using *stoichiometry*, which is pronounced "stoy-key-OM-eh-tree."

11.1 The Mole Method

Stoichiometry is defined as the *calculation of the quantities of substances involved in chemical reactions*. An important idea in stoichiometry is that a balanced equation tells us how many *moles* of the various substances take part in the reaction. For example, the balanced equation for the Haber process,

$$N_2 + 3H_2 \xrightarrow[\text{Pressure}]{\text{Heat}} 2NH_3$$

tells us that 1 *mole* of nitrogen and 3 *moles* of hydrogen react to form 2 *moles* of ammonia. (If you've forgotten what a mole is, refer back to Chapter 9.)

The reasoning behind this idea is as follows: The balanced chemical equation tells us the number of molecules or atoms of each substance that is involved in a single instance of the reaction. It also tells us in what *ratio* these molecules or atoms are involved. In the Haber process, the molecules are involved in the ratio one N_2 to three H_2 to two NH_3, or $1N_2 : 3H_2 : 2NH_3$. Using this ratio, we can also say that 1 *Avogadro's number* of nitrogen molecules and 3 *Avogadro's numbers* of hydrogen molecules produce 2 *Avogadro's numbers* of ammonia molecules. Because 1 Avogadro's number of molecules is 1 mole, we can say that 1 mole of nitrogen and 3 moles of hydrogen react to produce 2 moles of ammonia.

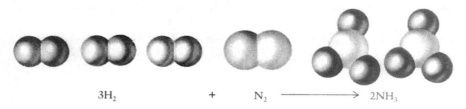

FIGURE 11.1 The Haber process.

CHAPTER 11: STOICHIOMETRY   207

This same sort of reasoning holds true for every balanced equation. *The coefficients in a balanced chemical equation indicate the number of moles of each substance that takes part in the reaction.*

Once we know the number of moles of each substance in a reaction, we can find out how many grams (or kilograms or pounds) of each substance are involved. For example, we know that the atomic mass of nitrogen is 14.0 and that the atomic mass of hydrogen is 1.0. Then we can say that since 1 mole of N_2 has a mass of 28.0 grams, 1 mole of H_2 has a mass of 2.0 grams, and 1 mole of NH_3 has a mass of 17.0 grams, 28.0 grams of nitrogen react with 6.0 grams of hydrogen to produce 34.0 grams of ammonia:

$$N_2 + 3H_2 \xrightarrow[\text{Pressure}]{\text{Heat}} 2NH_3$$

1 mole	3 moles	2 moles
28.0 grams	6.0 grams	34.0 grams

Note that we have satisfied the Law of Conservation of Mass. The mass of the product is equal to the sum of the masses of the reactants. In the remainder of this chapter, we shall apply this basic idea to several types of stoichiometric problems.

11.2 Quantities of Reactants and Products

The best way to learn how to solve stoichiometric problems is to work through a number of them. As we do this, remember that the balanced equation for a reaction contains the basic information about the relative quantities of substances involved. Also note that if we know the quantity of even a single reactant or product, we can obtain the quantities of all other reactants or products.

We can begin by asking how many moles of N_2 are needed to produce 5 moles of NH_3. The balanced equation is

$$N_2 + 3H_2 \rightarrow 2NH_3$$

1 mole		2 moles
?		5 moles

Set up the conversion:

$$5 \text{ moles NH}_3 \times \frac{1 \text{ mole N}_2}{2 \text{ moles NH}_3} = 2.5 \text{ moles N}_2$$

Next we can ask how many moles of H_2 are needed to produce 5 moles of NH_3. The balanced equation is

$$N_2 + 3H_2 \rightarrow 2NH_3$$

	3 moles	2 moles
	?	5 moles

Set up the conversion:

$$5 \text{ moles NH}_3 \times \frac{3 \text{ moles H}_2}{2 \text{ moles NH}_3} = 7.5 \text{ moles H}_2$$

Let us assume that we want to synthesize 136 grams of ammonia. How many grams of nitrogen and hydrogen do we need? The first thing we do is set up the balanced chemical equation.

$$N_2 + 3H_2 \rightarrow 2NH_3$$

Next we record what is given and what must be found.

$$N_2 + 3H_2 \rightarrow 2NH_3$$

? g	? g	136 g

208 ESSENTIAL CONCEPTS OF CHEMISTRY

The next step is to convert grams of ammonia into moles (because the chemical equation relates substances by moles, not grams).

$$? \text{ moles of } NH_3 = (136 \text{ g}) \left(\frac{1 \text{ mole}}{17.0 \text{ g}} \right) = 8.00 \text{ moles of } NH_3$$

$$N_2 + 3H_2 \rightarrow 2NH_3$$
$$?\text{g} \quad ?\text{g} \quad\quad\quad 136 \text{ g (which is 8.00 moles)}$$

Now we use the coefficients of the balanced equation to calculate the number of moles of hydrogen and nitrogen necessary to form 8 moles of ammonia. We do it in three steps. First, we determine the number of moles of nitrogen:

$$? \text{ moles of } N_2 = (8.00 \text{ moles } NH_3) \left(\frac{1 \text{ mole } N_2}{2 \text{ moles } NH_3} \right) = 4.00 \text{ moles of } N_2$$

We obtained the factor

$$\left(\frac{1 \text{ mole } N_2}{2 \text{ moles } NH_3} \right)$$

from the coefficients of the balanced equation. This factor tells us that 1 mole of nitrogen is necessary to produce 2 moles of ammonia. Now we are ready for the second step.

$$? \text{ moles of } H_2 = (8.00 \text{ moles } NH_3) \left(\frac{3 \text{ moles } H_2}{2 \text{ moles } NH_3} \right) = 12.0 \text{ moles of } H_2$$

We also obtained the factor

$$\left(\frac{3 \text{ moles } N_2}{2 \text{ moles } NH_3} \right)$$

from the balanced equation, which tells us that 3 moles of hydrogen are necessary to produce 2 moles of ammonia. We now have

$$N_2 \quad + \quad 3H_2 \quad \rightarrow \quad 2NH_3$$
$$?\text{g} \quad\quad\quad ?\text{g} \quad\quad\quad 136 \text{ g}$$
$$4.00 \text{ moles} \quad 12.0 \text{ moles} \quad 8.00 \text{ moles}$$

Another way to obtain the same information is to remember that the moles of reactants and products must be in the ratio of

$$N_2 : H_2 : NH_3$$
$$1 : 3 : 2$$

We could ask ourselves the question,

$$1 : 3 : 2 \quad \text{is the same as} \quad ? : ? : 8$$

To answer this, we would note that

$$1 : 3 : 2$$
$$\downarrow \times 4$$
$$? : ? : 8$$

or

$$1 : 3 : 2 \quad \text{is the same as} \quad 4 : 12 : 8$$

Our final step is to change *moles* of nitrogen and hydrogen into *grams*.

$$? \text{ grams of } N_2 = (4.00 \text{ moles}) \left(\frac{28.0 \text{ g}}{1 \text{ mole}} \right) = 112 \text{ g of } N_2$$

$$? \text{ grams of } H_2 = (12.0 \text{ moles}) \left(\frac{2.0 \text{ g}}{1 \text{ mole}} \right) = 24 \text{ g of } H_2$$

The problem is solved.

$$\begin{array}{ccccc}
N_2 & + & 3H_2 & \rightarrow & 2NH_3 \\
112 \text{ g} & & 24 \text{ g} & & 136 \text{ g} \\
4.00 \text{ moles} & & 12.0 \text{ moles} & & 8.00 \text{ moles}
\end{array}$$

We can check our mathematics by seeing whether this reaction obeys the Law of Conservation of Mass. The grams of nitrogen and hydrogen we start with should equal the grams of ammonia produced.

MASS OF REACTANTS	MASS OF PRODUCTS
Grams of N_2 = 112 g	Grams of NH_3 = 136 g
Grams of H_2 = 24 g	
Total mass = 136 g	Total mass = 136 g

To practice our stoichiometry, let's solve another problem. Many scientists have suggested using propane gas (C_3H_8) in motor vehicles. Because propane burns cleaner than today's gasolines, these scientists feel that the use of propane would drastically reduce the amount of automobile pollution. When propane reacts with sufficient oxygen, the resulting products are carbon dioxide and water, in the form of water vapor. Our problem is this: How many grams of oxygen are needed for combustion of 22 grams of propane? How many grams of carbon dioxide would be produced? How many grams of water would be produced? (Remember that C_3H_8 is the formula for propane.)

STEP 1

Write a balanced equation, listing what is given and what we want to find.

$$C_3H_8 + 5O_2 \rightarrow 3CO_2 + 4H_2O$$
$$22 \text{ g} \quad ? \text{ g} \quad ? \text{ g} \quad ? \text{ g}$$

STEP 2

Change the grams of C_3H_8 into moles.

$$? \text{ moles of } C_3H_8 = (22 \text{ g}) \left(\frac{1 \text{ mole}}{44.0 \text{ g}} \right) = 0.50 \text{ mole of } C_3H_8$$

$$C_3H_8 + 5O_2 \rightarrow 3CO_2 + 4H_2O$$
$$22 \text{ g} \quad ? \text{ g} \quad ? \text{ g} \quad ? \text{ g}$$
$$0.50 \text{ mole}$$

210 ESSENTIAL CONCEPTS OF CHEMISTRY

STEP 3

Use the coefficients of the equation to determine the moles of oxygen needed, as well as the moles of carbon dioxide and water produced.

$5O_2$

? moles of O_2 needed = $(0.50 \text{ mole } C_3H_8)\left(\dfrac{5 \text{ moles } O_2}{1 \text{ mole } C_3H_8}\right)$ = 5×.5 = 2.5/1 = 2.5

= 2.5 moles of O_2 needed

$3CO_2$

? moles of CO_2 produced = $(0.50 \text{ mole } C_3H_8)\left(\dfrac{3 \text{ moles } CO_2}{1 \text{ mole } C_3H_8}\right)$ = 3×.5 = 1.5/1 = 1.5

= 1.5 moles of CO_2 produced

$4H_2O$

? moles of H_2O produced = $(0.50 \text{ mole } C_3H_8)\left(\dfrac{4 \text{ moles } H_2O}{1 \text{ mole } C_3H_8}\right)$ = 5×.4 = 2/1 = 2

= 2.0 moles of H_2O produced

$$C_3H_8 \;+\; 5O_2 \;\rightarrow\; 3CO_2 \;+\; 4H_2O$$
| 22 g | ? g | ? g | ? g |
| 0.50 mole | 2.5 moles | 1.5 moles | 2.0 moles |

Note that you can obtain the same result by the other method.

$$1 : 5 : 3 : 4$$
$$0.50 : ? : ? : ?$$

or

$$1 : 5 : 3 : 4$$
$$0.50 : 2.5 : 1.5 : 2.0$$

STEP 4

Change moles of O_2, CO_2, and H_2O into grams.

$3 O_2$? grams of O_2 = $(2.5 \text{ moles})\left(\dfrac{32.0 \text{ g}}{1 \text{ mole}}\right)$ = $8\overline{0}$ g of O_2 32×2.5 = 80/1 = 80 g

? grams of CO_2 = $(1.5 \text{ moles})\left(\dfrac{44.0 \text{ g}}{1 \text{ mole}}\right)$ = 66 g of CO_2 12+32=44 × 1.5 = 66 = 66 g

? grams of H_2O = $(2.0 \text{ moles})\left(\dfrac{18.0 \text{ g}}{1 \text{ mole}}\right)$ = 36 g of H_2O ✓

$$C_3H_8 \;+\; 5O_2 \;\rightarrow\; 3CO_2 \;+\; 4H_2O$$
| 22 g | $8\overline{0}$ g | 66 g | 36 g |
| 0.50 mole | 2.5 moles | 1.5 moles | 2.0 moles |

STEP 5

Check your results.

MASS OF REACTANTS	MASS OF PRODUCTS
Grams of C_3H_8 = 22 g	Grams of CO_2 = 66 g
Grams of O_2 = $8\overline{0}$ g	Grams of H_2O = 36 g
Total mass = 102 g	Total mass = 102 g

Let's summarize the procedure for solving stoichiometric problems:

1. Write a balanced chemical equation for the reaction. Record the given number of grams under the chemical formula of that substance. Place question marks for the unknown numbers of grams under the corresponding chemical formulas.
2. Change the given number of grams into moles.
3. Use the coefficients of the balanced equation and the moles of the given amount of substance to determine the moles of the other substances involved in the reaction.
4. Change the moles of the other substances into grams.
5. Check the results to see whether the mass of the reactants equals the mass of the products.

Another way of summarizing this procedure is to think of it in the following sequence:

$$\text{Grams known} \rightarrow \text{moles known} \rightarrow \text{moles unknown} \rightarrow \text{grams unknown}$$

Before we continue we would like to show one more way to approach this problem. Some students find it easier to solve a problem such as this one using the mass-mass ratio from the balanced equation. We will find the grams of carbon dioxide and the grams of water produced using this method.

$$C_3H_8 + 5O_2 \rightarrow 3CO_2 + 4H_2O$$

1 mole		3 moles	4 moles
44.0 g		3(44.0 g)	4(18.0 g)
44.0 g		132.0 g	72.0 g

$$22 \text{ g } C_3H_8 \times \frac{132.0 \text{ g } CO_2}{44.0 \text{ g } C_3H_8} = 66 \text{ g } CO_2$$

$$22 \text{ g } C_3H_8 \times \frac{72.0 \text{ g } H_2O}{44.0 \text{ g } C_3H_8} = 36 \text{ g } H_2O$$

This procedure for solving mass-mass stoichiometric problems can be summarized as follows:

1. First write a balanced chemical equation for the reaction.
2. Underneath the chemical formulas, write the number of moles of the substance given and the number of moles of the substance asked for as indicated by the coefficients.
3. Convert these mole quantities into grams. This may be written below the formulas in the equation.
4. Use the gram-gram relationship as a conversion factor with the mass of the given substance to find the mass of the substance in question.

212 ESSENTIAL CONCEPTS OF CHEMISTRY

Practice solving stoichiometric problems by doing the following exercises. Work each problem in steps according to the procedure shown. You may also use the alternate strategy.

EXAMPLE 11.1

The *thermite* reaction is a chemical reaction that is important to railroads, because railroad workers use it to weld railroad tracks. In this reaction, a mixture of aluminum metal and iron(III) oxide (known as thermite) is heated to yield aluminum oxide and iron metal. Once begun, the reaction releases a tremendous amount of heat—so much that the molten iron produced fuses the railroad tracks (Figure 11.2).

How many grams of aluminum and iron(III) oxide are needed to produce 279 grams of iron metal? How many grams of aluminum oxide will be produced?

Solution

UNDERSTAND THE PROBLEM

We are told how many grams of product (iron metal) are formed, and we can use this information to calculate the grams of reactants [aluminum and iron(III) oxide] necessary to produce this amount of product.

DEVISE A PLAN

Use the strategy that was described previously.

CARRY OUT THE PLAN

The procedure may be followed in steps. [*Note:* Fe_2O_3 stands for iron(III) oxide, and Al_2O_3 stands for aluminum oxide.]

STEP 1

$$2Al + Fe_2O_3 \rightarrow Al_2O_3 + 2Fe$$
$$?g \quad ?g \quad\quad ?g \quad\quad 279g$$

STEP 2

$$? \text{ moles of Fe} = (279 \text{ g}) \left(\frac{1 \text{ mole}}{55.8 \text{ g}} \right) = 5.00 \text{ moles of Fe}$$

$$2Al + Fe_2O_3 \rightarrow Al_2O_3 + 2Fe$$
$$?g \quad ?g \quad\quad ?g \quad\quad 279\,g$$
$$\quad\quad\quad\quad\quad\quad\quad\quad\quad 5.00 \text{ moles}$$

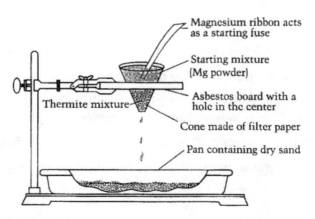

FIGURE 11.2 The thermite reaction in the lab.

STEP 3

$$? \text{ moles of Al} = (5.00 \text{ moles Fe})\left(\frac{2 \text{ moles Al}}{2 \text{ moles Fe}}\right)$$

$$= 5 \text{ moles of Al}$$

$$? \text{ moles of Fe}_2\text{O}_3 = (5.00 \text{ moles Fe})\left(\frac{1 \text{ mole Fe}_2\text{O}_3}{2 \text{ moles Fe}}\right)$$

$$= 2.50 \text{ moles of Fe}_2\text{O}_3$$

$$? \text{ moles of Al}_2\text{O}_3 = (5.00 \text{ moles Fe})\left(\frac{1 \text{ mole Al}_2\text{O}_3}{2 \text{ moles Fe}}\right)$$

$$= 2.50 \text{ moles of Al}_2\text{O}_3$$

2Al	+	Fe$_2$O$_3$	→	Al$_2$O$_3$	+	2Fe
?g		?g		?g		279 g
5.00 moles		2.50 moles		2.50 moles		5.00 moles

STEP 4

$$? \text{ grams of Al} = (5.00 \text{ moles})\left(\frac{27.0 \text{ g}}{1 \text{ mole}}\right) = 135 \text{ g of Al}$$

$$? \text{ grams of Fe}_2\text{O}_3 = (2.50 \text{ moles})\left(\frac{159.6}{1 \text{ mole}}\right) = 399 \text{ g of Fe}_2\text{O}_3$$

$$? \text{ grams of Al}_2\text{O}_3 = (2.50 \text{ moles})\left(\frac{102.0 \text{ g}}{1 \text{ mole}}\right) = 255 \text{ g of Al}_2\text{O}_3$$

2 Al	+	Fe$_2$O$_3$	→	Al$_2$O$_3$	+	2Fe
135 g		399 g		255 g		279 g
5.00 moles		2.50 moles		2.50 moles		5.00 moles

LOOK BACK

We check our results to be sure they satisfy the Law of Conservation of Mass.

MASS OF REACTANTS		MASS OF PRODUCTS	
Grams of Al	= 135 g	Grams of Al$_2$O$_3$	= 255 g
Grams of Fe$_2$O$_3$	= 399 g	Grams of Fe	= 279 g
Total mass	= 534 g	Total mass	= 534 g

Practice Exercise 11.1

Magnesium oxide is a compound that occurs in nature as the mineral periclase. It is composed of the elements magnesium and oxygen. In industry, magnesium oxide is used in the manufacture of refractory crucibles, firebricks, and casein glue. It is also used as a reflector in optical instruments. It combines with water to form magnesium hydroxide, which acts as an antacid in low concentrations. How many grams of magnesium and oxygen gas are needed to produce 4.032 grams of magnesium oxide?

EXAMPLE 11.2

Polyvinyl chloride (PVC) is one of the best-known and most widely used plastics in the world today. PVC is used to make food wrappings, luggage, plastic garbage bags, phonograph records, and automobile upholstery. It was originally thought that PVC was an extremely safe compound. But studies now indicate that there may be some hidden hazards. The problem is not with the PVC itself but with the compound vinyl chloride (C_2H_3Cl), from which the PVC is made. Vinyl chloride is a colorless gas, and studies show that it may cause liver damage and some forms of cancer in people who are in constant contact with it. Another problem with PVC and vinyl chloride is that when the compounds are burned, an extremely poisonous gas (hydrogen chloride, HCl) can be produced.

Our problem is this: Assume that the following reaction occurs when we burn vinyl chloride (C_2H_3Cl).

$$2C_2H_3Cl + 5O_2 \rightarrow 4CO_2 + 2H_2O + 2HCl$$

How many grams of hydrogen chloride (HCl) will be produced if 6.25 g of vinyl chloride are burned? How many grams of O_2 are needed, and how many grams of CO_2 and water are produced?

Solution

Follow the procedure we used before. Do it in steps.

STEP 1

$$2C_2H_3Cl + 5O_2 \rightarrow 4CO_2 + 2H_2O + 2HCl$$
$$6.25\,g \quad ?\,g \quad\quad ?\,g \quad\quad ?\,g \quad\quad ?\,g$$

STEP 2

$$? \text{ moles of } C_2H_3Cl = (6.25\,g)\left(\frac{1 \text{ mole}}{62.5\,g}\right) = 0.100 \text{ mole of } C_2H_3Cl$$

$$2C_2H_3Cl + 5O_2 \rightarrow 4CO_2 + 2H_2O + 2HCl$$
$$6.25\,g \quad ?\,g \quad\quad ?\,g \quad\quad ?\,g \quad\quad ?\,g$$
$$0.100 \text{ mole}$$

STEP 3

$$? \text{ moles of } O_2 = (0.100 \text{ mole } C_2H_3Cl)\left(\frac{5 \text{ moles } O_2}{2 \text{ moles } C_2H_3Cl}\right)$$
$$= 0.250 \text{ mole of } O_2$$

$$? \text{ moles of } CO_2 = (0.100 \text{ mole } C_2H_3Cl)\left(\frac{4 \text{ moles } CO_2}{2 \text{ moles } C_2H_3Cl}\right)$$
$$= 0.200 \text{ mole of } CO_2$$

$$? \text{ moles of } H_2O = (0.100 \text{ mole } C_2H_3Cl)\left(\frac{2 \text{ moles } H_2O}{2 \text{ moles } C_2H_3Cl}\right)$$
$$= 0.100 \text{ mole of } H_2O$$

$$? \text{ moles of } HCl = (0.100 \text{ mole } C_2H_3Cl)\left(\frac{2 \text{ moles } HCl}{2 \text{ moles } C_2H_3Cl}\right)$$
$$= 0.100 \text{ mole of } HCl$$

$$2C_2H_3Cl + 5O_2 \rightarrow 4CO_2 + 2H_2O + 2HCl$$

6.25 g	? g	? g	? g	? g
0.100 mole	0.250 mole	0.200 mole	0.100 mole	0.100 mole

STEP 4

$$? \text{ grams of } O_2 = (0.250 \text{ mole})\left(\frac{32.0 \text{ g}}{1 \text{ mole}}\right) = 8.00 \text{ g of } O_2$$

$$? \text{ grams of } CO_2 = (0.200 \text{ mole})\left(\frac{44.0 \text{ g}}{1 \text{ mole}}\right) = 8.80 \text{ g of } CO_2$$

$$? \text{ grams of } H_2O = (0.100 \text{ mole})\left(\frac{18.0 \text{ g}}{1 \text{ mole}}\right) = 1.80 \text{ g of } H_2O$$

$$? \text{ grams of } HCl = (0.100 \text{ mole})\left(\frac{36.5 \text{ g}}{1 \text{ mole}}\right) = 3.65 \text{ g of } HCl$$

STEP 5

MASS OF REACTANTS		MASS OF PRODUCTS	
Grams of C_2H_3Cl =	6.25 g	Grams of CO_2 =	8.80 g
Grams of O_2 =	8.00 g	Grams of H_2O =	1.80 g
		Grams of HCL =	3.65 g
Total mass =	14.25 g	Total mass =	14.25 g

Practice Exercise 11.2

Aluminum nitrate can be produced in a single-replacement reaction. Aluminum replaces zinc in the compound zinc nitrate, because aluminum is higher than zinc in the activity series. The reaction is as follows:

$$2Al + 3Zn(NO_3)_2 \rightarrow 2Al(NO_3)_3 + 3Zn$$

Aluminum nitrate is used in tanning leather, in antiperspirant preparations, and as a corrosion inhibitor. How many grams of aluminum nitrate are produced when 9.469 g of zinc nitrate react with aluminum? How many grams of aluminum are needed, and how many grams of zinc are produced?

11.3 The Limiting-Reactant Problem

Sometimes a chemist who wants to synthesize a compound finds that the quantity of product that can be formed is limited by the amount of one of the starting materials. To find out how much product can be formed, the chemist must first determine which one of the starting materials will be completely used up when the reaction is completed. The other will be in *excess*, so not all of it will be used up to form the product. In other words, the chemist must determine which one of the starting materials is the *limiting reactant*.

To clarify this, let's imagine an automobile manufacturer, the Ajax Automobile Company, which makes a model of a car that is called the Ajax. In order to make just *one* of these automobiles, the assembly line needs (among other things) the following parts: *one* body and *four* wheels. We can represent this statement as follows:

$$1 \text{ body} + 4 \text{ wheels} \rightarrow 1 \text{ automobile}$$

On a particular day, the assembly line has 50 bodies and 160 wheels. How many automobiles can be produced? If you use the number of bodies as the basis for your answer, you might think Ajax could produce

216 ESSENTIAL CONCEPTS OF CHEMISTRY

50 automobiles. However, if you use the number of wheels as the basis for your answer, you'll find that Ajax can produce only 40 automobiles. (Remember, it takes 4 wheels to produce an automobile—not counting spare tires. Therefore 160 ÷ 4 = 40.) So the company has 10 extra bodies but not enough wheels to use them. As a result, the most automobiles it can produce is 40. The "limiting reactant" in this case is the wheels, and the bodies are the "reactant" that is in excess.

Let's go through this entire procedure again, this time treating it as if it were a chemical problem.

STEP 1

We follow our usual procedure.

$$1 \text{ body} + 4 \text{ wheels} \rightarrow 1 \text{ automobile}$$
$$50 \text{ bodies} \quad 160 \text{ wheels} \quad ? \text{ automobiles}$$

STEP 2

In a chemical problem of this type, we would have to change grams into moles at this point. However, in our car problem, we already know the relationship of bodies to wheels. Therefore all we have to do is determine which substance gets used up first (bodies or wheels).

a. Determine how many bodies would be needed to use up 160 wheels.

$$? \text{ bodies needed} = (160 \text{ wheels}) \left(\frac{1 \text{ body}}{4 \text{ wheels}} \right) = 40 \text{ bodies}$$

b. Determine how many wheels would be needed to use up 50 bodies.

$$? \text{ wheels needed} = (50 \text{ bodies}) \left(\frac{4 \text{ wheels}}{1 \text{ body}} \right) = 200 \text{ wheels}$$

But the Ajax Company doesn't have 200 wheels—only 160. Therefore the wheels limit the number of automobiles it can produce.

STEP 3

We may use the number of wheels to calculate the number of automobiles Ajax can produce and the number of bodies it will use. (Note that we've already calculated the number of bodies that the company will use in step 2a.)

$$? \text{ automobiles} = (160 \text{ wheels}) \left(\frac{1 \text{ automobile}}{4 \text{ wheels}} \right) = 40 \text{ automobiles}$$

Now let's do a chemical problem. Assume that a chemist has 203 grams of $Mg(OH)_2$ (magnesium hydroxide) and 164 grams of HCl (hydrogen chloride). The chemist wants to make these substances react to form $MgCl_2$ (magnesium chloride), which can be used for fireproofing wood or in a disinfectant. How much $MgCl_2$ can the chemist produce, and how much water is formed?

STEP 1

We follow our usual procedure.

$$Mg(OH)_2 + 2HCl \rightarrow MgCl_2 + 2H_2O$$
$$203 \text{ g} \quad\quad 164 \text{ g} \quad\quad ? \text{ g} \quad\quad ? \text{ g}$$

STEP 2

Determine the number of moles of each starting substance.

$$? \text{ moles of Mg(OH)}_2 = (203 \text{ g}) \left(\frac{1 \text{ mole}}{58.3 \text{ g}} \right) = 3.48 \text{ moles of Mg(OH)}_2$$

$$? \text{ moles of HCl} = (164 \text{ g}) \left(\frac{1 \text{ mole}}{36.5 \text{ g}} \right) = 4.49 \text{ moles of HCl}$$

STEP 3

To find which starting substance is used up, use the following procedure.

a. Write the balanced equation, then place the number of moles of each substance below its formula.

$$\begin{array}{ccccc}
\text{Mg(OH)}_2 & + & 2\text{HCl} & \rightarrow & \text{MgCl}_2 + 2\text{H}_2\text{O} \\
203 \text{ g} & & 164 \text{ g} & & ?\text{g} \quad ?\text{g} \\
3.48 \text{ moles} & & 4.49 \text{ moles} & &
\end{array}$$

b. Determine how much HCl would be needed to use up all the Mg(OH)$_2$.

$$? \text{ moles of HCl needed} = (3.48 \text{ moles Mg(OH)}_2) \left(\frac{2 \text{ moles HCl}}{1 \text{ mole Mg(OH)}_2} \right)$$

$$= 6.96 \text{ moles of HCl needed}$$

But we have only 4.49 moles of HCl. Therefore:

c. Our limiting reactant must be HCl, and the Mg(OH)$_2$ must be the reactant in excess. We can check this by calculating the amount of Mg(OH)$_2$ needed to react with all the HCl.

$$? \text{ moles of Mg(OH)}_2 \text{ needed} = (4.49 \text{ moles HCl}) \frac{1 \text{ mole Mg(OH)}_2}{2 \text{ moles HCl}}$$

$$= 2.25 \text{ moles of Mg(OH)}_2 \text{ needed (or 131g)}$$

We have 3.48 moles of Mg(OH)$_2$, so we can expect to have 3.48 − 2.25 = 1.23 moles of Mg(OH)$_2$ left over, or in excess.

STEP 4

Using the number of moles of HCl that we have, calculate the moles of MgCl$_2$ and H$_2$O that will be formed.

$$? \text{ moles of MgCl}_2 = (4.49 \text{ moles HCl}) \left(\frac{1 \text{ mole of MgCl}_2}{2 \text{ moles HCl}} \right)$$

$$= 2.25 \text{ moles of MgCl}_2$$

$$? \text{ moles of H}_2\text{O} = (4.49 \text{ moles HCl}) \left(\frac{2 \text{ moles H}_2\text{O}}{2 \text{ moles HCl}} \right)$$

$$= 4.49 \text{ moles of H}_2\text{O}$$

218 ESSENTIAL CONCEPTS OF CHEMISTRY

STEP 5

Calculate the number of grams of $MgCl_2$ and H_2O formed.

$$? \text{ grams of } MgCl_2 = (2.25 \text{ moles}) \left(\frac{95.3 \text{ g}}{1 \text{ mole}} \right) = 214 \text{ g of } MgCl_2$$

$$? \text{ grams of } H_2O = (4.49 \text{ moles}) \left(\frac{18.0 \text{ g}}{1 \text{ mole}} \right) = 80.8 \text{ g of } H_2O$$

$$\begin{array}{ccccccc}
Mg(OH)_2 & + & 2HCl & \rightarrow & MgCl_2 & + & 2H_2O \\
131 \text{ g} & & 164 \text{ g} & & 214 \text{ g} & & 80.8 \text{ g} \\
2.25 \text{ moles} & & 4.50 \text{ moles} & & 2.25 \text{ moles} & & 4.50 \text{ moles}
\end{array}$$

STEP 6

Check your mathematics by seeing whether the reaction obeys the Law of Conservation of Mass.

MASS OF REACTANTS	MASS OF PRODUCTS
Grams of $Mg(OH)_2$ = 131 g	Grams of $MgCl_2$ = 214 g
Grams of HCl used = 164 g	Grams of H_2O = 80.8 g
Total mass used = 295 g	Total mass produced = 294.8 g (or 295)

As we expected, we had 72 g of extra $Mg(OH)_2$. This material remained unchanged during the course of the reaction.

EXAMPLE 11.3

Chloroform ($CHCl_3$), a quick-acting powerful anesthetic (often used in spy stories to incapacitate the victim), can decompose when it reacts with oxygen. The products formed are HCl and deadly phosgene gas ($COCl_2$), a substance so harmful when inhaled that it was used as a poison gas against enemy troops in World War I. How many grams of $COCl_2$ and HCl can be formed from 35.9 grams of $CHCl_3$ and 6.40 grams of O_2?

Solution

Follow the procedure we used before.

STEP 1

$$\begin{array}{ccccc}
2CHCl_3 & + O_2 & \rightarrow & 2COCl_2 & + 2HCl \\
35.9 \text{ g} & 6.40 \text{ g} & & ?\text{g} & ?\text{g}
\end{array}$$

STEP 2

$$? \text{ moles of } CHCl_3 = (35.9 \text{ g}) \left(\frac{1 \text{ mole}}{119.5 \text{ g}} \right) = 0.300 \text{ mole of } CHCl_3$$

$$? \text{ moles of } O_2 = (6.40 \text{ g}) \left(\frac{1 \text{ mole}}{32.0 \text{ g}} \right) = 0.200 \text{ mole of } O_2$$

STEP 3

Determine which of the substances is the limiting reactant.

$$2CHCl_3 \quad + \quad O_2 \quad \rightarrow \quad 2COCl_2 + 2HCl$$
$$35.9\,g \quad\quad\quad 6.40\,g \quad\quad\quad\quad ?\,g \quad\quad ?\,g$$
$$0.300 \text{ mole} \quad 0.200 \text{ mole}$$

$$? \text{ moles of } CHCl_3 \text{ needed} = (0.200 \text{ mole } O_2)\left(\frac{2 \text{ moles } CHCl_3}{1 \text{ mole } O_2}\right)$$
$$= 0.400 \text{ mole of } CHCl_3 \text{ needed}$$

We would need 0.400 mole of $CHCl_3$ to use all the oxygen. However, we do not have 0.400 mole of $CHCl_3$. Therefore the $CHCl_3$ must be the limiting reactant, and the O_2 must be the reactant in excess. Let's check this.

$$? \text{ moles of } O_2 \text{ needed} = (0.300 \text{ mole } CHCl_3)\left(\frac{1 \text{ mole } O_2}{2 \text{ moles } CHCl_3}\right)$$
$$= 0.150 \text{ mole of } O_2 \text{ needed (or } 4.80\,g)$$

We have more than enough oxygen present to use all the chloroform because we have 0.200 mole of O_2 and we need only 0.150 mole of it.

STEP 4

Using the moles of $CHCl_3$ that are present, calculate the moles of $COCl_2$, and HCl that will be formed.

$$? \text{ moles of } COCl_2 = (0.300 \text{ mole } CHCl_3)\left(\frac{2 \text{ moles } COCl_2}{2 \text{ moles } CHCl_3}\right)$$
$$= 0.300 \text{ moles of } COCl_2$$

$$? \text{ moles of HCl} = (0.300 \text{ mole } CHCl_3)\left(\frac{2 \text{ moles HCl}}{2 \text{ moles } CHCl_3}\right)$$
$$= 0.300 \text{ moles of HCl}$$

STEP 5

$$? \text{ grams of } COCl_2 = (0.300 \text{ mole})\left(\frac{99.0\,g}{1 \text{ mole}}\right) = 29.7\,g \text{ of } COCl_2$$

$$? \text{ grams of HCl} = (0.300 \text{ mole})\left(\frac{36.5\,g}{1 \text{ mole}}\right) = 11.0\,g \text{ of HCl}$$

$$2CHCl_3 \quad + \quad O_2 \quad \rightarrow \quad 2\,COCl_2 \quad + \quad 2HCl$$
$$35.9\,g \quad\quad\quad 4.80\,g \quad\quad\quad 29.7\,g \quad\quad\quad 11.0\,g$$
$$0.300 \text{ mole} \quad 0.150 \text{ mole} \quad 0.300 \text{ mole} \quad 0.300 \text{ mole}$$

STEP 6

Check your mathematics by seeing whether the reaction obeys the Law of Conservation of Mass.

MASS OF REACTANTS		MASS OF PRODUCTS	
Grams of $CHCl_3$ used =	35.9 g	Grams of $COCl_2$ =	29.7 g
Grams of O_2 used =	4.8 g	Grams of HCl =	11.0 g
Total mass used =	40.7 g	Total mass produced =	40.7 g

Note that in this reaction we had an excess of 0.050 mole (1.6 g) of O_2. This excess oxygen remained unchanged during the course of the reaction.

Practice Exercise 11.3

Potassium acetate, $KC_2H_3O_2$, is a salt with a molecular mass of 98.1. It can be produced by the reaction of potassium hydroxide with acetic acid. Potassium acetate has been used in veterinary medicine, in combating cardiac arrhythmia (irregular heart beat), and as an expectorant. It acts as a diuretic in animals. How many grams of potassium acetate can be produced from 28.0 g of potassium hydroxide and 120.0 g of acetic acid?

SUMMARY

Stoichiometry is the calculation of the quantities of elements or compounds involved in chemical reactions. Knowledge of stoichiometry enables chemists to determine the quantities of reactants needed and the quantities of products that are formed. It also permits chemists to produce industrial and commercial chemicals with little or no waste.

The coefficients in a balanced equation indicate the ratio in which molecules (and thus moles) of substances take part in a chemical reaction. The mole method for computing the quantities of reactants and products makes use of this fact. First we find the number of moles of each substance involved in the reaction from this ratio. Then, from the number of moles, we find the number of grams of each substance. The same method may be used to determine which of several reactants is the limiting reactant. It may also be used to determine the quantity of product that can be formed with the available amount of the limiting reactant.

KEY TERMS

Haber process (**Introduction**) synthesize (**Introduction**)
stoichiometry (**11.1**)

SELF-TEST EXERCISES

LEARNING GOAL 1

Simple Stoichiometry

◀ *1.* Sodium thiosulfate, $Na_2S_2O_3$, known as "hypo" in photography, is used to remove excess silver bromide, AgBr, in the film-developing process. The reaction is

$$Na_2S_2O_3 + AgBr \rightarrow Na_3Ag(S_2O_3)_2 + NaBr \text{ (unbalanced)}$$

(a) Balance the equation.
(b) How many grams of sodium thiosulfate are needed to react with 46.95 g of silver bromide?
(c) How many grams of $Na_3Ag(S_2O_3)_2$ are produced? How many grams of NaBr are produced?

2. A spectacular laboratory demonstration is the synthesis of sodium chloride (table salt) from its elements. In this reaction, pure sodium metal is heated in an enclosed glass container in the presence of green-colored chlorine gas. A violent reaction begins producing heat, light, and a white cloud of sodium chloride. The reaction can be represented by the following chemical equation:

$$Na + Cl_2 \rightarrow NaCl \text{ (unbalanced)}$$

(a) Balance the equation.
(b) How many grams of NaCl can be produced from 4.60 g of sodium metal?
(c) How many grams of chlorine gas are needed to react with the sodium metal?

◀ 3. Nitrous oxide (N_2O), commonly called laughing gas, can be prepared from the decomposition of ammonium nitrate:

$$NH_4NO_3 \xrightarrow{Heat} N_2O + H_2O$$

(a) Balance the equation.
(b) How many grams of ammonium nitrate are needed to prepare 2.2 g of nitrous oxide?
(c) How many grams of water are produced?

4. Lithium phosphate is a white crystalline powder that is soluble in dilute acids. It can be prepared by the reaction

$$H_3PO_4 + LiOH \rightarrow Li_3PO_4 + H_2O$$

(a) Balance the equation.
(b) How many grams of lithium phosphate can be produced from 168.0 g of lithium hydroxide?
(c) How many grams of phosphoric acid are needed, and how many grams of water are produced?

5. We can prepare the chemical explosive trinitrotoluene (TNT) by the reaction of toluene (C_7H_8) with nitric acid:

$$C_7H_8 + HNO_3 \rightarrow \underset{TNT}{C_7H_5N_3O_6} + H_2O$$

(a) Balance the equation.
(b) How many grams of toluene and nitric acid are necessary to manufacture $\overline{1,000}$ g of TNT?
(c) How many grams of water will be produced in the reaction?

◀ 6. Calcium hydroxide, also known as slaked lime, is used in mortar, plaster, cement, and other building and paving materials. When it reacts with phosphoric acid, it produces calcium phosphate and water.

(a) Write a balanced chemical equation for this reaction.
(b) How many grams of calcium phosphate can be produced from 148 g of calcium hydroxide?
(c) How many grams of phosphoric acid are needed, and how many grams of water are produced?

◀ 7. The compound ammonium sulfate can be used as a local analgesic (pain reliever). This compound can be prepared by a reaction of ammonium chloride and sulfuric acid.

(a) Write a balanced equation for this reaction.
(b) How many grams of sulfuric acid are needed to react with 15.9 g of ammonium chloride?
(c) How many grams of ammonium sulfate and hydrogen chloride are produced?

◀ 8. Barium chloride is used in manufacturing pigments and paints, in weighting and dyeing textile fabrics, and in tanning and finishing leather. When it reacts with potassium phosphate, the products formed are potassium chloride and barium phosphate.

(a) Write a balanced equation for this reaction.
(b) How many grams of potassium chloride can be produced from 0.208 g of barium chloride?
(c) How many grams of potassium phosphate are needed, and how many grams of barium phosphate are produced?

*9. Suppose you are a chemist who has been asked to analyze a sample containing potassium chlorate ($KClO_3$) and sodium chloride. You are asked to determine the percentage of potassium chlorate in the sample. You can do this by heating a weighed amount of the sample. The potassium chlorate will decompose to potassium chloride and oxygen gas. You can reweigh the sample after heating, to tell how much oxygen was lost. Then you can backtrack to find the number of grams of potassium chlorate that decomposed to yield this mass of oxygen. From this information, you can calculate the percentage of potassium chlorate in the sample.

Assume that the masses of your sample are as follows, and find the percentage of potassium chlorate in the mixture of potassium chlorate and sodium chloride:

$$\text{Grams of sample before heating} = 5.00 \text{ g}$$

$$\text{Grams of sample after heating} = 4.00 \text{ g}$$

10. A sample containing $NaClO_3$ and KCl must be analyzed to determine the percentage of $NaClO_3$. The sample is weighed and then heated, and the $NaClO_3$ decomposes to NaCl and O_2. The sample is weighed after heating to determine the amount of O_2 lost during decomposition. The number of grams of $NaClO_3$ that decomposed to yield the mass of oxygen is calculated. The percentage of $NaClO_3$ in the sample is then calculated. Perform this analysis using the following information:

$$\text{Grams of sample before heating} = 10.00 \text{ g}$$

$$\text{Grams of sample after heating} = 8.00 \text{ g}$$

11. Hydrogen peroxide (H_2O_2) is a compound with many uses. In dilute solutions (3% H_2O_2 in water), it can be used as an antiseptic for cuts. In a solution of 90% concentration, it can be used as a fuel for rockets. H_2O_2 is prepared in the following way:

$$\underset{\substack{\text{Barium} \\ \text{peroxide}}}{BaO_2} + H_3PO_4 \rightarrow H_2O_2 + Ba_3(PO_4)_2 \text{ (unbalanced)}$$

(a) Balance the equation.
(b) How many grams of hydrogen peroxide can be made from 338 g of barium peroxide?
(c) How much phosphoric acid is needed to react with 338 g of barium peroxide?

12. Lead(II) iodide is a chemical used in bronzing, printing, and photography, as well as in making gold pencils. It is composed of lead and iodine.

(a) Write a balanced equation for the formation of lead(II) iodide from its elements.
(b) How many grams of lead and iodine are needed to produce 1.38 g of lead(II) iodide?

*13. Suppose you have a solution containing 588 g of sulfuric acid, and you put $20\overline{0}$ g of magnesium into it. A reaction occurs. After the reaction is complete, 56 g of the original magnesium remain. How many grams of hydrogen were produced?

*14. Suppose you have a solution containing 216.8 g of HCl, and you put 100.0 g of magnesium into it. A reaction occurs. After the reaction is complete, 28.00 g of the original magnesium remain. How many grams of hydrogen were produced?

*15. How many grams of *air* are necessary for a complete combustion of 90.0 g of ethane (C_2H_6), producing carbon dioxide and water vapor? Assume that the air contains 23.0% oxygen by weight.

*16. How many grams of *air* are necessary for the complete combustion of 64.0 g of methane (CH_4), producing carbon dioxide and water vapor? Assume that the air contains 23.0% oxygen by weight.

◀ 17. Butane, C_4H_{10}, is an organic compound used as a fuel in some types of cigarette lighters. When butane is burned with oxygen, the following reaction takes place.

$$C_4H_{10} + O_2 \rightarrow CO_2 + H_2O \text{ (unbalanced)}$$

(a) Balance the equation.
(b) How many grams of oxygen are needed to burn 23.2 g of butane?
(c) How many grams of carbon dioxide and water are produced?

◀ 18. Ethane, C_2H_6, is an organic compound that together with 90% propane and 5% butane composes bottled gas. When ethane is burned with oxygen, the following reaction takes place:

$$C_2H_6 + O_2 \rightarrow CO_2 + H_2O \text{ (unbalanced)}$$

(a) Balance the equation.
(b) How many grams of oxygen are needed to burn $18\overline{0}$ g of ethane?
(c) How many grams of carbon dioxide and water are produced?

19. Ozone, O_3, is a pale-blue gas that can be formed in the atmosphere from O_2. The reaction is

$$3O_2 \rightarrow 2O_3$$

How many grams of ozone can be formed from 64.0 g of O_2?

◀ 20. Ozone (O_3) is formed from oxygen gas in the reaction

$$3O_2 \rightarrow 2O_3$$

How many grams of oxygen gas are needed to form 0.048 g of ozone?

21. Aluminum sulfate has been used to tan leather, to treat sewage, and as an antiperspirant. The compound can be formed by the acid-base reaction

$$Al(OH)_3 + H_2SO_4 \rightarrow Al_2(SO_4)_3 + H_2O \text{ (unbalanced)}$$

(a) Balance the equation.
(b) How many grams of aluminum hydroxide and sulfuric acid are needed to produce $5\overline{00}$ g of aluminum sulfate?
(c) How many grams of water are produced?

22. Zinc sulfide is used as a pigment for paints, oilcloths, linoleum, and leather. It is composed of the elements zinc and sulfur.

(a) Write a balanced equation for the formation of zinc sulfide from its elements.
(b) How many grams of zinc and sulfur are needed to produce 2.44 g of zinc sulfide?

23. In a blast furnace, iron(III) oxide is converted to iron metal. The overall reaction for this process is

$$Fe_2O_3 + CO \rightarrow Fe + CO_2 \text{ (unbalanced)}$$

(a) Balance the equation.
(b) How many grams of iron(III) oxide are needed to produce 454.0 g of iron metal (about one pound)?
(c) How many grams of CO are needed, and how many grams of CO_2 are produced?

24. Silver nitrate, a topical anti-infective agent, reacts with copper to form copper(II) nitrate and silver metal.

(a) Write a balanced equation for this reaction.
(b) How many grams of copper are needed to react with 1.70 g of silver nitrate?
(c) How many grams of silver metal and copper(II) nitrate are formed from the reaction?

224 ESSENTIAL CONCEPTS OF CHEMISTRY

25. Nitric acid is prepared commercially by the *Ostwald process*. The first part of this process involves the reaction of ammonia with oxygen to produce nitric oxide, using platinum as a catalyst:

$$NH_3 + O_2 \xrightarrow{Pt} NO + H_2O \text{ (unbalanced)}$$

(a) Balance the equation.
(b) How many grams of oxygen are needed to react with 425 g of ammonia?
(c) How many grams of nitric oxide and water are produced?

26. Glucose, $C_6H_{12}O_6$, reacts with oxygen to produce carbon dioxide and water.

(a) Write a balanced equation for this reaction.
(b) How many grams of glucose are needed to react with 3.20 g of oxygen?
(c) How many grams of carbon dioxide and water are produced?

*27. Suppose that you have a solution of $50\overline{0}$ g of dissolved $Cu(NO_3)_2$ and you add $30\overline{0}$ g of Zn metal to it. A reaction occurs. After the reaction is complete, 127 g of Zn remain. How many grams of $Zn(NO_3)_2$ are produced?

28. Hydrogen gas reacts with iodine to produce hydrogen iodide. Hydrogen iodide is a strong irritant that is used in the manufacture of hydroiodic acid.

(a) Write a balanced equation for the reaction between hydrogen and iodine.
(b) How many grams of hydrogen gas and iodine are needed to produce $32\overline{0}$ g of hydrogen iodide?

LEARNING GOAL 2

Limiting-Reactant Problems

◀ 29. The compound acetylene (C_2H_2) is used as a fuel for welding (for example, in the oxyacetylene torch). In years gone by, acetylene was used as a surgical anesthetic. The compound results from the reaction

$$\underset{\text{Calcium carbide}}{CaC_2} + H_2O \rightarrow Ca(OH)_2 + \underset{\text{Acetylene}}{C_2H_2} \text{ (unbalanced)}$$

(a) Balance this equation.
(b) If you have 128 g of calcium carbide and 144 g of water, how many grams of calcium hydroxide and acetylene can you produce?

◀ 30. In the Ostwald process, which is used to produce nitric acid, ammonia gas (NH_3) is initially converted to nitric oxide (NO) according to the following reaction:

$$NH_3 + O_2 \rightarrow NO + H_2O \text{ (unbalanced)}$$

(a) Balance the equation.
(b) How many grams of NO and how many grams of H_2O are produced from the reaction of 102.0 g of NH_3 and 320.0 g of O_2?

31. Pure iron is obtained by extracting it from its most abundant ore, hematite, Fe_2O_3. The process is carried out in a device known as a blast furnace. The chemical reaction can be represented by the following equation:

$$Fe_2O_3 + CO \rightarrow Fe + CO_2$$

(a) Balance the equation.
(b) How many grams of iron and how many grams of carbon dioxide can be obtained from the reaction of 399.8 g of Fe_2O_3 and 168.0 g of CO?

32. Nitrogen gas reacts with hydrogen gas to form ammonia gas (NH_3). How many grams of ammonia gas can be formed from 0.014 g of N_2 and 0.020 g of H_2?

◀ 33. Water can be produced from its elements by the reaction

$$H_2 + O_2 \xrightarrow{\text{Electric spark}} H_2O \text{ (unbalanced)}$$

(a) Balance the equation.
(b) How many grams of water can be produced from $1\overline{0}$ g of H_2 and 64 g of O_2?

34. Propane, C_3H_8, reacts with oxygen gas to form carbon dioxide and water according to the equation

$$C_3H_8 + 5O_2 \rightarrow 3CO_2 + 4H_2O$$

How many grams of carbon dioxide and water can be formed from 11.0 g of propane and 32 g of oxygen gas?

35. When calcium phosphide is placed in water, it forms phosphine gas (PH_3). In air, PH_3 usually bursts into flames. In fact, this substance (PH_3) may be responsible for people observing faint flickers of light in marshes. The PH_3 could be produced by the reduction of naturally occuring phosphorus compounds. The reaction for the formation of PH_3 from calcium phosphide is

$$Ca_3P_2 + H_2O \rightarrow PH_3 + Ca(OH)_2 \text{ (unbalanced)}$$

(a) Balance the equation.
(b) How many grams of PH_3 can be produced from 515 g of calcium phosphide and 216 g of water?
(c) How many grams of calcium hydroxide are produced?

◀ 36. Aluminum reacts with iron(III) oxide to produce iron and aluminum oxide.

(a) Write a balanced equation for this reaction.
(b) How many grams of aluminum oxide and iron can be produced from 135 g of Al and $8\overline{00}$ g of iron(III) oxide?

37. Elementary phosphorus can be produced by the reaction

$$Ca_3(PO_4)_2 + 3SiO_2 + 5C \rightarrow 3CaSiO_3 + 5CO + P_2$$

Determine the number of moles of each product formed from 8.0 moles of $Ca_3(PO_4)_2$, $2\overline{0}$ moles of SiO_2, and 45 moles of C.

38. Barium metal reacts with oxygen gas to form barium oxide, a chemical used for drying gases and solvents.

(a) Write a balanced equation for this reaction.
(b) How many grams of barium oxide can be produced from $1\overline{00}$ g of barium and $1\overline{00}$ g of oxygen gas?

39. An inexpensive way of preparing pure hydrogen gas involves the following reaction between iron and steam:

$$Fe + H_2O \rightarrow Fe_3O_4 + H_2 \text{ (unbalanced)}$$

(a) Balance the equation.
(b) How many grams of hydrogen gas can be prepared from 225 g of Fe and 225 g of H_2O?
(c) How many grams of Fe_3O_4 are produced?

40. Sulfur dioxide is a chemical used in preserving fruits and vegetables and as a disinfectant in breweries and food factories. It is produced from the reaction of sulfur and oxygen gas.

    (a) Write a balanced equation for this reaction.
    (b) How many grams of sulfur dioxide can be prepared from 100.0 g of sulfur and 200.0 g of oxygen gas?

EXTRA EXERCISES

◀ 41. (a) Write the balanced equation for the reaction in which chlorine, Cl_2, reacts with potassium metal, K, to produce potassium chloride. How many moles of chlorine are needed to react with 10.0 g of potassium?
    (b) How many grams of potassium chloride are produced?

◀ 42. Consider the reaction

$$Cu(NO_3)_2 \rightarrow CuO + NO_2 + O_2 \text{ (unbalanced)}$$

    (a) Balance the equation.
    (b) What mass of O_2 is produced if $10\overline{0}$ g of copper(II) nitrate react?
    (c) How much copper(II) nitrate must react if 54.0 g of NO_2 are formed?

43. Assume that $10\overline{0}$ g of carbon and $15\overline{0}$ g of oxygen are available for the reaction

$$C + O_2 \rightarrow CO_2$$

    (a) What is the limiting reactant?
    (b) How much CO_2 is formed?

44. A reaction is carried out with 8.70 g of hydrogen and 64.0 g of oxygen to produce water.

    (a) Write a balanced equation for this reaction.
    (b) What amount of each substance will be on hand when the reaction is complete?

45. Calculate the number of grams of oxygen gas produced by the decomposition of 1.00 kg of potassium chlorate. [*Hint*: $KClO_3 \rightarrow KCl + O_2$ (unbalanced)]

46. Write the equation for the reaction between zinc metal and sulfuric acid. What mass of zinc is required to produce $60\overline{0}$ g of hydrogen gas?

47. Two experiments are performed. In the first, a 50.0-g plate of silver is immersed in a solution of copper(II) chloride. In the second experiment, a 50.0-g plate of lead is immersed in the same solution.

    (a) Determine in which experiment a reaction occurs, and write a balanced equation for this reaction.
    (b) Assuming that the metal is the limiting reactant, calculate the masses of all products formed in the reaction that occurs.

48. Aluminum reacts with sulfuric acid to produce aluminum sulfate and hydrogen gas. How many grams of hydrogen are produced from the reaction of 60.0 g of aluminum with excess sulfuric acid?

49. Hydrochloric acid reacts with sodium hydroxide to produce sodium chloride and water. Suppose that 36.5 g of HCl react with 80.0 g of NaOH. How many grams of sodium chloride are produced? How many grams of the excess reactant are left at the end of the reaction?

50. Carbon reacts with oxygen under certain conditions to produce carbon monoxide. Suppose that 48.0 g of carbon react with 64.0 g of oxygen gas to produce *only* carbon monoxide. How many grams of carbon monoxide are produced?

CUMULATIVE REVIEW Chapters 9–11

1. How many moles are there in 3.43 g of $Al_2(SO_4)_3$?
2. How many moles are there in 4.8 g of C atoms?

3. Determine the number of grams in each of the following:

   (a) 0.750 mole of H_2SO_4
   (b) 5.00 moles of H_2O
   (c) 0.00700 mole of H_2CO_3
   (d) 50.0 moles of $HC_2H_3O_2$

4. After completing an experiment, you find 4.6 g of unreacted Na in the vessel. How many atoms does this represent?

5. After a reaction is complete, it is determined that 5.328 g of Sn combined with 1.436 g of O. Calculate the empirical formula of this compound.

6. The empirical formula for the antibiotic nonactin is $C_{10}H_{16}O_3$. If the molecular mass of nonactin is 736, what is its molecular formula?

7. The empirical formula for fructose, the natural sugar found in fruit juice, fruits, and honey is CH_2O. If the molecular mass of fructose is $18\overline{0}$, what is its molecular formula?

8. The molecular formula for aspirin is $C_9H_8O_4$. You have 36.0 g of aspirin.

   (a) How many moles of aspirin are there?
   (b) How many molecules of aspirin are there?
   (c) How many moles of carbon atoms are there?
   (d) How many moles of hydrogen atoms are there?
   (e) How many moles of oxygen atoms are there?

9. A 6.08-g sample of nitrogen combines with 13.90 g of oxygen to produce a compound whose molecular mass is 92.0. What is the molecular formula of this compound?

10. Calculate the molecular mass of each of the following compounds:

    (a) $Ca_3(PO_4)_2$ (b) CH_4 (c) NO_2 (d) $Ba(C_2H_3O_2)_2$

11. Calculate the percentage composition of each of the following compounds:

    (a) H_2S (b) $NaCl$ (c) $Fe(C_2H_3O_2)_3$ (d) K_2SO_4

12. Determine the molecular formula of a compound that has a molecular mass of 86.0 and is composed of 83.7% carbon and 16.3% hydrogen by mass.

13. Find the percentage by mass of carbon, hydrogen, and nitrogen in the compound nicotine, given that its empirical formula is C_5H_7N and that its molecular mass is 162.0.

14. How many water molecules are there in a 0.720-g sample of H_2O?

15. A patient has been told to take a $1,\overline{000}$ mg supplement of elemental calcium each day. How many grams of $CaCO_3$ must she take in order to receive $1,\overline{000}$ mg of elemental calcium?

16. Write a balanced chemical equation for the reaction in which calcium bromide + sulfuric acid forms hydrogen bromide + calcium sulfate.

17. When magnesium metal reacts with oxygen, magnesium oxide is formed. Write a balanced equation for this reaction.

18. A flask found in a chemistry laboratory was coated with a thin film that had a cloudy appearance. Chemical analysis of the film showed it to be ammonium chloride formed by reaction of ammonia gas and hydrogen chloride gas. Write the balanced equation for this reaction.

19. Identify each of the following reactions as a combination, decomposition, single-replacement, or double-replacement reaction:

    (a) $C_{12}H_{22}O_{11} \rightarrow 12C + 11H_2O$
    (b) $H_2 + Br_2 \rightarrow 2HBr$
    (c) $2KBr + Cl_2 \rightarrow 2KCl + Br_2$
    (d) $2HCl + Ca(OH)_2 \rightarrow CaCl_2 + 2H_2O$

228   ESSENTIAL CONCEPTS OF CHEMISTRY

20. Complete and balance each of the following equations. Use the activity series and solubility table when necessary.

(a) $H_2 + Cl_2 \rightarrow ?$

(b) $NaCl(aq) + Pb(NO_3)_2(aq) \rightarrow ? + ?$

(c) $Al + H_2SO_4 \rightarrow ? + ?$

(d) $SrCO_3 \xrightarrow{Heat} ? + ?$

21. Complete and balance the following equation:

$$CaCO_3 \xrightarrow{Heat} ? + ?$$

22. Balance the following equations:

(a) $TiCl_4 + H_2O \rightarrow TiO_2 + HCl$

(b) $P_4O_{10} + H_2O \rightarrow H_3PO_4$

23. Complete and balance the following equations. If necessary, check the solubility table to be sure that the reaction can take place.

(a) $Fe(OH)_3 + H_3PO_4 \rightarrow ? + ?$

(b) $Pb(OH)_2 + HNO_3 \rightarrow ? + ?$

24. Acid rain is formed in the atmosphere by the reaction of sulfur trioxide (an air pollutant) and water, forming sulfuric acid. Complete and balance the equation describing this process.

25. In an automobile engine, fuel that contains a chemical called octane (C_8H_{18}) reacts with oxygen to form carbon dioxide and water. Write a balanced equation for this reaction.

26. Isopropyl alcohol (C_3H_7OH), or rubbing alcohol, reacts with oxygen to produce carbon dioxide and water. Write a balanced equation for this reaction.

27. Hard water containing $CaSO_4$ can be softened by adding washing soda (Na_2CO_3). Write a balanced chemical equation for this reaction. Which chemical precipitates out, causing the water to soften?

28. When the sugar sucrose ($C_{12}H_{22}O_{11}$) is heated, it decomposes to form carbon and water. Write a balanced chemical equation for this reaction.

29. In the following reaction, the sum of the coefficients is 9. True or false? (*Hint*: Balance the equation before answering.)

$$Al + H_2SO_4 \rightarrow H_2 + Al_2(SO4)_3$$

30. In the following reaction, the sum of the coefficients is 4. True or false? (*Hint*: Balance the equation before answering.)

$$NaNO_3 \rightarrow NaNO_2 + O_2$$

31. In the following reaction, metallic zinc is oxidized and hydrogen ions are reduced. True or false?

$$Zn + H_2SO_4 \rightarrow ZnSO_4 + H_2$$

32. In the following reaction, sodium is oxidized and chlorine is reduced. True or false?

$$Na + Cl_2 \rightarrow NaCl$$

33. In the following reaction, tin is the reducing agent and HNO_3 is the oxidizing agent. True or false?

$$Sn + HNO_3 \rightarrow SnO_2 + NO_2 + H_2O$$

34. The sum of the coefficients for the following oxidation-reduction reaction is 38. True or false?

$$K_2Cr_2O_7 + FeCl_2 + HCl \rightarrow$$
$$CrCl_3 + KCl + FeCl_3 + H_2O \text{ (unbalanced)}$$

35. Balance the following redox reaction:

$$C + H_2SO_4 \rightarrow CO_2 + SO_2 + H_2O$$

36. When copper(II) oxide reacts with hydrogen, copper and water are produced. How many moles of copper oxide are needed to completely react with 5 moles of hydrogen?

37. Consider the following chemical equation:

$$2Na + 2H_2O \rightarrow 2NaOH + H_2$$

How many moles of sodium hydroxide are produced when 0.46 mole of sodium reacts with 0.20 mole of water?

38. Calculate the number of grams of propane (C_3H_8) needed to produce 6.0 moles of carbon dioxide by the following reaction:

$$C_3H_8 + 5O_2 \rightarrow 3CO_2 + 4H_2O$$

39. How many grams of hydrochloric acid are needed to produce $34\overline{0}$ g of hydrogen sulfide by the following reaction?

$$FeS + HCl \rightarrow FeCl_2 + H_2S \text{ (unbalanced)}$$

40. A 30.0-g sample of ethane (C_2H_6) is allowed to react with 16.0 g of oxygen to produce carbon dioxide and water. How many grams of carbon dioxide will this reaction produce?

41. How many grams of phosphoric acid are needed to react with 1,360 g of lithium hydroxide in the following reaction?

$$H_3PO_4 + 3LiOH \rightarrow Li_3PO_4 + 3H_2O$$

42. Calcium hydroxide reacts with phosphoric acid to produce calcium phosphate and water. How many grams of calcium phosphate can be produced from 49.0 g of phosphoric acid? How many grams of calcium hydroxide are needed?

43. When the compound Epsom salts (magnesium sulfate heptahydrate, $MgSO_4 \cdot 7H_2O$) is heated, 1 mole of magnesium sulfate and 7 moles of water are produced. How many moles of magnesium sulfate are produced when 15.0 moles of Epsom salts are heated? How many grams is this?

44. When hydrogen peroxide (H_2O_2) decomposes, water and oxygen are produced. How many grams of hydrogen peroxide are needed to produce 63.0 g of water?

45. When sodium chloride and lead(II) nitrate react, lead(II) chloride solid and sodium nitrate result. When 66.4 g of lead(II) nitrate react with 5.85 g of sodium chloride, how many grams of lead(II) chloride form?

46. Carbon dioxide gas reacts with lithium hydroxide to form lithium carbonate and water. How many grams of lithium hydroxide are needed to produce 148 g of lithium carbonate?

47. Magnesium and oxygen react to form magnesium oxide. How many grams of magnesium are required to react with 48.0 g of oxygen?

48. Barium and water form barium hydroxide and hydrogen gas upon reaction. When 137.3 g of barium are mixed with 72.0 g of water, how much barium hydroxide and how much hydrogen are produced? Which reactant is in excess and by how much?

CHAPTER 12

THE CHEMISTRY OF SOLUTIONS

This is the story of Jim Lind, a 30-year-old accountant who suffered a Kidney-stone attack and decided to ensure that he wouldn't suffer other one in the future. Kidney stones are formed in the kidneys or urinary tract when certain salts precipitate out of the urine. Two types of stones may form: uric acid stones and calcium oxalate stones. A combination of hereditary factors and diet make some people stone formers. A kidney-stone attack can be very painful if the stone is large enough to get trapped in the kidneys or urinary tract and impede the flow of urine.

Some kidney stones pass through the urinary tract on their own. However, if the stones are too large, there are several methods that physicians may use to facilitate their removal. If the stone is very large and the kidney is damaged, surgical removal of the kidney may be necessary. If the stone is lodged in the urinary tract, a method called cystoscopy may be used to remove the stone. This method involves the insertion of a device into the urethra to catch the stone and pull it out. A new technique that employs sound waves to pulverize the stone in the kidney or urinary tract can also be used. Once the stone is pulverized, smaller pieces flow out with the urine.

Jim's stone was removed using cystoscopy, a procedure requiring an overnight hospital stay. The stone was analyzed and found to be composed of calcium oxalate. Jim asked his doctor how he could help prevent future stones from forming. The doctor instructed him to drink several liters of water a day to help keep his urine dilute. The doctor also told Jim that there was no guarantee that he would not suffer another attack.

Unhappy with his doctor's prognosis, Jim decided to do some research of his own. He had some background in chemistry and remembered that certain salts, like calcium oxalate, are alkaline. He wondered if there was a way to keep his urine acidic, so that the alkaline calcium oxalate stones would stay dissolved. Jim spoke to several faculty members in the food-science department of a nearby university, who told him that there were ways to keep his urine acidic by following an appropriate diet. They also told him that there were diets that could lower his calcium intake and the amount of oxalate in his body. Jim shared this information with his doctor and they decided to try this nutritional approach. Ten years have passed without Jim suffering another kidney-stone attack.

LEARNING GOALS

After you have studied this chapter, you should be able to:

1. Define the terms *solution, solute*, and *solvent*.
2. Predict whether substances will form solutions using the idea that "like dissolves like."
3. Define the terms *saturated solution, unsaturated solution*, and *supersaturated solution*.

232 ESSENTIAL CONCEPTS OF CHEMISTRY

4. Calculate concentrations of solutions in percent by mass, percent by volume, and percent by mass-volume.
5. Calculate the molarity, number of moles, or volume of a solution when you are given the other two quantities.
6. Explain the concept of equivalents for an acid or base and calculate the number of equivalents when you are given the number of grams or moles.
7. Calculate the normality, number of equivalents, or volume of a solution when you are given the other two quantities.
8. Calculate concentrations of normal and molar solutions after they have been diluted.
9. Explain what electrolytic and noneletrolytic solutions are and give examples of each.
10. Define the terms *ion* and *ionization*.
11. Explain what is meant by the molality of a solution.
12. Calculate the molality of a solution, given mass of solute and mass of solvent.
13. Define the term *colligative properties* and explain how a solute affects the boiling point and freezing point of a solution.
14. Define the terms *freezing-point depression constant* (K_f) and *boiling point elevation constant* (K_b).
15. Calculate the boiling and freezing points of a solution when you are given the molality of the solution, its K_f, and its K_b.
16. Define the following terms: *diffusion, osmosis, osmotic pressure,* and *osmometry*.

INTRODUCTION

Solutions are all around us. We drink them, we breathe them, we swim in them, we are even composed of them. Every time you drink a cup of tea or a soda, you swallow a solution. Each time you breathe, you inhale a solution—air. When you swim in the ocean, you're swimming in a solution of salt in water. Even your blood is a solution.

There is a lot more to solutions than just two things put together. In this chapter, we look at the types of solutions, some of their properties, and some special units that apply to solutions.

12.1 Like Dissolves Like

In Chapter 3, we defined a *solution* as a homogeneous mixture. A solution can contain two substances, three substances, or more. The most common types of solutions are made by dissolving a solid in a liquid—for example, a salt in water. However, solutions can be made up from combinations of all three states of matter. Table 12.1 shows eight possible types of solutions and gives an example of each.

Because solutions are always composed of at least two substances, we need to be able to identify the role that each substance plays. The **solute** is *the substance that is being dissolved,* whereas the **solvent** is

TABLE 12.1 Examples of Different Types of Solutions

	SOLVENT		
SOLUTE	*Solid*	*Liquid*	*Gas*
Solid	Copper metal dissolved in sliver metal (for example, coins)	Salt dissolved in water	—
Liquid	Mercury in silver (dental fillings)	Ethyl alcohol dissolved in water	Water vapor in air
Gas	Hydrogen dissolved in platinum metal	carbon dioxide dissolved in water (soda water)	Oxygen gas dissolved in nitrogen gas

the substance that is doing the dissolving. For example, in a solution of salt water, salt is the solute and water is the solvent.

When a solution is composed of two substances in the same state (as is a liquid–liquid solution), it is difficult to establish which substance is the solute and which is the solvent. One rule of thumb is that the substance present in the larger amount is the solvent. In a solution of 10 mL of ethyl alcohol and 90 mL of water, the ethyl alcohol is the solute and the water is the solvent.

EXAMPLE 12.1

Determine which is the solute and which is the solvent in each of the following solutions:

(a) sugar and water

(b) hydrogen chloride gas and water

(c) 75.0 mL of ethyl alcohol and 25.5 mL of water

(d) 80 mL of nitrogen and 20 mL of hydrogen

(e) soda water

Solution

(a) Sugar is the solute; water is the solvent.

(b) Hydrogen chloride is the solute; water is the solvent.

(c) Water is the solute; ethyl alcohol is the solvent.

(d) Hydrogen is the solute; nitrogen is the solvent.

(e) Carbon dioxide is the solute; water is the solvent.

Practice Exercise 12.1

Which is the solute and which is the solvent in each of the following solutions?

(a) salt and water

(b) sulfur trioxide gas and water

(c) $3\overline{0}$ mL of isopropyl alcohol and 25 mL of water

(d) 40 mL of hydrogen gas and 50 mL of oxygen gas

(e) carbonated mineral water

Before we go on, we should note that some pairs of liquids do not form solutions when mixed; such liquids are said to be **immiscible**. For example, gasoline and water do not readily form a solution. The same holds true for oil and vinegar, as in salad dressing (Figure 12.1). There are also substances that do mix, but only to a slight degree. (The opposite of immiscible is miscible. Pairs of liquids that form solutions when mixed are said to be *miscible*.)

FIGURE 12.1   Two immiscible liquids.

How can we determine whether two substances will form a solution? The answer is found in the statement "like dissolves like." To understand its significance, recall that a molecule may be either polar or nonpolar. "Like dissolves like" means that polar substances dissolve other polar substances and that nonpolar substances dissolve other nonpolar substances. But a polar and a nonpolar substance generally do not form a solution. Alcohol and water mix because both are composed of polar molecules. Gasoline and water do not mix, because gasoline is composed of nonpolar molecules and water is composed of polar molecules.

EXAMPLE 12.2

State whether the following pairs of substances form solutions. (*Hint:* Refer to Chapter 7.)

(a) methane (CH_4) and water

(b) HBr and water

(c) *trans*-dichloroethene and carbon tetrachloride

Solution

(a) Methane is nonpolar; water is polar. They do not form a solution.

(b) Both HBr and water are polar. They do form a solution.

(c) *trans*-dichloroetriene is nonpolar, and so is carbon tetrachloride. They do form a solution.

Practice Exercise 12.2

State whether the following pairs of substances form solutions. (*Hint:* You may want to refer to Chapter 7.)

(a) oil and water

(b) hydrogen chloride gas and water

(c) benzene and carbon tetrachloride

12.2 Saturated, Unsaturated, and Supersaturated Solutions

Suppose we take a glass of water and decide to add salt. We begin with a teaspoonful of salt. As the salt and water mix, the salt begins to dissociate. Here's how we think the process happens. We know that water molecules are polar, and the negative end of the water molecule (the oxygen portion) can align itself toward the positive sodium ions. The positive end of the water molecule (the hydrogen portion) can align itself toward the negative chloride ions. Thus, the polar water molecules surround ions on the surface of the sodium chloride crystal, forcing them to break away from the actual surface of the crystal. The first ions to break away become hydrated sodium ions and hydrated chloride ions, meaning that each ion is surrounded by a group of water molecules. A newly exposed surface is formed and is ready to be acted on by other water molecules. The process continues until all of the sodium chloride crystals become hydrated sodium ions and hydrated chloride ions. The salt is then said to have dissolved in the water. We can stir the salt water to increase the rate at which the salt dissolves. Stirring helps distribute the hydrated ions uniformly throughout the solution.

We can add another teaspoonful of salt, stir, and watch the salt crystal disappear into solution. As the number of hydrated ions in solution increase the chance of ions colliding with each other increases. Besides colliding with each other, ions can collide with solvent molecules, and they can collide with any remaining undissolved solid. When they collide with the undissolved solute, they leave the solution. If we continue to repeat the process we will eventually reach the point at which, no matter how much we stir, no more salt will dissolve. For any given temperature there is a point at which no more solute can dissolve in a particular quantity of solvent. Once this point is reached, the solution is said to be **saturated**. When a solution is saturated, the rate at which solid dissolves into the solution is the same as the rate at which dissolved solid crystallizes out of the solution. Ions leave the solution to return to the undissolved solute at the same rate at which ions enter the solution.

An **unsaturated** solution contains less solute than can be dissolved at a particular temperature. In an unsaturated solution, more solute can be dissolved into the solvent without changing the temperature. This happens when the number of ions in the solution is low, making the chance of collision between hydrated ions and the undissolved solute low. The result is that the undissolved solute won't have much opportunity to recapture ions, so they stay in solution.

Although it appears as if nothing is happening once a solution is saturated, we know that a dynamic process is occurring at the molecular level. The undissolved solute is in equilibrium with the dissolved solute. This dynamic equilibrium can be represented as follows:

$$\text{Undissolved solute} \rightleftharpoons \text{dissolved solute}$$

The double arrow means the reaction is reversible.

Table 12.2 gives the solubility of several substances in water. **Solubility** is the amount of a solute that dissolves in a particular amount of solvent to give a saturated solution. Note that solubility varies according to the temperature, that is, a particular solution that is saturated at one temperature is not necessarily saturated at another temperature. In general, an increase in temperature causes more solute to be dissolved.

It is possible to prepare solutions that contain a greater concentration of solute than is needed for saturation. Suppose we start with a teaspoonful of sugar and a glass of water at room temperature. We add the sugar to the water and stir. The sugar dissolves in the water. We repeat the process until no more sugar will dissolve. At this point we find sugar resting on the bottom of the glass, and we know that the solution is saturated. We decide to heat he solution. The sugar that was resting on the bottom of the glass now dissolves. We add more sugar, and it also dissolves when stirred. We allow the solution to cool slowly to room temperature. The solid does not precipitate out of the solution, and what we have is a **supersaturated** solution. Such a solution is unstable and will revert to a saturated solution if disturbed. Such action will cause the excess solute to crystallize rapidly from the solution. If we were to tie a string around a tiny sugar crystal and suspend the string in the supersaturated sugar–water solution, we could grow rock candy, which is made of large sugar crystals.

EXAMPLE 12.3

Determine whether each of the following solutions is unsaturated, saturated, or supersaturated (see Table 12.2).

(a) 200.0 g of $AgNO_3$ dissolved in 200.0 g of water at 0°C
(b) 50.0 g of $CuSO_4$ dissolved in 150.0 g of water at 50°C
(c) 220.0 g sucrose ($C_{12}H_{22}O_{11}$) dissolved in 75.00 g of water at 50°C

Solution

UNDERSTAND THE PROBLEM

In each case, we are given an amount of substance dissolved in a given amount of water at a certain temperature. We are asked whether the resulting solution is saturated (that is, there is just enough solute to achieve dynamic equilibrium), unsaturated (more of the solute could dissolve in the solution), or supersaturated (the solution contains more dissolved substance than a saturated solution).

TABLE 12.2 Solubility of Various Substances in Water at 0°C and 50°C

	SOLUBILITY (g SOLUTE/ 100 g SOLVENT)	
SOLUTE	*0°C*	*50°C*
$CuSO_4$	14.3	33.3
NaCl	35.7	37.0
$AgNO_3$	122.0	455.0
CsCl	161.4	218.5
$C_{12}H_{22}O_{11}$	203.9 (20°C)	260.4

236 ESSENTIAL CONCEPTS OF CHEMISTRY

DEVISE A PLAN

For each example use the factor-unit method and calculate the grams, of solute per 100 g of water. Compare this value to the value in Table 12.2.

CARRY OUT THE PLAN

(a) $?\text{g AgNO}_3 = (100.0 \text{ g } \cancel{H_2O}) \left(\dfrac{200.0 \text{ g AgNO}_3}{200.0 \text{ g } \cancel{H_2O}} \right)$

$= 100.0 \text{ g AgNO}_3 \text{ (at } 0°\text{C)}$

This solution is *unsaturated;* Table 12.2 states that 122.0 g $AgNO_3$ can be dissolved in 100.0 g H_2O at 0°C.

(b) $?\text{g CuSO}_4 = (100.0 \text{ g } \cancel{H_2O}) \left(\dfrac{50.0 \text{ g CuSO}_4}{150.0 \text{ g } \cancel{H_2O}} \right)$

$= 33.3 \text{ g CuSO}_4 \text{ (at } 50°\text{C)}$

This solution is *saturated;* Table 12.2 states that 33.3 g of $CuSO_4$ can be dissolved in 100.0 g H_2O at 50°C.

(c) $?\text{g sucrose} = (100.0 \text{ g } \cancel{H_2O}) \left(\dfrac{220.0 \text{ g sucrose}}{75.00 \text{ g } \cancel{H_2O}} \right)$

$= 293.3 \text{ g sucrose (at } 50°\text{C)}$

This solution is *supersaturated;* Table 12.2 states that 260.4 g of sucrose can be dissolved in 100.0 g H_2O at 50°C.

LOOK BACK

Recheck your calculations and see if your answers are reasonable.

Practice Exercise 12.3

Determine whether each of the following solutions is unsaturated, saturated, or supersaturated (see Table 12.2).

(a) 60.0 g of CsCl dissolved in 25.0 g of water at 50°C
(b) 75.0 g of CsCl dissolved in 50.0 g of water at 0°C
(c) 185.0 g of NaCl dissolved in 500.0 g of water at 50°C

12.3 Concentrations of Solutions by Percent

Because solutions are mixtures, their components can be present in different ratios. For example, we can make many different salt-and-water solutions, each with a different **concentration**, or ratio of solute to solvent. The concentrations or relative quantities of solutions can vary widely, so we must have a way of describing them. (It is important to remember, however, that at any given temperature there is a limit to the amount of solute that can be dissolved in a solution. This concentration, called the **saturation point**, is the point at which no more solute dissolves to form a stable solution.)

One method of defining the concentrations of solutions is based on the percent of solute in the solution. This method can cause confusion, because there can be three types of percent concentrations:

1. Percent by mass
2. Percent by volume
3. Percent by mass–volume

A 30%-by-mass solution of alcohol and water is not the same as a 30%-by-volume solution of alcohol and water. Therefore a label that reads "30% alcohol in water" tells us nothing. The solution could be a 30%-by-mass solution, a 30%-by-volume solution, or a 30%-by-mass–volume solution. Chemists seldom use the percent method of defining concentrations; however, biologists and medical workers do use percent by mass–volume, so we will discuss all three types.

Percent-by-Mass Solutions

To find percent by mass, we divide the *mass of the solute* by the *total mass of the solution* and multiply the result by 100. (The mass of the solution equals the mass of the solute plus the mass of the solvent.) For example, suppose we made a solution of alcohol in water by mixing 30 g of alcohol with enough water to make 100 g of solution (Figure 12.2). The concentration of our solution would be 30% by mass.

$$\text{Percent alcohol by mass} = \frac{\text{mass of alcohol}}{\text{total mass of solution}} \times 100$$

$$= \frac{30 \text{ g}}{100 \text{ g}} \times 100 = 30\%$$

EXAMPLE 12.4

Find the concentrations of the following solutions in percent by mass.

(a) $2\overline{0}$ g of salt with enough water to make $6\overline{0}$ g of solution

(b) $5\overline{0}$ g of sugar with enough water to make $35\overline{0}$ g of solution

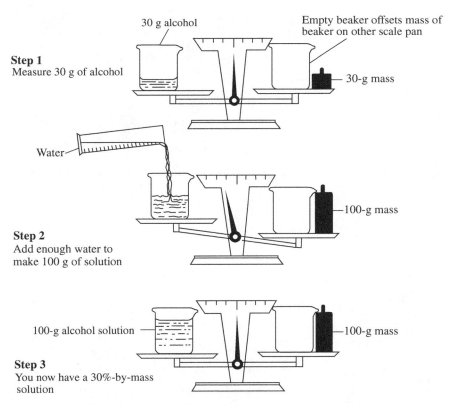

FIGURE 12.2 Preparing a 30%-by-mass alcohol–water solution.

Solution

(a) Percent salt by mass = $\dfrac{\text{mass of salt}}{\text{total mass of solution}} \times 100$

$= \dfrac{2\bar{0}\text{ g}}{6\bar{0}\text{ g}} \times 100 = 33\%$

(b) Percent sugar by mass = $\dfrac{\text{mass of sugar}}{\text{total mass of solution}} \times 100$

$= \dfrac{5\bar{0}\text{ g}}{35\bar{0}\text{ g}} \times 100 = 14\%$

Practice Exercise 12.4

Find the concentrations of the following solutions in percent by mass:

(a) 40.0 g of salt with enough water to make 90.0 g of solution

(b) 75.0 g of sugar with enough water to make 250.0 g of solution

Percent-by-Volume Solutions

To find percent by volume, we divide the *volume of the solute* by the *total volume of the solution* and multiply the result by 100. For example, suppose we mixed $3\bar{0}$ mL of alcohol with enough water to make $1\bar{0}0$ mL of solution (Figure 12.3). The concentration of the solution would be 30% alcohol by volume.

$$\text{Percent alcohol by volume} = \dfrac{\text{volume of alcohol}}{\text{total volume of solution}} \times 100$$

$$= \dfrac{3\bar{0}\text{ mL}}{1\bar{0}0\text{ mL}} \times 100 = 3\bar{0}\%$$

EXAMPLE 12.5

Find the concentrations of the following solutions in percent by volume:

(a) 50.0 mL of alcohol with enough water to make $4\bar{0}0$ mL of solution

(b) $1\bar{0}$ mL of benzene is added to $4\bar{0}$ mL of carbon tetrachloride to make $5\bar{0}$ mL of solution

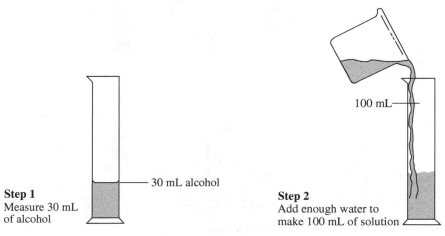

FIGURE 12.3 Preparing a 30%-by-volume alcohol–water solution.

Solution

(a) Percent alcohol by volume $= \dfrac{\text{volume of alcohol}}{\text{total volume of solution}} \times 100$

$= \dfrac{50.0 \text{ mL}}{400 \text{ mL}} \times 100 = 12.5\%$

(b) Percent benzene by volume $= \dfrac{\text{volume of benzene}}{\text{total volume of solution}} \times 100$

$= \dfrac{1\overline{0} \text{ mL}}{(1\overline{0} \text{ mL} + 4\overline{0} \text{ mL})} \times 100$

$= \dfrac{1\overline{0} \text{ mL}}{5\overline{0} \text{ mL}} \times 100 = 2\overline{0}\%$

Practice Exercise 12.5

Find the concentrations of the following solutions in percentage by volume:

(a) 60.0 mL of alcohol with enough water to make 300.0 mL of solution

(b) 40.0 mL of carbon tetrachloride in 50.0 mL of benzene

Percent-by-Mass–Volume Solutions

To find percent by mass–volume, we divide the *mass of the solute* in grams by the *volume of the solution* in milliliters and multiply the result by 100. For example, suppose that we mix $3\overline{0}$ g of alcohol with enough water to make $1\overline{00}$ mL of solution (Figure 12.4). The concentration of the solution would be 30% alcohol by mass-volume.

$$\text{Percent alcohol by mass–volume} = \dfrac{\text{mass of alcohol}}{\text{volume of solution (mL)}} \times 100$$

$$= \dfrac{3\overline{0} \text{ g}}{1\overline{00} \text{ mL}} \times 100 = 3\overline{0}\%$$

EXAMPLE 12.6

Find the concentrations of the following solutions in percent by mass–volume:

(a) 2.0 g of iodine in enough carbon tetrachloride to make $8\overline{0}$ mL of solution

(b) $13\overline{0}$ g of sugar in enough water to make $65\overline{0}$ mL of solution

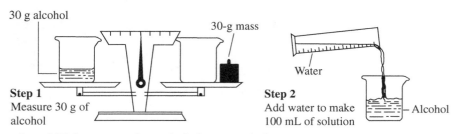

FIGURE 12.4 Preparing a 30%-by-mass–volume alcohol–water solution.

Solution

(a) Percent iodine by mass–volume = $\dfrac{\text{mass of iodine}}{\text{volume of solution}} \times 100$

$= \dfrac{2.0 \text{ g}}{8\overline{0} \text{ mL}} \times 100 = 2.5\%$

(b) Percent sugar by mass–volume = $\dfrac{\text{mass of sugar}}{\text{volume of solution}} \times 100$

$= \dfrac{13\overline{0} \text{ g}}{65\overline{0} \text{ mL}} \times 100 = 20.0\%$

Practice Exercise 12.6

Find the concentrations of the following solutions in percent by mass–volume:

(a) 4.5 g of *trans*-dichloroethene in enough carbon tetrachloride to make 100.0 mL of solution

(b) 150.0 g of sucrose in enough water to make 750.0 mL of solution

12.4 Molarity

Chemists run many reactions in solutions. Because most of these reactions take place between the solute *particles*, we need a concentration unit that expresses the *number of particles of solute* present in a given amount of solution. The concentration unit chemists use is called *molarity*. The **molarity** of a solution is *the number of moles of solute per liter of solution*. Molarity (abbreviated M) is equal to

$$\dfrac{\text{number of moles of solute}}{\text{number of liters of solution}} \quad \text{so that} \quad M = \dfrac{\text{moles}}{\text{liter}} \quad \text{or} \quad M = \dfrac{n}{V}$$

There can be no confusion when the concentrations of solutions are expressed in these terms. A solution made by dissolving 117 g (2.00 moles) of NaCl in enough water to obtain 1 L of solution (Figure 12.5) would be a

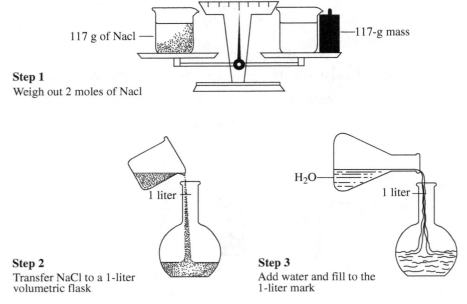

Step 1
Weigh out 2 moles of NaCl

Step 2
Transfer NaCl to a 1-liter volumetric flask

Step 3
Add water and fill to the 1-liter mark

FIGURE 12.5 Preparing 1 L of a 2 M NaCl solution.

2-molar NaCl solution (usually written as 2 M NaCl). (For a review of moles, see Chapter 9.) To calculate the molarity of this solution, we would write

$$M = \frac{\text{moles}}{\text{liters}} \qquad ?M = \frac{2 \text{ moles}}{1 \text{ liter}} = 2\,M$$

Chemists often need to prepare solutions that have particular concentrations. *Volumetric flasks* are helpful in preparing such solutions. These flasks are calibrated to contain a specific volume of a solution at a particular temperature (usually 20°C), as shown in Figure 12.5.

EXAMPLE 12.7

Calculate the molarity of each of the following solutions:

(a) 3.65 g of HCl in enough water to make $5\overline{00}$ mL of solution

(b) 160 g of NaOH in enough water to make 6.0 L of solution

(c) 36.8 g of ethyl alcohol (C_2H_6O) in enough water to make $1,6\overline{00}$ mL of solution

Solution

(a) First determine the number of moles of HCl.

$$3.65 \text{ g of HCl} \times \frac{1 \text{ mole of HCl}}{36.5 \text{ g of HCl}} = 0.100 \text{ mole of HCl}$$

Then convert the volume of solutions to liters.

$$5\overline{00} \text{ mL} \times \frac{1 \text{ L}}{1000 \text{ mL}} = 0.500 \text{ L}$$

Then,

$$M = \frac{n}{V}$$

$$= \frac{0.100 \text{ mole}}{0.500 \text{ L}}$$

$$= 0.200\,M$$

(b) We have 6.0 L of solution, and we first calculate that

$$160 \text{ g of NaOH} = 4.0 \text{ moles}$$

Then,

$$M = \frac{\text{moles}}{\text{liter}} = \frac{n}{V} \qquad ?M = \frac{4.0 \text{ moles}}{6.0 \text{ L}} = 0.67\,M$$

(c) We first calculate that

$$36.8 \text{ g of } C_2H_6O = 0.800 \text{ mole}$$
$$1,6\overline{00} \text{ mL of Solution} = 1.600 \text{ L}$$

Then,

$$M = \frac{\text{moles}}{\text{liter}} = \frac{n}{V} \qquad ?\ M = \frac{0.800 \text{ mole}}{1.600 \text{ L}} = 0.500\ M$$

Practice Exercise 12.7

Calculate the molarity of each of the following solutions:

(a) 9.80 g of H_2SO_4 in enough water to make 0.500 L of solution

(b) 4.60 g of KOH in enough water to make 5.00 L of solution

(c) 73.6 g of ethyl alcohol (C_2H_6O) in enough water to make 1.500 L of solution

EXAMPLE 12.8

Suppose that you have a supply of 3.0 M NaOH solution in your laboratory. You take $20\overline{0}$ mL of this solution and evaporate it to dryness. How many grams of *solid* NaOH remain?

Solution

First solve the molarity formula,

$$M = \frac{n}{V}$$

for n, the number of moles:

$$n = MV$$

Then substitute the values.

$$n = \frac{3.0 \text{ mole}}{\text{liter}} \times 0.200 \text{ liter}$$

$$= 0.60 \text{ mole of NaOH}$$

Convert the moles of NaOH to grams.

$$0.60 \text{ mole NaOH} \times \frac{40.0 \text{ g NaOH}}{1 \text{ mole NaOH}} = 24 \text{ g of NaOH}$$

Practice Exercise 12.8

Suppose that you have a supply of $2.00 \times 10^{-3}\ M$ $Ca(OH)_2$ solution in your laboratory. You evaporate 250.0 mL of this solution to dryness. How many grams of *solid* $Ca(OH)_2$ do you obtain?

EXAMPLE 12.9

Suppose that you have a supply of 0.500 M $Ca(NO_3)_2$ solution in your laboratory. How many milliliters of this solution must be evaporated to obtain 32.8 g of solid $Ca(NO_3)_2$?

Solution

Change the 32.8 g of $Ca(NO_3)_2$ to moles and then solve for the number of liters of 0.50 M $Ca(NO_3)_2$ solution needed to obtain this number of moles.

$$32.8 \text{ g of } Ca(NO_3)_2 \times \frac{1 \text{ mole of } Ca(NO_3)_2}{164.1 \text{ g of } Ca(NO_3)_2} = 0.200 \text{ mole of } Ca(NO_3)_2$$

Solve the molarity formula,

$$M = \frac{n}{V}$$

for V, the number of liters:

$$V = \frac{n}{M}$$

Then substitute the values.

$$V = \frac{0.200 \text{ mole}}{0.500 \text{ mole}/L}$$

$$= 0.400 \text{ L} \quad \text{or} \quad 4\overline{00} \text{ mL}$$

You would need to evaporate 0.400 L ($4\overline{00}$ mL) of a 0.500 M Ca(NO$_3$)$_2$ solution to obtain 32.8 g of solid Ca(NO$_3$)$_2$.

Practice Exercise 12.9

Suppose that you have a supply of 0.200 M Mg(NO$_3$)$_2$ solution in your laboratory. How many milliliters of this solution must be evaporated to obtain 1.48 g of solid Mg(NO$_3$)$_2$?

12.5 Normality

Chemists have developed another important concentration concept: *normality*. The definition of normality is similar to that of molarity, but normality is easier to use in certain chemical calculations. It is most often applied to solutions of acids and bases. The **normality** of a solution *is the number of equivalents of solute per liter of solution*. Normality (abbreviated N) is equal to

$$\frac{\text{number of equivalents of solute}}{\text{number of liters of solution}} \quad \text{so that} \quad N = \frac{\text{equivalents}}{\text{liter}}$$

$$\text{or } N = \frac{\text{Eq}}{\text{L}}$$

Of course, the question "What is an equivalent?" immediately comes to mind. Because we'll use normality for calculations involving acids and bases, we will define an equivalent in terms of an acid or a base. An **equivalent of an acid** is *the mass of the acid that contains Avogadro's number of hydrogen ions*. An **equivalent of a base** is *the mass of the base that contains Avogadro's number of hydroxide ions*.

To find the number of grams in 1 equivalent of an acid or a base, first obtain the molecular mass of the substance. Then divide the molecular mass in grams by the number of replaceable hydrogen ions (for an acid) or the number of replaceable hydroxide ions (for a base). This is called the equivalent mass, For example, sulfuric acid (H$_2$SO$_4$) has a molecular mass of 98.1. Because H$_2$SO$_4$ contains two replaceable hydrogen ions, the equivalent mass, or the grams per equivalent, is as follows:

$$\text{Equivalent mass of H}_2\text{SO}_4 = \frac{\text{molecular mass}}{\text{number of replaceable hydrogen ions}}$$

$$= \frac{98.1 \text{ g}}{2 \text{ Eq}} = \frac{49.1 \text{ g}}{\text{Eq}}$$

244 ESSENTIAL CONCEPTS OF CHEMISTRY

TABLE 12.3 The Mass of 1 Mole and of 1 Equivalent for Some Common Acids and Bases

SUBSTANCE	MASS OF 1 MOLE	MASS OF 1 EQUIVALENT	NUMBER OF REPLACEABLE H^+ OR OH^- IONS
HCl (hydrochloric acid)	$\dfrac{36.5\text{ g}}{1\text{ mole}}$	$\dfrac{36.5\text{ g}}{1\text{ equivalent}}$	1
$HC_2H_3O_2$ (acetic acid)	$\dfrac{60.0\text{ g}}{1\text{ mole}}$	$\dfrac{60.0\text{ g}}{1\text{ equivalent}}$	1
H_2SO_4 (sulfuric acid)	$\dfrac{98.1\text{ g}}{1\text{ mole}}$	$\dfrac{49.1\text{ g}}{1\text{ equivalent}}$	2
H_3PO_4 (phosphoric acid)	$\dfrac{98.0\text{ g}}{1\text{ mole}}$	$\dfrac{32.7\text{ g}}{1\text{ equivalent}}$	3
NaOH (sodium hydroxide)	$\dfrac{40.0\text{ g}}{1\text{ mole}}$	$\dfrac{40.0\text{ g}}{1\text{ equivalent}}$	1
$Ca(OH)_2$ (calcium hydroxide)	$\dfrac{74.1\text{ g}}{1\text{ mole}}$	$\dfrac{37.1\text{ g}}{1\text{ equivalent}}$	2

This means that a 1 N solution of H_2SO_4 contains 49.1 g of H_2SO_4 per liter of solution. Table 12.3 lists the masses of 1 mole and 1 equivalent for some common acids and bases.

EXAMPLE 12.10

What is the normality of the solution formed by dissolving 196 g of H_2SO_4 in enough water to make $8\overline{00}$ mL of solution? What is the molarity of the solution?

Solution

First convert the grams of H_2SO_4 to equivalents and then use the normality formula. H_2SO_4 has

$$\frac{49.1\text{ g}}{1\text{ equivalent}} \quad \text{or} \quad \frac{1\text{ equivalent}}{49.1\text{ g}}$$

Therefore

$$?\text{ equivalent of }H_2SO_4 = (196\text{ g})\left(\frac{1\text{ equivalent}}{49.1\text{ g}}\right)$$

$$= 3.99\text{ equivalents of }H_2SO_4$$

$$N = \frac{Eq}{L} = \frac{\text{equivalents}}{\text{liter}} \qquad ?\,N = \frac{3.99\text{ equivalents}}{0.800\text{ L}} = 4.99\,N$$

To find the molarity of this solution, we first convert grams of H_2SO_4 to moles and then find the number of moles per liter. (The molecular mass of H_2SO_4 is 98.1.)

$$? \text{ moles of } H_2SO_4 = (196 \text{ g})\left(\frac{1 \text{ mole}}{98.1 \text{ g}}\right) = 2.00 \text{ moles of } H_2SO_4$$

$$M = \frac{\text{moles}}{\text{liter}} \qquad ? M = \frac{2.00 \text{ moles}}{0.800 \text{ L}} = 2.50 \, M \text{ solution of } H_2SO_4$$

Practice Exercise 12.10

What is the normality of the solution formed by dissolving 60.0 g of $HC_2H_3O_2$ in enough water to make 500.0 mL of solution? What is the molarity of the solution?

EXAMPLE 12.11

What are the normality and the molarity of the solution formed by dissolving 19.6 g of H_3PO_4 in enough water to make $\overline{300}$ mL of solution?

Solution

Change the 19.6 g of H_3PO_4 to equivalents and then use the normality formula. H_3PO_4 has

$$\text{Equivalent mass of } H_3PO_4 = \frac{\text{moecular mass}}{\text{number of replaceable } H^+ \text{ions}}$$

$$= \frac{98.0 \text{g}}{3 \text{Eq}}$$

$$= \frac{32.7 \text{g}}{\text{Eq}}$$

$$\frac{32.7 \text{g}}{1 \text{ equivalent}} \quad \text{or} \quad \frac{1 \text{ equivalent}}{32.7 \text{g}}$$

Therefore

$$? \text{ equivalent of } H_3PO_4 = (19.6 \text{ g})\left(\frac{1 \text{ equivalent}}{32.7 \text{ g}}\right)$$

$$= 0.599 \text{ equivalent of } H_3PO_4$$

$$N = \frac{\text{equivalents}}{\text{liter}} \qquad ? N = \frac{0.599 \text{ equivalent}}{0.300 \text{ L}} = 2.00 \, N$$

To find the molarity of the solution, change grams of H_3PO_4 to moles and then find the number of moles per liter.

$$? \text{ moles of } H_3PO_4 = (19.6 \text{ g})\left(\frac{1 \text{ mole}}{98.0 \text{ g}}\right) = 0.200 \text{ mole of } H_3PO_4$$

$$M = \frac{\text{moles}}{\text{liter}} = \frac{n}{V} \qquad ? M = \frac{0.200 \text{ mole}}{0.300 \text{ L}} = 0.667 \, M \text{ solution of } H_3PO_4$$

Practice Exercise 12.11

What are the normality and the molarity of the solution formed by dissolving 7.23 g of HCl in enough water to make 450.0 mL of solution?

EXAMPLE 12.12

How many grams of $Ca(OH)_2$ are there in $5\overline{0}$ mL of a 3.0 N solution?

Solution

Solve the normality formula,

$$N = \frac{Eq}{L}$$

for Eq, the number of equivalents:

$$Eq = N \times L$$

Then substitute the values.

$$Eq = 3.0 Eq / L \times 0.050 L$$
$$= 0.15 Eq \text{ of } Ca(OH)_2$$

Next, find the equivalent mass of $Ca(OH)_2$.

$$\text{Equivalent mass of } Ca(OH)_2 = \frac{\text{molecular mass}}{\text{number of replaceable } OH^{1-} \text{ ions}}$$

$$= \frac{74.1 g}{2 Eq}$$

$$= 37.1 \frac{g}{Eq}$$

Convert the 0.15 Eq of $Ca(OH)_2$ to grams.

$$0.15 \text{ Eq } Ca(OH)_2 \times \frac{37.1 \text{ g } Ca(OH)_2}{1 \text{ Eq } Ca(OH)_2} = 5.6 \text{ g } Ca(OH)_2$$

Practice Exercise 12.12

How many grams of NaOH are there in 75.0 mL of a 4.00 N NaOH solution?

12.6 Dilution of Solutions

Solutions of acids and bases are usually purchased from chemical supply houses. For the sake of economy, the suppliers furnish most of these solutions in highly concentrated form. (This is like buying concentrated soups and adding water to them.) Chemists have to know how to dilute concentrated solutions to the strengths they want. If the concentration is expressed as a normality, the formula for doing this is

$$N_c V_c = N_d V_d$$

where N_c is the normality of the concentrated solution, V_c is the volume of the concentrated solution, N_d is the normality of the dilute solution, and V_d is the volume of the dilute solution. If the concentration is expressed in terms of molarity, the formula becomes

$$M_c V_c = M_d V_d$$

The volume V can be expressed in any convenient unit. But the units used, such as liters or milliliters, must be the same on both sides of the equation. This formula can be applied when solving practical laboratory problems like those in the next two examples.

EXAMPLE 12.13

How would you prepare 2 L of 3 M HCl from a 12 M HCl solution?

Solution

You need to find the amount V_c of concentrated HCl that, when diluted with water, will give 2 L of 3 M HCl.

$$M_c V_c = M_d V_d \quad \text{so} \quad V_c = \frac{M_d V_d}{M_c}$$

$$= \frac{(3\,M)\,(2\,L)}{12\,M} = 0.5\,L$$

Therefore, to make the final solution, measure 0.5 L of concentrated HCl into a flask and dilute with water until you have 2 L. (But see the cautionary note in Example 12.14.)

Practice Exercise 12.13

How would you prepare 3.00 L of a 4.00 M HCl solution from a stock solution of 6.00 M HCl?

EXAMPLE 12.14

How would you prepare 300 mL of a 0.6 N H$_2$SO$_4$ solution from a 36 N solution?

Solution

You need the amount of V_c of concentrated H$_2$SO$_4$ that, when diluted with water, will give 300 mL of 0.6 N H$_2$SO$_4$.

$$N_c V_c = N_d V_d \quad \text{so} \quad V_c = \frac{N_d V_d}{N_c}$$

$$= \frac{(0.6\,N)(300\,\text{mL})}{36\,N} = 5\,\text{mL}$$

Therefore, to make the final solution, measure 5 mL of concentrated H$_2$SO$_4$ into a flask and dilute with water to 300 mL. This will be a 0.6 N H$_2$SO$_4$ solution. (A cautionary note: *Always add acid to water, never the reverse. In the case of H$_2$SO$_4$, adding water to the concentrated acid will make the acid react violently. In the problem we just solved, it would be wise to put most of the water in the flask before adding the H$_2$SO$_4$.*)

Practice Exercise 12.14

How would you prepare 500.0 mL of a 0.400 N H$_3$PO$_4$ solution from a stock solution of 12.0 N H$_3$PO$_4$?

12.7 Ionization in Solutions

For centuries, scientists have tried to find out what happens when one substance dissolves in another. The chemists of the 1800s were puzzled by the fact that some solutions would conduct electric current and others would not. The English physicist Michael Faraday tried to explain this phenomenon. He classified solutions generally in the following way: **Electrolytic solutions** are *solutions that conduct electric current*. A solute that produces ions in solution is an electrolyte. **Nonelectrolytic solutions** are *solutions that do not conduct electric current*. A solute that does not produce ions in solution is a nonelectrolyte.

Faraday said that electrolytic solutions—such as sodium chloride in water—contain charged particles. He called these particles *ions* (from a Greek word that may be translated as "wanderer"). He said that ions wander through the solution carrying electric current (Figure 12.6). But questions arose about the nature of these ions.

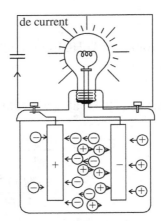

FIGURE 12.6 An electrolytic solution conducts electricity.

For example, why are there ions in a salt-and-water solution but no ions in a sugar-and-water solution? Another scientist, the Swedish chemist Svante Arrhenius, came up with the answer.

In 1884, in his Ph.D. thesis, Arrhenius advanced his ideas about ions. He suggested that Faraday's ions were really simple atoms (or groups of atoms) carrying a positive or a negative charge. He said that when some substances dissolve in solution, they break up into ions, for example:

$$NaCl(s) \xrightarrow{H_2O} Na^{1+}(aq) + Cl^{1-}(aq)$$

He called this process **ionization**. Actually, the solid sodium chloride **dissociates** when placed in water, because it is composed of ions (Figure 12.7). When placed in water, the ions separate from each other and are free to move about in the solution. Solutions that contain ions are electrolytic because they contain charged particles that can carry electric current. Substances in solution that conduct electricity are called *electrolytes*. A solution of sodium chloride actually contains a 1:1 mixture of sodium *ions* and chloride *ions*. A bottle label reading "1 *M* NaCl" refers to how the solution was made and not what is really in the bottle.

On the other hand, some substances dissolve in solution and *do not ionize*. They simply break up into their neutral molecules and become surrounded by the molecules of solvent:

$$C_6H_{12}O_6(s) \xrightarrow{H_2O} C_6H_{12}O_6(aq)$$
(Glucose)

These solutions are nonelectrolytic because they contain no charged particles to carry electric current (Figure 12.8). Substances that do not ionize in solution are called *nonelectrolytes*.

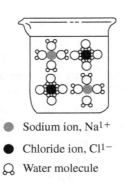

● Sodium ion, Na^{1+}
● Chloride ion, Cl^{1-}
⚛ Water molecule

FIGURE 12.7 Dissociation of NaCl in water.

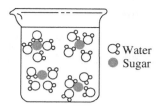

FIGURE 12.8 Sugar dissolves in water without ionizing.

Electrolytes and nonelectrolytes acting as solutes have some very interesting effects on the boiling points and freezing points of solutions. However, before we can examine these effects, we must discuss another way of expressing the concentration of solutes in solutions.

12.8 Molality

We can express the concentration of a solute in a solution in terms of the *number of moles of solute per kilogram of solvent*. This concentration unit is known as **molality** (m).

$$\text{Molality } (m) = \frac{\text{number of moles of solute}}{\text{kilogram of solvent}} = \frac{n}{\text{kg}}$$

We can determine the molality of a solution if we know either the number of moles or the mass of a solute dissolved in a given mass of solvent.

EXAMPLE 12.15

Calculate the molality of each of the following solutions:

(a) 36 g of glucose ($C_6H_{12}O_6$) dissolved in $5\overline{0}0$ g of water

(b) 1.03 g of sodium bromide (NaBr) dissolved in $25\overline{0}$ g of water

(c) 3.40 g of ammonia (NH_3) dissolved in $3\overline{0}0$ g of water

Solution

(a) First change the 36 g of $C_6H_{12}O_6$ (molecular mass = 180.0) to moles.

$$? \text{ moles} = (36 \text{ g}) \left(\frac{1 \text{ mole}}{180.0 \text{ g}} \right) = 0.20 \text{ mole}$$

Next change $5\overline{0}0$ g of water to kilograms.

$$? \text{ Kg} = (5\overline{0}0 \text{ g}) \left(\frac{1 \text{ kg}}{1,000 \text{ g}} \right) = 0.500 \text{ kg}$$

Finally, substitute these numbers into the molality formula.

$$m = \frac{\text{number of moles of solute}}{\text{Kilogram of solvent}}$$

$$? \, m = \frac{0.20 \text{ mole of glucose}}{0.500 \text{ kg of water}} = 0.40 \, m$$

(b) First change 1.03 g of NaBr (molecular mass = 102.9) to moles.

$$? \text{ moles} = (1.03 \text{ g})\left(\frac{1 \text{ mole}}{102.9 \text{ g}}\right) = 0.0100 \text{ mole}$$

Next change $25\overline{0}$ g of water to kilograms.

$$? \text{ kg} = (25\overline{0} \text{ g})\left(\frac{1 \text{ kg}}{1{,}000 \text{ g}}\right) = 0.250 \text{ kg}$$

Finally, substitute these numbers into the molality formula.

$$m = \frac{\text{number of moles of solute}}{\text{kilogram of solvent}}$$

$$? \, m = \frac{0.0100 \text{ mole of NaBr}}{0.250 \text{ kg of H}_2\text{O}} = 0.0400 \, m$$

(c) First change the 3.40 g of NH_3 (molecular mass = 17.0) to moles.

$$? \text{ moles} = (3.40 \text{ g})\left(\frac{1 \text{ mole}}{17.0 \text{ g}}\right) = 0.200 \text{ mole}$$

Next change 300 g of water to kilograms.

$$? \text{ kg} = (3\overline{00} \text{ g})\left(\frac{1 \text{ kg}}{1{,}000 \text{ g}}\right) = 0.300 \text{ kg}$$

Finally, substitute these numbers into the molality formula.

$$m = \frac{\text{number of moles of solute}}{\text{kilogram of solvent}}$$

$$? \, m = \frac{0.200 \text{ mole of NH}_3}{0.300 \text{ kg of H}_2\text{O}} = 0.667 \, m$$

Practice Exercise 12.15

Calculate the molality of each of the following solutions:

(a) 72 g of glucose ($C_6H_{12}O_6$) dissolved in $4\overline{00}$ g of water

(b) 1.17 g of NaCl dissolved in $5\overline{00}$ g of water

(c) 0.680 g of NH_3 dissolved in $25\overline{0}$ g of water

Chemists find that molality is a useful concept when they must deal with the effect of a solute on the boiling point and freezing point of a solution. The next section explains what this means.

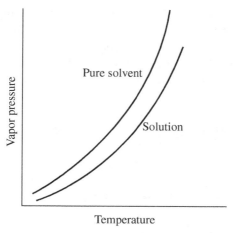

FIGURE 12.9 At a given temperature, the vapor pressure of the pure solvent is always greater than that of the solution of a non-volatile solute.

12.9 Colligative (Collective) Properties of Solutions

Certain properties of solutions depend more on the number of solute particles present than on the type of solute particles. In other words, these properties are the same regardless of what substance is the solute. *These properties called* **colligative properties**, *depend on the number of solute particles present in a given quantity of solvent.* For example, a mole of sugar molecules will have the same effect on the physical properties of a solution as will a mole of soluble starch or glycerol molecules.

One important colligative property of solutions is related to the vapor pressure of the solution. When a solute is dissolved in a solvent, the vapor pressure of the solution *decreases*. In other words, if we compared the vapor pressure of pure water at 50°C with the vapor pressure of a sugar–water solution at 50°C, we would find that the sugar–water solution had a lower vapor pressure than the pure water (Figure 12.9).

Two other colligative properties of solutions result from this vapor-pressure–reduction property.

1. The boiling point of a solution is *hig*her than the boiling point of the pure solvent alone. This is called the **boiling-point-elevation** property.

2. The freezing point of a solution is *lower* than the freezing point of the pure solvent alone. This is called the **freezing-point-depression** property.

For example, when salt is added to water, the boiling point of the water is raised, and food cooks in the water at a higher temperature. And, by adding salt to the ice that is used to freeze ice cream, we lower its freezing point. We both raise the boiling point and lower the freezing point of automobile radiator water when we add antifreeze/coolant.

12.10 Boiling-Point Elevation, Freezing-Point Depression, and Molality

The amount by which the boiling point of a solution is elevated or the freezing point lowered depends on (1) the solvent and (2) the concentration of the solute particles. For water, the **boiling-point–elevation constant** K_b is 0.52°C per mole of solute particles per kilogram of water. That is, a 1-molal (1 m) solution of sugar in water raises the boiling point of the water from 100°C to 100.52°C. The **freezing-point–depression constant** K_f for water is 1.86°C per mole of solute particles per kilogram of water. So a 1 m solution of sugar in water would have a freezing point of –1.86°C. Other solvents have different values for the constants K_b and K_f.

The values of K_f and K_b for water are the same for all non-ionic solutes. Thus a 1 m solution of glycerin in water also raises the boiling point of water to 100.52°C and lowers the freezing point to –1.86°C. (However, if we dissolve 1 mole of NaCl in 1,000 g of water, the effect on the boiling point and freezing point would be about double, because the NaCl ionizes into Na^{1+} ions and Cl^{1-} ions. In other words, 1 mole of NaCl yields

252 ESSENTIAL CONCEPTS OF CHEMISTRY

2 moles of solute particles. However, here we'll discuss only non-ionizing solutes.) For a solute that does not ionize in water, we can make this statement: *A 1-molal solution of molecular solute lowers the freezing point of 1 kg of water by 1.86°C and raises the boiling point by 0.52°C.*

Each additional mole of solute per kilogram of solvent lowers the freezing point of the solution by an additional 1.86°C and raises the boiling point of the solution by an additional 0.52°C. We can state this mathematically as

$$\Delta t = mK_f \quad \text{(for freezing-point depression)}$$

$$\Delta t = mK_b \quad \text{(for boiling-point elevation)}$$

where Δt is the boiling-point elevation or freezing-point depression in Celsius degrees.

EXAMPLE 12.16

Calculate the boiling point of each of the following solutions:

(a) 36 g of glucose dissolved in $50\overline{0}$ g of water

(b) 9.2 g of glycerol ($C_3H_8O_3$) in 250 g of water

Solution

(a) First we need to determine the molality of a solution of 36 g of glucose in $50\overline{0}$ g of water. If you look back at Example 12.15(a), you'll see that we calculated this molality to be 0.40 m.

Now use the formula for boiling-point elevation to calculate the change in temperature Δt.

$$\Delta t = mK_b = (0.40\,m)(0.52°C/m) = 0.21°C$$

If pure water boils at 100°C, then this 0.40 m solution boils at 100.21°C.

(b) First determine the molality of a solution of 9.2 g of $C_3H_8O_3$ (molecular mass = 92.0) in $25\overline{0}$ g of water.

$$?\text{ moles } C_3H_8O_3 = (9.2\,g)\left(\frac{1\text{ mole}}{92.0\,g}\right) = 0.10 \text{ mole}$$

$$?\text{ kg water} = (25\overline{0}\,g)\left(\frac{1\,Kg}{1{,}000\,g}\right) = 0.250 \text{ kg}$$

$$?\,m = \frac{0.10 \text{ mole}}{0.250 \text{ kg}} = 0.40\,m$$

Now use the formula for boiling-point elevation to calculate the change in temperature Δt.

$$\Delta t = mK_b = (0.40\,m)(0.52°C/m) = 0.21°C$$

If pure water boils at 100°C, then this 0.40 *m* solution boils at 100.21°C.

Practice Exercise 12.16

Calculate the boiling point of each of the following solutions:

(a) 72.0 g of glucose ($C_6H_{12}O_6$) dissolved in $25\overline{0}$ g of water

(b) 1.84 g of glycerol ($C_3H_8O_3$) dissolved in $50\overline{0}$ g of water

CHAPTER 12: THE CHEMISTRY OF SOLUTIONS

EXAMPLE 12.17

Calculate the freezing point of each of the solutions in Example 12.16.

Solution

(a) We already know that the molality of this solution is 0.40 m. We also know that K_f for water is 1.86°C. Therefore

$$\Delta t = mK_f = (0.40\ m)(1.86°C/m) = 0.74°C$$

Because pure water freezes at 0°C, this 0.40 m solution freezes at –0.74°C.

(b) We already know that the molality of this solution is 0.40 m. Therefore

$$\Delta t = mK_f = (0.40\ m)(1.86°C/m) = 0.74°C$$

This 0.40 m solution freezes at –0.74°C.

Practice Exercise 12.17

Calculate the freezing point of each of the solutions in Practice Exercise 12.16.

12.11 The Processes of Diffusion and Osmosis

Diffusion and osmosis are important colligative properties in living systems and in the chemistry laboratory as well. In this section, we'll take a brief look at each of these processes.

If you place a few drops of dye in a glass of water, what happens? Very quickly the dye disperses through the water. Given enough time, the dye will distribute itself evenly throughout the water (Figure 12.10). This process is known as **diffusion**. It occurs because the dye moves from an area of high dye concentration to an area of low dye concentration. In other words, there is a *concentration gradient*, or gradually changing difference in concentration, between the place in the water where the dye is dropped and the rest of the water. The diffusion process stops when the dye is evenly distributed throughout the water, so that the concentration gradient no longer exists.

It is this process of diffusion that moves most of the substances in our body. For example, the breathing process is based on the diffusion of gases from areas of high concentration to areas of low concentration. And the process of diffusion is the system by which electrolytes in the body are carried to where they are needed.

A special type of diffusion process called **osmosis** involves the *passage of water through a semipermeable membrane, like cellophane or cell walls*. In this process, *water* moves from an area of low solute concentration to an area of high solute concentration. The semipermeable membrane allows only the water, not the solute, to pass through it.

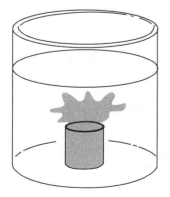
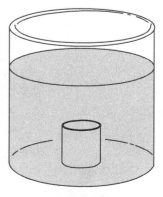

FIGURE 12.10 A dye will diffuse through a solution from areas of high dye concentration to areas of low dye concentration, Eventually, the dye will distribute itself evenly throughout the solution.

For example, consider a beaker of water having two compartments separated by a semipermeable membrane (Figure 12.11). Compartment A is filled with a 5% sucrose solution. Compartment B is filled with an equal volume of a 20% sucrose solution. Soon after the solutions are placed in the beaker, we notice a shift in the water levels of the compartments. Water moves from the compartment containing the 5% sucrose solution to the compartment containing the 20% sucrose solution, in an attempt to equalize the sucrose concentrations in both compartments (Figure 12.12). This movement of water is the process of osmosis.

You may be thinking, "Why doesn't the sucrose in the 20% solution move through the semipermeable membrane to equalize the concentrations of the two solutions?" This doesn't happen because an osmotic membrane does not allow solute particles to pass through it but does allow solvent particles to pass freely.

Osmotic Pressure

Consider the two sucrose solutions that we've just discussed. After the process of osmosis has proceeded for a while, it finally stops. Osmosis ends when the rate of water molecules passing from compartment A to compartment B equals the rate of water molecules passing from compartment B to compartment A (Figure 12.13). In other words, a dynamic equilibrium is in progress. But why does this occur?

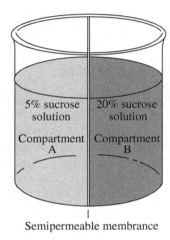

FIGURE 12.11 A beaker with two compartments seperated by a semipermeable membrane.

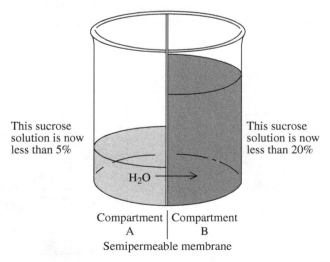

FIGURE 12.12 This is what happens to our two solutions after osmosis has proceeded for a while. Water has left compartment A, increasing the concentration of that sucrose solution. The water has gone into compartment B, diluting that sucrose solution.

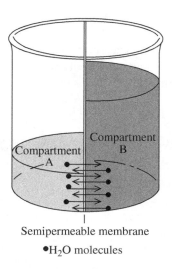

FIGURE 12.13 The process of osmosis stops when the number of water molecules moving from compartment A to compartment B equals the number of water molecules moving from compartment B to compartment A.

The answer lies in looking at the height of the water in each compartment. The height of the water in compartment B is greater than that in A (Figure 12.14). At first, water molecules moved from A to B in an attempt to equalize the concentrations of the two sucrose solutions. But eventually the height of the water in compartment B caused a pressure to be exerted on the water molecules in that compartment, pushing them back into compartment A. When these two opposing forces became equal, the process of osmosis stopped. The pressure exerted by the water in compartment B at this point is called the *osmotic pressure* and is related to the concentration of the solute in the solution. **Osmotic pressure** can be defined as *the amount of pressure that must be applied to prevent the flow of water through a membrane*. This concept is extremely important in the transport of body fluids.

Like other colligative properties, osmotic pressure can be used to determine the molecular mass of a solute in a solution. A Dutch chemist named van't Hoff developed a mathematical expression that relates the osmotic pressure to the molar concentration of the solution. A technique based on this equation, called **osmometry**, has proved to be far more sensitive for molecular mass determinations than freezing-point depression or boiling, point elevation. In this method, the difference in the height of the water columns in the two compartments is used to obtain the osmotic pressure of the system. From this information the molecular mass of the solute is determined. Although we won't pursue this calculation here, chemists have found this technique quite useful in the laboratory.

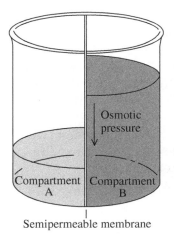

FIGURE 12.14 Osmotic pressure is due to the height of the water column in compartment B.

EXAMPLE 12.18

Two solutions of equal volume are placed in a beaker similar to the one in Figure 12.11. Compartment A of the beaker contains a 10% NaCl solution. Compartment B of the beaker contains a 5% NaCl solution. Which way does the water move, from A to B or from B to A?

Solution

The water moves from compartment B to compartment A. The solvent moves from the side of low solute concentration to the side of high solute concentration.

Practice Exercise 12.18

Cells in the body have intracellular and extracellular fluid. The intracellular fluid has an osmotic pressure equivalent to 5% glucose. If an individual is given an intravenous solution that has an osmotic pressure equivalent to 10% glucose, would water move into or out of the cell? (*Hint:* The intravenous solution replaces the extracellular fluid.)

SUMMARY

A solution is a homogeneous mixture composed of at least two substances. The solute is the substance that is dissolved, and the solvent is the substance in which it is dissolved. Solutions may be saturated, unsaturated, or supersaturated, depending on the quantity of solute dissolved in a given quantity of solvent. Pairs of liquids that do not form solutions when mixed are said to be immiscible, whereas pairs of liquids that form solutions are said to be miscible. In general, polar substances dissolve polar substances, and nonpolar substances dissolve nonpolar substances.

Because solutions are mixtures, they may have various ratios of solvent to solute. These ratios, or concentrations, may be measured as percentage by mass, percentage by volume, or percentage by mass–volume. In addition, the concentration of a solution may be measured as its molarity: the number of moles of solute per liter of solution. Concentration may also be measured as normality: the number of equivalents of solute per liter of solution.

An equivalent of any acid is the mass of that acid that contains Avogadro's number of hydrogen ions. An equivalent of a base is the mass of that base that contains Avogadro's number of hydroxide ions. The number of grams in one equivalent of an acid or a base is computed by dividing the molecular mass of the substance by the number of replaceable hydrogen ions (for an acid) or replaceable hydroxide ions (for a base) in one molecule of the substance.

Electrolytic solutions are solutions that conduct electricity, and nonelectrolytic solutions are solutions that do not conduct electricity. Electrolytic solutions are capable of conducting electricity because they contain ions, which are charged atoms or groups of atoms. Substances that ionize, or break up into ions, when they dissolve are called electrolytes.

The molality of a solution is the number of moles of solute per kilogram of solvent. Molality is useful in dealing with colligative properties of solutions, which are properties that depend mainly on the number of solute particles present (in a given quantity of solvent) rather than on the type of solute particles. Vapor-pressure reduction, boiling-point elevation, and freezing-point depression are examples of colligative properties. In general, a solution exhibits a lower vapor pressure, a higher boiling point, and a lower freezing point than the pure solvent alone.

Diffusion is the movement of a solute in a solution from an area of high solute concentration to an area of low solute concentration. Osmosis is the passage of water through a semipermeable membrane from the less concentrated to the more concentrated solution.

KEY TERMS

boiling-point elevation (**12.9**)
boiling-point–elevation
constant (**12.10**)
colligative properties (**12.9**)
concentration (**12.3**)
diffusion (**12.11**)
dissociation (**12.7**)
electrolytic solution (**12.7**)
equivalent(of an acid or a base) (**12.5**)
freezing-point depression (**12.9**)
saturated solution (**12.2**)
saturation point (**12.3**)
solubility (**12.2**)
solute (**12.1**)

freezing-point–depression
constant (**12.10**)
immiscible (**12.1**)
ionization (**12.7**)
molality (**12.8**)
molarity (**12.4**)
nonelectrolytic solution (**12.7**)
normality (**12.5**)
osmometry (**12.11**)
osmosis (**12.11**)
osmotic pressure (**12.11**)
solvent (**12.1**)
supersaturated solution (**12.2**)
unsaturated solution (**12.2**)

SELF-TEST EXERCISES

LEARNING GOAL 1

Definitions of Solution, Solute, and Solvent

◀ 1. Match the word on the left with its definition on the right.

(a) Solute
(b) Solvent
(c) Solution
(d) Immiscible

1. A homogeneous mixture
2. The substance in a solution that gets dissolved
3. Substances that do not form solutions when mixed
4. The substance in a solution that causes another substance to dissolve

2. When a solution is composed of two or more substances in the same state, how do chemists determine which is the solute and which is the solvent?

◀ 3. Which substance is the solute and which is the solvent in each of the following solutions?

(a) 10%-by-weight hydrochloric acid
(b) Seltzer water (also known as club soda or soda water)
(c) 4%-by-volume benzene in carbon tetra-chloride

4. A coin composed of two metals that were melted, mixed together, and solidified is an example of a solution. True or false?

5. Give an example of

(a) A solid–solid solution   (b) A solid–liquid solution   (c) A solid–gas solution
(d) A liquid–liquid solution

◀ 6. Which is the solute and which the solvent in each of the following solutions?

(a) 10%-by-mass sodium chloride in water
(b) 5%-by-mass benzene in carbon tetrachloride
(c) 10%-by-mass–volume methyl alcohol in ethyl alcohol

LEARNING GOAL 2

Substances That Form Solutions

7. "Like dissolves like." True or false? What does this statement refer to in terms of molecular structure?
◀ 8. You are served a cup of coffee in a restaurant. Assuming that no sugar has been added to the coffee, you add 2 teaspoons of sugar, stir well, and notice a layer of sugar at the bottom of the cup that won't dissolve. What type of solution have you produced?
9. The point at which no more solute dissolves in a solution is called the _____ point.
10. What are the meanings of the terms *miscible* and *immiscible*?

LEARNING GOAL 3

Saturated, Unsaturated, and Supersaturated Solutions

11. (a) A solution that contains less solute than can be dissolved in a given quantity of solvent at a particular temperature is a(n) _____ solution.
    (b) A solution that contains more solute than is needed for saturation for a given quantity of solvent at a particular temperature is a(n) _____ solution.

12. Determine whether each of the following solutions is unsaturated, saturated, or supersaturated. (See Table 12.2.)

    (a) 15.0 g of NaCl in 50.0 g of water at 50°C
    (b) 45.5 g of $AgNO_3$ in 50.0 g of water at 50°C

◀ 13. You are given a saturated solution that contains undissolved solute. The solute is constantly dissolving, yet the concentration of the solute in the solution remains the same. Explain.
14. You are given a solution by your instructor who asks you to determine whether or not it is supersaturated. Explain the procedure you would use to test for supersaturation.

LEARNING GOAL 4

Percent Concentrations

◀ 15. What is the concentration in percent by mass of the following solutions?

   (a) 5.00 g of magnesium nitrate in enough water to make 80.0 g of solution.
   (b) 60.0 g of ethanol in enough water to make 300.0 g of solution.
   (c) 2.00 g of sodium chloride *plus* 23.0 g of water.

◀ 16. What is the concentration in percent by mass of each of the following solutions?

   (a) 80.0 g of sodium chloride with enough water to make $25\overline{0}$ g of solution
   (b) 7.50 g of methyl alcohol with enough water to make $40\overline{0}$ g of solution
   (c) 50.0 g of potassium nitrate *plus* $15\overline{0}$ g of water

17. How many grams of KOH are there in $20\overline{0}$ g of $1\overline{0}$%-by-mass KOH solution?
18. How many grams of NaOH are there in 500.0 g of a 15.0%-by-mass NaOH solution?
19. What is the percent by volume of the solute in each of the following solutions?

   (a) $20\overline{0}$ mL of ether in enough carbon tetrachloride to make 1.0 L of solution
   (b) 5 mL of alcohol in enough water to make 80 mL of solution

20. What is the percent by volume of the solute in each of the following solutions?

   (a) 300.0 mL of ether in enough carbon tetrachloride to make 2.00 L of solution
   (b) 25.0 mL of alcohol in enough water to make 100.0 mL of solution

*21. Find the volume of alcohol present in $10\overline{0}$ g of 10.0%-by-mass alcohol–water solution. (*Hint:* The density of alcohol is 0.800 g/mL.)

*22. Find the volume of alcohol present in 250.0 g of a10.0%-by-mass alcohol–water solution. (*Hint:* The density of alcohol is 0.800 g/mL.)

23. How many milliliters of chloroform are there in $5\overline{00}$ mL of a 4.0%-by-volume chloroform–ether solution?

24. How many milliliters of carbon tetrachloride are there in 250.0 mL of a 5.00%-by-mass–volume carbon tetrachloride–benzene solution?

◀ 25. What are the percent by mass–volume of the solutes in the following solutions?

   (a) $5\overline{0}$ g of sodium chloride in 1.5 L of solution
   (b) $1\overline{0}$ g of ammonia dissolved in $2\overline{00}$ g of solution (Assume that the density of the solution is 1.00 g/cm³.)

◀ 26. What is the percent by mass–volume of the solute in each of the following solutions?

   (a) 25.0 g of sodium nitrate in 1.00 L of solution
   (b) 15.0 g of sodium hydroxide dissolved in 100.0 g of solution (Assume that the density of the solution is 1.00 g/cm³.)

27. How many grams of sugar are there in $10\overline{0}$ mL of a 15%-by-mass–volume sugar–water solution?

28. How many grams of lactose (milk sugar) are there in 150.0 mL of a 10%-by-mass–volume lactose–water solution?

*29. If a person's blood-sugar level falls below 60 mg per 100 mL, insulin shock can occur. What is the percent by mass of sugar in the blood at this level? (Assume that the density of blood is 1.2 g/mL.)

◀ 30. A person's blood-sugar level is calculated to be 120.0 mg per $1\overline{00}$ mL blood. What is the percentage by mass of sugar in the blood at this level? (The density of blood is 1.20 g/mL.)

31. What are the concentrations in percent by mass of the following solutions?

   (a) 25.0 g of NaCl in enough water to make $2\overline{00}$ g of solution
   (b) 8.00 g of sugar in enough water to make $16\overline{0}$ g of solution

32. What is the concentration in percent by mass of each of the following solutions?

   (a) 55.0 of Na_2CO_3 in enough water to make 400.0 g of solution
   (b) 14.5 g of sugar in enough water to make 290.0 g of solution

33. How many grams of KCl are there in $5\overline{00}$ g of a 6.0%-by-mass KCl solution?

34. How many grams of KNO_3 are there in 750.0 g of a 7.50%-by-mass KNO_3 solution?

35. What are the percents by volume of the solutes in the following solutions?

   (a) $15\overline{0}$ mL of ethyl alcohol in enough water to make 3.0 L of solution
   (b) $5\overline{00}$ mL of methyl alcohol (dry gas) in enough gasoline to make 60.0 L of solution (This represents a can of dry gas in a nearly full tank of gasoline in a standard-size automobile.)

36. What is the percent by volume of the solute in each of the following solutions?

   (a) 175 mL of isopropyl alcohol in enough water to make 4.00 L of solution
   (b) 600.0 mL of methyl alcohol (dry gas) in enough gasoline to make 60.0 L of solution

◀ 37. Determine the volume of ether present in $5\overline{00}$ g of a 4.0%-by-mass ether–carbon tetrachloride solution. (*Hint:* The density of ether is 0.714 g/cm³.)

*38. Determine the volume of ether in 750.0 g of a 10.0%-by-mass ether–carbon tetrachloride solution. (*Hint:* The density of ether is 0.714 g/cm³.)

39. How many milliliters of ethyl alcohol are there in 2.00 L of a 10.0%-by-volume ethyl alcohol–water solution?

40. How many milliliters of isopropyl alcohol are there in 4.50 L of a 20.0%-by-volume isopropyl alcohol–water solution?

260 ESSENTIAL CONCEPTS OF CHEMISTRY

<u>41</u>. What are the percents by mass–volume of the solutes in the following solutions?

  (a) 25.0 g of sodium nitrate in $8\overline{00}$ mL of solution
  (b) 0.800 g of potassium chloride in 16.0 mL of solution

42. What is the percent by mass–volume of the solute in each of the following solutions?

  (a) 50.0 g of sodium carbonate in 500.0 mL of solution
  (b) 0.160 g of potassium nitrate in 80.0 mL of solution

<u>43</u>. How many grams of sodium chloride are there in $25\overline{0}$ mL of a 5.00%-by-mass–volume solution of sodium chloride and water?

44. How many grams of sodium chloride are there in 100.0 mL of a 9.00%-by-mass–volume solution of sodium chloride and water?

LEARNING GOAL 5

Calculations Involving Molarity

◄ <u>45</u>. What is the molarity of each of the following solutions?

  (a) 5.4 g of HCl in enough water to make $5\overline{00}$ mL of solution
  (b) 117 g of sodium chloride in enough water to make 4 L of solution

◄ 46. What is the molarity of each of the following solutions?

  (a) 1.18 g of HCl in enough water to make 2.00 L of solution
  (b) 23.40 g of sodium chloride in enough water to make 5.00 L of solution

*<u>47</u>. What is the molarity of a 10.0%-by-mass HCl solution? (Assume that the density of the solution is 1.00 g/mL.)

*48. What is the molarity of a 5.00%-by-mass HCl solution? (Assume that the density of the solution is 1.00 g/cm³.)

<u>49</u>. How many grams of KOH are obtained by evaporating to dryness $5\overline{0}$ mL of a 3.0 M KOH solution?

50. How many grams of NaOH are obtained by evaporating to dryness 25.0 mL of a 2.00 M NaOH solution?

*<u>51</u>. The *Merck Manual* reports that a blood-alcohol level of $15\overline{0}$ to $2\overline{00}$ mg per $1\overline{00}$ mL produces intoxication and that a level of $3\overline{00}$ to $4\overline{00}$ mg per $1\overline{00}$ mL produces unconsciousness. At a blood-alcohol level above $5\overline{00}$ mg per $1\overline{00}$ mL, a person may die. What is the molarity of the blood with respect to alcohol at the level of $5\overline{00}$ mg per $1\overline{00}$ mL? (*Hint:* The molecular mass of ethyl alcohol is 46.0.)

◄ *52. A person has been drinking ethyl alcohol. What is the molarity of the alcohol in this person's blood if the alcohol concentration is $4\overline{00}$ mg per $1\overline{00}$ mL of blood?

◄ <u>53</u>. Salt (NaCl) is a necessary ingredient of our diets, and the body has a system to maintain a delicate balance of salt. In certain illnesses, this salt balance can be lost, and a physician or nurse must administer salt intravenously, using a 0.85% solution. What is the molarity of a 0.85%-by-mass–volume sodium chloride solution?

◄ 54. How many milliliters of water are needed to make a 2.00 M NaCl solution if 5.85 g of NaCl are present?

<u>55</u>. What is the molarity of each of the following solutions?

  (a) 24.5 g of H_2SO_4 in enough water to make 1.50 L of solution
  (b) 10.1 g of KNO_3 in enough water to make $5\overline{00}$ mL of solution

56. What is the molarity of each of the following solutions?

  (a) 49.0 g of H_2SO_4 in enough water to make 2.00 L of solution
  (b) 2.02 g of KNO_3 in enough water to make 200.0 mL of solution

*57. What is the molarity of a 5.00%-by-mass magnesium chloride solution? (Assume that the density of the solution is 1.00 g/cm³.)

*58. What is the molarity of a 7.50%-by-mass sodium nitrate solution? (Assume that the density of the solution is 1.00 g/cm³.)

59. How many grams of Na_3PO_4 are obtained by evaporating to dryness $25\overline{0}$ mL of a 0.400 M Na_3PO_4 solution?

60. How many grams of $NaC_2H_3O_2$ are obtained by evaporating to dryness 400.0 mL of a 0.500 M $NaC_2H_3O_2$ solution?

LEARNING GOAL 6

Equivalents of Acids and Bases

◀ 61. Determine the number of equivalents in each of the following samples:

(a) 4.9 g of H_2SO_4 (b) 7.2 g of HCl (c) 32.1 g of $Fe(OH)_3$

62. Determine the number of equivalents in each of the following samples:

(a) 9.8 g of H_2SO_4 (b) 3.6 g of HCl (c) 6.42 g of $Fe(OH)_3$

63. For each of the following acids and bases, determine the number of grams per mole and the number of grams per equivalent:

(a) H_2SO_3 (b) $Fe(OH)_3$ (c) $Fe(OH)_2$ (d) H_3BO_3

64. For each of the following acids and bases, determine the number of grams per mole and the number of grams per equivalent:

(a) $HC_2H_3O_2$ (b) NaOH (c) $Mg(OH)_2$ (d) HNO_3

65. Determine the number of equivalents in each of the following:

(a) 1.23 g of H_2SO_3 (b) 26.7 g of $Fe(OH)_3$

66. Determine the number of equivalents in each of the following:

(a) 246 g of H_2SO_3 (b) 4.00 g of NaOH

◀ 67. Determine the number of grams in each of the following:

(a) 25.0 equivalents of $Fe(OH)_2$ (b) 0.450 equivalent of H_3BO_3

◀ 68. Determine the number of grams in each of the following:

(a) 0.250 equivalent of $Mg(OH)_2$ (b) 2.00 equivalents of $HC_2H_3O_2$

LEARNING GOAL 7

Calculations Involving Normality

◀ 69. Determine the normality of each of the following solutions:

(a) $97\overline{0}$ g of H_3PO_4 in enough water to make 10.0 L of solution
(b) 3.7 g of $Ca(OH)_2$ in enough water to make $8\overline{00}$ mL of solution
(c) Calculate the molarities of the solutions in (a) and (b).

◀ 70. Determine the normality of each of the following solutions:

(a) 465 g of H_3PO_4 in enough water to make 8.00 L of solution
(b) 740.0 g of $Ca(OH)_2$ in enough water to make 740.0 mL of solution
(c) Calculate the molarities of the solutions in (a) and (b).

71. How many grams of Ca(OH)$_2$ are there in $8\overline{00}$ mL of a 2.0 N Ca(OH)$_2$ solution?
72. How many grams of Mg(OH)$_2$ are there in $75\overline{0}$ mL of a 3.00 N Mg(OH)$_2$ solution?
73. How many milliliters of a 0.60 N H$_3$PO$_4$ solution do you need to get 4.9 g of H$_3$PO$_4$?
74. How many milliliters of a 0.450 N H$_3$PO$_4$ solution do you need to get 98.0 g of H$_3$PO$_4$?
75. Determine the normality of each of the following solutions:

    (a) 11.23 g of Fe(OH)$_2$ in enough water to make $25\overline{0}$ mL of solution
    (b) 4.12 g of H$_3$BO$_3$ in enough water to make $4\overline{00}$ mL of solution
    (c) Calculate the molarities of the solutions in (a) and (b).

76. Determine the normality of each of the following solutions:

    (a) 2.246 g of Fe(OH)$_2$ in enough water to make 500.0 mL of solution
    (b) 16.48 g of H$_3$BO$_3$ in enough water to make 500.0 mL of solution
    (c) Calculate the molarities of the solutions in (a) and (b).

◀ 77. How many grams of Al(OH)$_3$ are there in $25\overline{0}$ mL of a 0.500 N Al(OH)$_3$ solution?
◀ 78. How many grams of Fe(OH)$_3$ are there in 500.0 mL of a 0.250 N Fe(OH)$_3$ solution?
79. How many milliliters of 0.100 N H$_2$SO$_4$ do you need to obtain 2.94 g of H$_2$SO$_4$?
80. How many milliliters of a 0.200 N H$_3$PO$_4$ solution do you need to obtain 29.40 g of H$_3$PO$_4$?

LEARNING GOAL 8

Calculations for Dilutions of Solutions

81. How would you make $5\overline{00}$ mL of 4.0 N H$_2$SO$_4$ solution from a 16 N solution?
◀ 82. How would you make 250.0 mL of 8.00 N H$_3$PO$_4$ solution from a 10.0 N solution?
83. How would you make 3 L of a 0.1 N NaOH solution from a 6 N solution?
84. How would you make 4.00 L of 0.200 N Ca(OH)$_2$ solution from 2.00 N solution?
85. How would you prepare $25\overline{0}$ mL of a 5.00 M HCl solution from a 12.0 M solution?
86. How would you prepare 450.0 mL of a 7.50 M HNO$_3$ solution from an 11.0 M solution?
87. How would you prepare 5.00 L of 0.0500 N H$_2$SO$_4$ solution from a 3.00 M solution?
88. How would you prepare 4.00 L of a 0.0200 N HC$_2$H$_3$O$_2$ solution from a 2.00 N solution?

LEARNING GOAL 9 AND 10

Electrolytic and Nonelectrolytic Solutions; Ions and Ionization

◀ 89. Define the terms *ion* and *ionization* and give examples of each.
90. True or false:

    (a) Ions are groups of charged atoms.   (b) Charged elements are not ions.

◀ 91. Explain why some solutions conduct electric current and others do not.
◀ 92. Solution A conducts electricity, but solution B does not. If solutions A and B are mixed and no chemical reaction occurs, will the resulting solution conduct electricity?
◀ 93. Define and give an example of an electrolytic solution and of a nonelectrolytic solution.
94. Which will conduct electricity, a glass of salt water or a cup of tea with sugar? Explain.

LEARING GOAL 11

Definition of Molality

◀ 95. Define *molality*. What information is necessary to calculate the molality of a solution?
◀ 96. How is *molality* different from *molarity*?

LEARNING GOAL 12

Calculation of Molality

*__97__. Calculate the molality of the intravenous salt solution in Self-Test Exercise 53. (Assume that there is 0.85 g of NaCl per 100 g of water.)

◀ 98. A solution contains 0.373 g of KCl per 100 g of water. What is its molality?

◀ __99__. Determine the molality of each of the followin of solutions:

(a) 6.4 g of methanol (CH_3OH) in 250 g of water
(b) 90.0 g of glucose ($C_6H_{12}O_6$) in $1,500$ g of water

◀ 100. Determine the molality of each of the following solutions:

(a) 0.128 g of CH_3OH in 500 g of water
(b) 450 g of $C_6H_{12}O_6$ in $2,000$ g of water

LEARNING GOAL 13

Definition of Colligative Properties and Effect of Solute on Boiling and Freezing Points

◀ 101. Explain how a solute affects the boiling point and the freezing point of a solution.

◀ 102. What would happen to the engine of an automobile if antifreeze were not added to the water in the radiator during a cold winter?

LEARNING GOAL 14

Definition of K_f and K_b

◀ 103. Define the following terms and give an example of each:

(a) K_f (b) K_b

◀ 104. What are the values of K_f and K_b for water? What do they mean?

LEARNING GOAL 15

Calculations Involving Boiling and Freezing Points of Solutions

◀ __105__. You can winterize a typical car radiator system by adding 1.00 gal of antifreeze to 1.00 gal of water. At what temperature does this water–antifreeze solution begin to freeze? (Assume that the antifreeze is pure ethylene glycol, $C_2H_6O_2$, which has a density of 1.10 g/cm³.) (*Hint:* You must change the 1.00 gal of ethylene glycol solute to grams and then to moles. Also, the 1.00 gal of water must be changed to kilograms.)

106. Calculate the boiling point of a solution prepared by dissolving 5.40 g of glucose in 500 g of water. Glucose is $C_6H_{12}O_6$.

__107__. Calculate the boiling point of the solution in Self-Test Exercise 105.

◀ 108. Calculate the boiling point of a solution prepared by dissolving 0.184 g of glycerol ($C_3H_8O_3$) in $1,000$ g of water.

__109__. Calculate the boiling point and the freezing point of a solution prepared by dissolving 27.0 g of glucose, $C_6H_{12}O_6$, in 100 g of water.

110. Calculate the freezing point of the solution in Self-Test Exercise 108.

LEARNING GOAL 16

Diffusion and Osmosis

111. (a) The passage of water through a semipermeable membrane is the process of _____.
(b) The movement of a solute in a solvent from an area of high solute concentration to an area of low solute concentration is the process of _____.

◀ 112. A 50.0-g sample of a freshly peeled raw potato is placed in a beaker containing a saturated NaCl solution. The potato is left in the beaker for two hours. It is then removed, dried, and weighed. Will the mass of the potato be lower, higher, or the same? Explain.

◀ 113. Two solutions of equal volume are placed in a beaker similar to the one in Figure 12.11. Compartment A of the beaker contains a 5% KCl solution. Compartment B of the beaker contains a 20% KCl solution. Which way does the water move, from A to B or from B to A?

114. Cells in the body have both intracellular and extra-cellular fluid. The intracellular fluid has an osmotic pressure equivalent to 0.9% sodium chloride. If an individual is given an intravenous solution that has an osmotic pressure equivalent to 2.0% sodium chloride, would water move into or out of the cell? (*Hint:* The intravenous solution replaces the extracellular fluid.)

EXTRA EXERCISES

◀ *115. Use the kinetic theory to explain how and why a solute affects the boiling and freezing points of a solution.

116. Calculate the molarity of a solution of HCl that has a hydrogen-ion concentration of 10^{-2} M.

117. Calculate the molarity and the normality of a solution of sulfuric acid that has a [H^{1+}] of 10^{-2} M.

118. How many grams of H_2SO_4 are necessary to make 10.0 L of a 5.00%-by-mass sulfuric acid solution? (Assume that the density of the solution is 1.00 g/mL.)

119. Find the molarity and the normality of the solution in Extra Exercise 118.

120. How does a solution differ from a mixture?

◀ 21. In Chicago, salt is sometimes used on roads to keep the roads free of ice. In northern Minnesota, however, people use sand to avoid slipping on ice. Why don't the Minnesota residents try to use salt to get rid of their ice?

◀ *122.* Which of the following would lower the freezing point of $1,\overline{000}$ g of water by the greatest amount? Explain your choice.

(a) $1\overline{00}$ g of glucose ($C_6H_{12}O_6$)
(b) $1\overline{00}$ g of sucrose ($C_{12}H_{22}O_{11}$)
(c) $1\overline{00}$ g of ethanol (C_2H_6O)

◀ *123. Concentrated hydrochloric acid is delivered as 37.00% HCl by mass. The density of this solution is 1.19 g/mL. What volume of this acid is needed to produce 1.00 L of a 0.500 M HCl solution?

124. You dissolve a 15.0-g sample of an unknown substance (nonelectrolyte) in $1\overline{00}$ g of water. The solution freezes at −1.86°C. What is the molecular mass of the substance?

125. A chemist prepares a solution by taking 1.00 L of water and adding 2.00 moles of solute. Is this solution a 2.00 M solution or a 2.00 m solution? Explain.

*126. You add a 10.0-g sample of a substance to $1,\overline{000}$ g of water. You find that the freezing point of the solution is −5.00°C. What is the molecular mass of the substance?

127. You add 5.85 g of NaCl to $1,\overline{000}$ g of water. What is the approximate boiling point of the solution?

128. Determine the number of equivalents in each of the following:

(a) 73.0 g HCl (b) 98.1 g of H_2SO_4

129. Determine the number of equivalents in each of the following:

(a) 1.00 mole of HCl (b) 1.00 mole of H_2SO_4

APPENDIX A

BASIC MATHEMATICS FOR CHEMISTRY

LEARNING GOALS

After you have studied this appendix, you should be able to:

1. Add and subtract algebraically.
2. Multiply and divide fractions.
3. Multiply and divide exponential numbers.
4. Use the factor-unit method in solving problems.
5. Add and subtract decimals.
6. Multiply and divide decimals.
7. Solve simple algebraic equations.
8. Set up and solve proportions.
9. Solve simple density problems.
10. Solve problems involving percentages.
11. Perform mathematical operations using a calculator.

INTRODUCTION

Full appreciation of many ideas in chemistry requires some basic mathematics. (There are also concepts in chemistry that require absolutely *no* mathematics.) This appendix is here so that you can review and master the mathematical skills you need to be successful in chemistry. Whenever you are in doubt about a mathematical problem as you work your way through the book, turn to this appendix for guidance.

You'll find that chemical mathematics is not very difficult, as long as you can perform some basic operations. Work your way through this appendix carefully and try your hand at all the sample problems. Then do the exercises at the end. Remember that the knowledge you gain here will greatly increase your success in chemistry.

A.1 Adding and Subtracting Algebraically

In chemistry we deal with both positive and negative numbers, so it is important to know how to add and subtract both. The process is called algebraic addition and subtraction. Here are the rules to follow.

Addition

1. *When the signs are alike:* When the signs of two numbers are the same, add the numbers and keep the same sign.

$$+4 + 2 = +6$$
$$-4 - 2 = -6$$

2. *When the signs are different:* When the signs of two numbers are not the same, subtract the numbers and keep the sign of the larger number.

$$-4 + 2 = -2$$
$$+4 - 2 = +2$$

Subtraction

Change the sign of the *number being subtracted* and follow the rules for addition. (*Note:* Plus signs before positive numbers are usually left out, but minus signs before negative numbers are always written.)

$4 - (-2)$	becomes	$4 + 2 = 6$
$4 - (+2)$	becomes	$4 - 2 = 2$
$-4 - (+2)$	becomes	$-4 - 2 = -6$
$-4 - (-2)$	becomes	$-4 + 2 = -2$

A.2 Fractions

A ratio of two numbers is called a **fraction**. Some examples of fractions are $\frac{1}{4}, \frac{1}{2}, \frac{2}{5}, \frac{1}{11},$ and $\frac{2}{1}$. A fraction has two parts: the **numerator**, which is the top number, and the **denominator**, which is the bottom number.

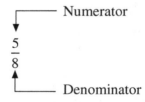

Here are some basic rules to follow in dealing with fractions.

Multiplication

Multiply numerators; then multiply denominators.

$$\frac{1}{2} \times \frac{3}{4} = \frac{3}{8} \qquad \frac{5}{9} \times \frac{9}{5} = \frac{45}{45} = 1$$

Division

Invert (turn upside down) the fraction to the right of the division sign, and then follow the rules for multiplication.

$$\frac{1}{2} \div \frac{1}{4} \quad \text{becomes} \quad \frac{1}{2} \times \frac{4}{1} = \frac{4}{2} = 2$$

$$\frac{3}{4} \div \frac{5}{9} \quad \text{becomes} \quad \frac{3}{4} \times \frac{9}{5} = \frac{27}{20}$$

A.3 Exponents

Exponents are numbers written to the right of and above another number (called the **base number**). A positive exponent tells you to multiply the base number by itself. The exponent tells you how many times the base number is taken. For example, $(4)^2$ means take the number 4 twice: 4×4. Here 4 is the base number and 2 is the exponent.

$$(4)^2 \leftarrow \text{Exponent}$$
$$\uparrow \text{Base number}$$

Sometimes exponents are negative. For example, we may have

$$(4)^{-2} \quad \text{which means} \quad \frac{1}{(4)^2}$$

$$\frac{1}{(4)^{-2}} \quad \text{which means} \quad \frac{(4)^2}{1}$$

Example A.1 shows how exponents are used.

EXAMPLE A.1

Perform the indicated operation: (a) $(3)^2$ (b) $(4)^3$ (c) $(3)^1$ (d) $(3)^{-2}$ (e) $\dfrac{1}{2^{-3}}$

Solution

(a) $(3)^2 = 3 \times 3 = 9$

(b) $(4)^3 = 4 \times 4 \times 4 = 64$

(c) $(3)^1 = 3$

(d) $(3)^{-2} = \dfrac{1}{(3)^2} = \dfrac{1}{3 \times 3} = \dfrac{1}{9}$

(e) $\dfrac{1}{(2)^{-3}} = \dfrac{(2)^3}{1} = 2 \times 2 \times 2 = 8$

Sometimes it is necessary to *multiply* exponential numbers that have the *same base*. This is done as shown in Example A.2.

EXAMPLE A.2

Perform the indicated operation:

(a) $(4)^8(4)^2$ (b) $(3)^2(3)^{-1}$ (c) $(8)^2(8)^1$ (d) $(2)^{-2}(2)^{-3}$ (e) $(3)^{-2}(3)^2$

Solution

Keep the same base and *add the exponents algebraically.*

(a) $(4)^8(4)^2 = (4)^{10}$

(b) $(3)^2(3)^{-1} = (3)^1 = 3$

(c) $(8)^2(8)^1 = (8)^3$

(d) $(2)^{-2}(2)^{-3} = (2)^{-5}$

(e) $(3)^{-2}(3)^2 = (3)^0 = 1$ (Any number to the zero power is equal to 1.)

268 ESSENTIAL CONCEPTS OF CHEMISTRY

Sometimes it is necessary to *divide* exponential numbers that have the *same base*. This is done as shown in Example A.3.

EXAMPLE A.3

Perform the indicated operation: (a) $\dfrac{(2)^5}{(2)^2}$ (b) $\dfrac{(2)^5}{(2)^{-2}}$ (c) $\dfrac{(2)^{-5}}{(2)^2}$ (d) $\dfrac{(2)^{-5}}{(2)^{-2}}$

Solution

Keep the same base and subtract the exponent in the denominator from the exponent in the numerator *algebraically*.

(a) $\dfrac{(2)^5}{(2)^2} = (2)^{5-2} = (2)^3$

(b) $\dfrac{(2)^5}{(2)^{-2}} = (2)^{5-(-2)} = (2)^{5+2} = (2)^7$

(c) $\dfrac{(2)^{-5}}{(2)^2} = (2)^{-5-(+2)} = (2)^{-5-2} = (2)^{-7}$

(d) $\dfrac{(2)^{-5}}{(2)^{-2}} = (2)^{-5-(-2)} = (2)^{-5+2} = (2)^{-3}$

In Examples A.2 and A.3, only two numbers have been multiplied or divided. However, sometimes there are more than two numbers in the calculation, as in Example A.4.

EXAMPLE A.4

Perform the indicated operation:

(a) $(5)^3(5)^3(5)^2$ (b) $(a)^5(a)^3(a)^{-4}$ (c) $\dfrac{(b)^3(b)^6}{(b)^2}$ (d) $\dfrac{(2)^5(2)^{-6}}{(2)^{-9}}$ (e) $\dfrac{(a)^{-3}(a)^4(a)^6}{(a)^{-2}(a)^7}$

Solution

(a) $(5)^3(5)^3(5)^2 = (5)^8$ The exponents were added algebraically.

(b) $(a)^5(a)^3(a)^{-4} = (a)^4$ The exponents were added algebraically.

(c) $\dfrac{(b)^3(b)^6}{(b)^2} = \dfrac{(b)^9}{(b)^2} = (b)^{9-2} = (b)^7$ When you work out a problem one step at a time, it becomes easy.

(d) $\dfrac{(2)^5(2)^{-6}}{(2)^{-9}} = \dfrac{(2)^{-1}}{(2)^{-9}} = (2)^8$ (e) $\dfrac{(a)^{-3}(a)^4(a)^6}{(a)^{-2}(a)^7} = \dfrac{(a)^7}{(a)^5} = (a)^2$

A.4 Working with Units

Almost all numbers, in chemistry or anywhere else, are accompanied by units. It is important to be able to work with these units as well as with the numbers. Study the following examples carefully.

EXAMPLE A.5

How many feet are there in 36 inches?

Solution

We are going to use what is known as the **factor-unit method** of analysis. Here's how it works. Because there are 12 inches in 1 foot, we can express this relationship mathematically as

$$\frac{12 \text{ inches}}{1 \text{ foot}} \quad \text{(reads "12 inches per 1 foot")}$$

or as

$$\frac{1 \text{ foot}}{12 \text{ inches}} \quad \text{(reads "1 foot per 12 inches")}$$

These are called factor units. We set the problem up so that when we multiply the quantity in question (36 inches) by the appropriate factor unit, we end up with what we want (feet). Here, we have

$$? \text{ feet} = (36 \text{ inches})\left(\frac{1 \text{ foot}}{12 \text{ inches}}\right) = 3 \text{ feet}$$

Notice how the unit *inches* cancels out in Example A.5 when the numbers are multiplied. You may wonder how you'd know that you must use the factor unit

$$\left(\frac{1 \text{ foot}}{12 \text{ inches}}\right) \text{ and not the factor unit } \left(\frac{12 \text{ inches}}{1 \text{ foot}}\right)$$

That's easy. If you used the wrong factor unit, the term *inches* wouldn't cancel out. You would arrive at

$$? \text{ feet} = (36 \text{ inches})\left(\frac{12 \text{ inches}}{1 \text{ foot}}\right) = \frac{432 (\text{inches})^2}{1 \text{ foot}}$$

which has no meaning to you. Try the problems in the following examples for practice in using the factor-unit method.

EXAMPLE A.6

How many inches are there in 5 feet?

Solution

Set the problem up this way:

$$? \text{ inches} = (5 \text{ feet})\left(\frac{12 \text{ inches}}{1 \text{ foot}}\right) = 60 \text{ inches}$$

Note how the unit *feet* cancels out when the numbers are multiplied with their units.

EXAMPLE A.7

How many minutes are there in 4 hours? How many seconds are there in 4 hours?

Solution

There are 60 minutes in 1 hour. So the factor unit is

$$\frac{60 \text{ minutes}}{1 \text{ hour}}$$

The problem is solved this way:

$$? \text{ minutes} = (4 \cancel{\text{ hours}})\left(\frac{60 \text{ minutes}}{1 \cancel{\text{ hour}}}\right) = 240 \text{ minutes}$$

There are 60 seconds in 1 minute. Our factor unit for this part of the problem is

$$\frac{60 \text{ seconds}}{1 \text{ minute}}$$

So the problem is solved as follows:

$$? \text{ seconds} = (240 \cancel{\text{ minutes}})\left(\frac{60 \text{ seconds}}{1 \cancel{\text{ minute}}}\right) = 14{,}400 \text{ seconds}$$

EXAMPLE A.8

How many gallons are there in 60 quarts?

Solution

There are 4 quarts in 1 gallon. Therefore

$$? \text{ gallons} = (60 \cancel{\text{ quarts}})\left(\frac{1 \text{ gallon}}{4 \cancel{\text{ quarts}}}\right) = 15 \text{ gallons}$$

EXAMPLE A.9

Change 60 miles per hour to feet per second.

Solution

Here you can really see the value of the factor-unit method. However, we have to know that 1 mile equals 5,280 feet and that 1 hour equals 3,600 seconds. Now, 60 miles per hour means

$$60\frac{\text{miles}}{\text{hour}} \quad \text{which is the same as} \quad \frac{60 \text{ miles}}{1 \text{ hour}}$$

So the complete calculation reads

$$? \frac{\text{feet}}{\text{seconds}} = \left(\frac{60 \cancel{\text{ miles}}}{1 \cancel{\text{ hour}}}\right)\left(\frac{5{,}280 \text{ feet}}{1 \cancel{\text{ mile}}}\right)\left(\frac{1 \cancel{\text{ hour}}}{3{,}600 \text{ seconds}}\right)$$

$$= 88 \frac{\text{feet}}{\text{second}}$$

A.5 Decimals

Chemistry is such an exact science that the measurements we take usually consist of a string of numbers with a decimal point—for example, 2.7135 grams. When we are working with decimal numbers, we have to keep certain rules in mind.

Addition and Subtraction

In adding or subtracting numbers, it is important to line up the columns correctly. Keep the decimal points stacked one over the other.

APPENDIX A: BASIC MATHEMATICS FOR CHEMISTRY

EXAMPLE A. 10

Perform the indicated operation: (a) 25.8 + 107.09 + 88.004 + 0.011 (b) 342.78 − 14.99

Solution

Line up the numbers properly.

```
(a)    25.8        (b)    342.78
      107.09              − 14.99
       88.004             327.79
        0.011
      220.905
```

Multiplication

When we multiply 2.4 × 1.6, where does the decimal point go? First we write the numbers and perform the multiplication, ignoring the decimal point.

```
         2.4
       × 1.6
         144
          24
         384
```

To find out where the decimal point goes, we count the number of digits to the right of the decimal point in each term. (There is one digit to the right of the decimal point in each term.) Then we add these numbers of digits (one plus one equals two) and count off this new number from the right-hand side of the answer. In our example, we count off two digits from the right, and we place the decimal point between the 3 and the 8. This gives us an answer of 3.84.

EXAMPLE A. 11

Perform the indicated operation: (a) (4.25)(5) (b) (7.12)(3.64) (c) (5.222)(4.11)

Solution

```
(a)   4.25     (b)    7.12      (c)    5.222
       × 5          × 3.64            × 4.11
     21.25           2848              5222
                     4272              5222
                     2136             20888
                    25.9168           21.46242
```

Division

Where does the decimal point go when we divide 11 by 0.56? First we write the problem in a more familiar way.

$$0.56\overline{)11}$$

(Here 0.56 is called the **divisor**, and 11 is called the **dividend**.) Our next step is to move the decimal point in the divisor so that all digits are to the left of the decimal point: 0.56 becomes 56. We must also move the decimal point in the dividend the same number of places: 11 becomes 1,100.

$$0.56\overline{)11.00} \quad \text{becomes} \quad 56\overline{)1{,}100.00}$$

We now perform the division, locating the decimal point in the answer directly above the decimal point in the dividend.

```
           19.64
   56)1,100.00
       56xxx
        540
        504
         360
         336
          240
          224
           16
```

EXAMPLE A.12

Perform the indicated operation: (a) $\dfrac{9.25}{2.5}$ (b) $\dfrac{24.138}{7.45}$

Solution

(a) $2.5\overline{)9.25}$ (b) $7.45\overline{)24.138}$

```
         3.7                    3.24
   25)92.5              745)2,413.80
      75 x                   2235 xx
      175                     1788
      175                     1490
        0                     2980
                              2980
                                 0
```

A.6 Solving Algebraic Equations

Very often in chemistry, we need to solve algebraic equations ("find the unknown"). So study the following examples carefully.

EXAMPLE A. 13

Solve each of the following equations for the unknowns a, y, and z:

(a) $3a = 9$ (b) $4y + 5 = 37$ (c) $9z - 3 = 22$

Solution

To solve an algebraic equation, we try to isolate the unknown.

(a) In $3a = 9$, we can isolate the a by dividing both sides of the equation by 3. This operation removes the 3 from the left-hand side of the equation. Remember, what we do to one side of the equation we *must* do to the other side, in order to maintain the equality.

$$\frac{\cancel{3}a}{\cancel{3}} = \frac{9}{3}$$

After simplifying both sides of the equation, we get the equality

$$a = 3$$

(b) In $4y + 5 = 37$, we want to isolate the y. Our first step is to remove the 5 from the left-hand side of the equation. We do this by subtracting 5 from both sides of the equation.

$$4y + 5 - 5 = 37 - 5 \quad \text{so} \quad 4y = 32$$

We now divide both sides of the equation by 4.

$$\frac{\cancel{4}y}{\cancel{4}} = \frac{32}{4}$$

After simplifying both sides of the equation, we get the equality

$$y = 8$$

(c) In $9z - 3 = 22$, the first step is to remove the 3 from the left-hand side of the equation. We do this by adding 3 to both sides of the equation.

$$9z - 3 + 3 = 22 + 3 \quad \text{so} \quad 9z = 25$$

Now divide both sides of the equation by 9.

$$\frac{\cancel{9}z}{\cancel{9}} = \frac{25}{9}$$

After simplifying both sides of the equation, we have the equality

$$z = \frac{25}{9}$$

EXAMPLE A.14

Solve each of the following equations for the unknown: (a) $\dfrac{2a}{3} = \dfrac{12}{9}$ (b) $5y + 3 = 2y - 42$

Solution

(a) In the equation

$$\frac{2a}{3} = \frac{12}{9}$$

We can remove the 3 on the left-hand side by multiplying both sides of the equation by 3.

$$(3)\left(\frac{2a}{3}\right) = \left(\frac{12}{9}\right)(3) \quad 2a = \frac{36}{9} \quad 2a = 4$$

Now divide both sides of the equation by 2.

$$\frac{\cancel{2}a}{\cancel{2}} = \frac{4}{2}$$

After simplifying both sides of the equation, we have the equality

$$a = 2$$

274 ESSENTIAL CONCEPTS OF CHEMISTRY

(b) In the equation
$$5y + 3 = 2y - 42$$
we first gather all the terms containing the unknown on one side. Place all y terms on the left-hand side of the equation by subtracting $2y$ from each side of the equation.
$$5y + 3 - 2y = 2y - 42 - 2y \quad \text{so} \quad 3y + 3 = -42$$
Then subtract 3 from each side of the equation.
$$3y + 3 - 3 = -42 - 3 \quad \text{so} \quad 3y = -45$$
Now divide each side of the equation by 3.
$$\frac{3y}{3} = \frac{-45}{3}$$
After simplifying both sides of the equation, we have the equality
$$y = -15$$

A.7 Ratios and Proportions

There are four wheels on a car. This relationship can be expressed mathematically in the following ways:

$$4 \text{ wheels} : 1 \text{ car} \quad \text{or} \quad \frac{4 \text{ wheels}}{1 \text{ car}}$$

Both statements express the fact that there are four wheels on one car. Such statements are known as **ratios** because they reveal the numerical relationship between different things. A ratio is just like a factor unit, and it can be used as an alternative to the factor-unit method to solve many types of problems. But to use ratios in problem solving, you must understand the concept of proportions. A **proportion** is *an equality between two ratios*, as in

$$\frac{1}{2} = \frac{5}{10} \quad \text{or} \quad 1:2 = 5:10$$

(When the second expression is used, the symbol : is often read "is to" and the symbol = is read "as.") In a problem involving a proportion, you are usually given three of the four numbers and are asked to determine the missing number. Example A.15 shows how this works.

EXAMPLE A.15

If there are 4 wheels on 1 car, how many wheels are there on 5 cars?

Solution

The proportion is set up in the following way:

$$\frac{4 \text{ wheels}}{1 \text{ car}} = \frac{y \text{ wheels}}{5 \text{ cars}}$$

To solve for y, multiply each side of the equation by 5 cars.

$$(5 \text{ cars})\left(\frac{4 \text{ wheels}}{1 \text{ car}}\right) = \left(\frac{y \text{ wheels}}{5 \text{ cars}}\right)(5 \text{ cars})$$

$$20 \text{ wheels} = y \text{ wheels}$$

In other words, $y = 20$ wheels, so there are 20 wheels on 5 cars.

A.8 Solving Word Equations

Many of the mathematical equations we use in chemistry are word equations, such as

$$\text{Density} = \frac{\text{mass}}{\text{volume}} \quad \text{sometimes abbreviated as } D = \frac{m}{V}$$

As written, the equation is set up so that we can solve for the density of a substance if we are given its mass and volume. In some problems, however, we are given the density and volume of a substance and are asked to solve for its mass. We do so as shown in Example A.16.

EXAMPLE A.16

Given the following information, solve for the mass m:

$$D = 0.8 \, \frac{\text{g}}{\text{mL}} \quad V = 30 \text{ mL}$$

Solution

First write the formula.

$$D = \frac{m}{V}$$

Then isolate the m by multiplying both sides of the equation by V.

$$(D)(V) = \left(\frac{m}{\cancel{V}}\right)(\cancel{V}) \quad \text{so} \quad DV = m$$

Now substitute the numbers given in the problem and solve for the mass in grams.

$$\left(0.8 \, \frac{\text{g}}{\cancel{\text{mL}}}\right)(30 \, \cancel{\text{mL}}) = m \quad \text{so} \quad 24 \text{ g} = m$$

In some problems, we are given the density and mass of a substance and are asked to solve for its volume.

EXAMPLE A.17

Given the following information, solve for the volume V:

$$D = 2 \, \frac{\text{g}}{\text{mL}} \quad m = 20 \text{ g}$$

Solution

Write the density formula in its original form.

$$D = \frac{m}{V}$$

We must get the V into the numerator by itself. This can be done in two steps:

1. Multiply both sides of the equation by V.

$$(D)(V) = \left(\frac{m}{\cancel{V}}\right)(\cancel{V}) \quad \text{so} \quad DV = m$$

2. Divide both sides of the equation by D.

$$\frac{\cancel{D}V}{\cancel{D}} = \frac{m}{D} \quad \text{so} \quad V = \frac{m}{D}$$

Now substitute the numbers given in the problem and solve for the volume in millimeters.

$$V = \frac{m}{D} \quad \text{so} \quad V = \frac{20 \text{ g}}{2 \text{ g/mL}} = 10 \text{ mL}$$

Do you understand how the unit *gram* cancels out to give milliliters in the answer? We actually have

$$V = 20 \text{ g} \div \frac{2 \text{ g}}{1 \text{ mL}}$$

When we apply the *rules for division of fractions* discussed in Section A.2, this expression becomes

$$20 \text{ g} \times \frac{1 \text{ mL}}{2 \text{ g}} = 10 \text{ mL}$$

and the unit *gram* cancels out.

A.9 Calculating and Using Percentages

A **percentage** is *the number of parts of something out of 100 parts*. For example, if a chemistry class has 100 students and 40 of the students are women, the percentage of women is 40 percent (also written as 40%).

$$40 \text{ women per } 100 \text{ students} = 40\%$$

We can state this mathematically (using the factor-unit method) as follows:

$$\text{Percentage women} = \left(\frac{40 \text{ women}}{100 \text{ students}}\right)(100 \text{ students})$$
$$= 40 \text{ percent women}$$

However, we usually set the formula up as follows:

$$\textbf{Percentage} = \frac{\text{number of items of interest}}{\text{total number of items}} \times 100$$

Suppose that we have a chemistry class with 60 students and that 48 of them are women. What is the percentage of women in the class? By the formula,

$$\text{Percentage} = \frac{\text{number of items of interest}}{\text{total number of items}} \times 100$$
$$= \frac{48 \text{ women}}{60 \text{ students}} \times 100 = 80\% \text{ women}$$

EXAMPLE A.18

A box of mixed vegetables has 25 carrots, 30 tomatoes, and 95 heads of lettuce. What is the percentage of each vegetable in the box?

Solution

First obtain the total number of vegetables in the box.

$$25 \text{ carrots} + 30 \text{ tomatoes} + 95 \text{ lettuces} = 150 \text{ vegetables}$$

Now calculate the percentage of each vegetable, using our formula.

$$\% \text{ carrots} = \frac{25 \text{ carrots}}{150 \text{ vegetables}} \times 100 = 16.7\% \text{ carrots}$$

$$\% \text{ tomatoes} = \frac{30 \text{ tomatoes}}{150 \text{ vegetables}} \times 100 = 20\% \text{ tomatoes}$$

$$\% \text{ lettuces} = \frac{95 \text{ lettuces}}{150 \text{ vegetables}} \times 100 = 63.3\% \text{ lettuces}$$

Note that the total of the percentages equals 100%. In other words, the whole is the sum of the parts.

Sometimes we know the percentage of something and we want to solve for the particular number of items. For example, say that 30% of the people in a particular community own a car. If this community has 800 people, how many people own cars? We can solve this problem using the factor-unit method if we remember that 30% car ownership means

$$\frac{30 \text{ people own cars}}{100 \text{ people in town}}$$

Therefore

$$? \text{ people owning cars} = (800 \text{ people}) \left(\frac{30 \text{ people owning cars}}{100 \text{ people}} \right)$$
$$= 240 \text{ people own cars}$$

However, a simpler method of solving this problem is to move the decimal point on the percent number two places to the left and multiply it by the number of people in town. In this method, 30% becomes 0.30. Therefore

$$800 \text{ people} \times 0.30 = 240 \text{ people own cars}$$

EXAMPLE A.19

A box of fruit contains apples, oranges, and bananas. There are 500 pieces of fruit in the box; 20% are apples, 30% are oranges, and 50% are bananas. How many pieces of each kind of fruit are in the box?

Solution

Turn each percent into its decimal equivalent and multiply it by the total number of fruit in the box.

$$20\% \text{ becomes } 0.20 \text{ (for apples)}$$
$$30\% \text{ becomes } 0.30 \text{ (for oranges)}$$
$$50\% \text{ becomes } 0.50 \text{ (for bananas)}$$

Therefore

$$500 \text{ fruit} \times 0.20 = 100 \text{ apples}$$
$$500 \text{ fruit} \times 0.30 = 150 \text{ oranges}$$
$$500 \text{ fruit} \times 0.50 = 250 \text{ bananas}$$

Note that the total number of pieces of fruit is 500.

A.10 Using the Calculator

If they are used correctly, electronic calculators speed up the process of making mathematical computations. One common error is the simple mistake of pushing the buttons incorrectly (in the wrong order, for instance). Many such errors can be detected by first estimating the answer so you can tell whether the calculated answer makes sense.

You should follow the specific directions given for your particular calculator, of course, but the following rules apply to most calculators:

1. Enter the operations in the order in which they are written.

$$\boxed{4} \boxed{\times} \boxed{5} \boxed{=} \boxed{20}$$

First press the "4" key, then the "×" key, then the "5" key, followed by the "=" key. The answer "20" will appear in the electronic display position.

278 ESSENTIAL CONCEPTS OF CHEMISTRY

EXAMPLE A.20

Perform the following operations:

(a) $\boxed{10}\boxed{\times}\boxed{5}\boxed{=}$

(b) $\boxed{10}\boxed{+}\boxed{5}\boxed{=}$

(c) $\boxed{10}\boxed{\div}\boxed{5}\boxed{=}$

(d) $\boxed{10}\boxed{-}\boxed{5}\boxed{=}$

Solution

(a) 50
(b) 15
(c) 2
(d) 5

2. Once the calculator is directed to add, subtract, multiply, or divide two numbers, the results of the mathematical operation are stored in its memory. Another operation key (+, −, /, or ÷) can then be pressed and another calculation carried out.

EXAMPLE A.21

Perform the following operations:

(a) $\boxed{10}\boxed{\times}\boxed{5}\boxed{=}\boxed{+}\boxed{4}\boxed{=}$

(b) $\boxed{20}\boxed{\div}\boxed{5}\boxed{-}\boxed{4}\boxed{=}$

(c) $\boxed{6}\boxed{-}\boxed{4}\boxed{\times}\boxed{20}\boxed{=}$

(d) $\boxed{8}\boxed{+}\boxed{6}\boxed{\times}\boxed{2}\boxed{=}$

Solution

(a) 54
(b) 0
(c) 40
(d) 28

3. Your calculator probably has a floating decimal point. Once you reach the proper place for the decimal point, press the $\boxed{\cdot}$ key to position the decimal point.

4. Final zeros to the right of the decimal point are not shown in the electronic display after a calculation is complete.

SELF-TEST EXERCISES

LEARNING GOAL 1

Adding and Subtracting Algebraically

1. *Add* the following numbers algebraically:

(a) 5 + 12   (b) −5 − 10   (c) 5 + 10   (d) −5 + 10

2. *Subtract* the following numbers algebraically:

(a) 3 − (+2)   (b) 3 − (− 2)   (c) −3 − (+2)   (d) −3 − (− 2)

3. *Add* the following numbers algebraically:

(a) 8 + 25   (b) 8 − 25   (c) − 8 + 25   (d) −8 − 25

APPENDIX A: BASIC MATHEMATICS FOR CHEMISTRY 279

<u>4</u>. *Subtract* the following numbers algebraically:
(a) $55 - (+44)$ (b) $55 - (-44)$ (c) $-55 - (+44)$ (d) $-55 - (-44)$

LEARNING GOAL 2

Multiplying and Dividing Fractions

<u>5</u>. Perform the following operations:
(a) $\dfrac{1}{8} \times \dfrac{1}{2}$ (b) $\dfrac{1}{8} \div \dfrac{1}{2}$ (c) 8^3 (d) $(2)^{-5}$

<u>6</u>. Perform the following operations:
(a) $\dfrac{3}{5} \times \dfrac{2}{7}$ (b) $\dfrac{20}{50} \times \dfrac{9}{5}$ (c) $\dfrac{5}{6} \div \dfrac{10}{3}$ (d) $\dfrac{25}{35} \div \dfrac{7}{5}$

LEARNING GOAL 3

Multiplying and Dividing Exponential Numbers

<u>7</u>. Perform the following operations: (a) $(4)^{-2}(4)^{-5}$ (b) $(4)^2(4)^3$ (c) $(6)^4(6)^{-2}$ (d) $(5)^{-8}(5)^9$

<u>8</u>. Perform the following operations: (a) $\dfrac{(4)^5}{(4)^2}$ (b) $\dfrac{(4)^5}{(4)^{-3}}$ (c) $\dfrac{(a)^3(a)^2}{(a)^4}$ (d) $\dfrac{(b)^6(b)^4}{(b)^3(b)^5}$

<u>9</u>. Perform the following operations:

(a) $(25)^2$ (b) $(10)^3$ (c) $(99)^1$ (d) $(8)^{-3}$ (e) $\dfrac{1}{(4)^{-2}}$

<u>10</u>. Perform the following operations: (a) $(5)^2(5)^3$ (b) $(12)^3(12)^9$ (c) $(10)^2(10)^{-2}$ (d) $(4)^{-8}(4)^{-2}$

<u>11</u>. Perform the following operations:

(a) $\dfrac{(4)^5}{(4)^3}$ (b) $\dfrac{(12)^3}{(12)^{-2}}$ (c) $\dfrac{(20)^{-4}}{(20)^3}$ (d) $\dfrac{(20)^{-4}}{(20)^{-3}}$

<u>12</u>. Perform the following operations:

(a) $(4)^4(4)^3(4)^2$ (b) $(x)^2(x)^3(x)^{-4}$ (c) $\dfrac{(y)^2(y)^8}{(y)^3}$ (d) $\dfrac{(p)^2(p)^{-9}}{(p)^{-8}}$ (e) $\dfrac{(10)^{-3}(10)^5(10)^7}{(10)^{-8}(10)^9}$

<u>13</u>. Perform the following operations: (a) $(2)^6$ (b) $(3)^{-4}$ (c) $\dfrac{1}{(4)^3}$ (d) $\dfrac{1}{(4)^{-3}}$

<u>14</u>. Perform the following operations: (a) $(10)^5(10)^{-2}$ (b) $(2)^{-3}(2)^{-2}$ (c) $(3)^{-5}(3)^2$ (d) $(6)^{-8}(6)^{10}$

<u>15</u>. Perform the following operations:

(a) $\dfrac{(3)^7}{(3)^5}$ (b) $\dfrac{(2)^4(2)^{-6}}{(2)^7(2)^{-12}}$ (c) $\dfrac{(a)^{12}(a)^{-10}}{(a)^{-7}(a)^{-6}}$ (d) $\dfrac{(b)^{-20}(b)^{-10}}{(b)^{-9}(b)^{-8}}$

LEARNING GOAL 4

Solving Problems Using the Factor-Unit Method

<u>16</u>. $(4 \text{ feet})\left(\dfrac{12 \text{ inches}}{1 \text{ foot}}\right) = ?$

<u>17</u>. $A = l \times w$; therefore $w = ?$ Solve the equation for w in terms of A and l.
<u>18</u>. How many yards are there in 18 feet?
<u>19</u>. How many eggs are there in 4 dozen eggs?
<u>20</u>. (a) How many hours are there in 360 minutes?
(b) How many hours are there in 7,200 seconds?

280 ESSENTIAL CONCEPTS OF CHEMISTRY

21. How many pints are there in 8 quarts?
22. Change 1 mile per hour to feet per second.
23. $(72 \text{ inches})\left(\dfrac{1 \text{ foot}}{12 \text{ inches}}\right) = ?$
24. $(15 \text{ feet})\left(\dfrac{12 \text{ inches}}{1 \text{ foot}}\right) = ?$
25. How many feet are there in 100 yards?
26. How many inches are there in 5 yards?
27. How many dozen eggs do you have if you have 288 eggs?
28. How many minutes are there in 24 hours? How many seconds are there in 24 hours?
29. How many gallons are there in 40 quarts?
30. A rocket is traveling at 22,000 feet per second. How many miles per hour is it traveling?

LEARNING GOAL 5

Adding and Subtracting Decimals

31. Perform the following operations:
    (a) $15.4 + 117.33 + 16.909 + 0.044$ (b) $171.82 - 30.41$
32. Perform the following operations:
    (a) $25.431 + 0.761 + 0.325 + 0.008$
    (b) $123.25 - 19.54$
    (c) $98.77 - 38.25 + 45.62$
    (d) $254.37 - 68.26 - 38.33$

LEARNING GOAL 6

Multiplying and Dividing Decimals

33. Perform the following operations:
    (a) $(7.33)(4)$ (b) $(9.01)(4.28)$ (c) $(3.111)(8.7)$
34. Perform the following operations:
    (a) $\dfrac{6.324}{3.1}$ (b) $\dfrac{25.30}{5.06}$
35. Perform the following operations:
    (a) $(9.54)(3.27)$ (b) $(6.5)(7.1)$ (c) $\dfrac{80.4}{2.02}$
    (d) $\dfrac{640.75}{8.5}$

LEARNING GOAL 7

Solving Simple Algebraic Equations

36. Solve each of the following equations for the unknown:
    (a) $25a = 100$ (b) $40y + 13 = 26$ (c) $12z - 12 = 24$

37. Solve each of the following equations for the unknown: (a) $\dfrac{6x}{3} = \dfrac{36}{12}$ (b) $12y + 6 = 6y + 12$

38. Solve the equation $PV = nRT$ for (a) P, (b) n, (c) T, (d) n/V.

39. Solve each of the following equations for the unknown:

(a) $4t = 64$ (b) $6y + 3 = 45$ (c) $2x - 7 = 33 - 3x$ (d) $40k + 8 = 96 + 51k$

40. Solve the equation $A = bcd$ for (a) b, (b) c, (c) A/b.

LEARNING GOAL 8

Solving Proportions

41. Solve each of the following equations for the unknown:

(a) $\dfrac{x}{25} = \dfrac{4}{5}$ (b) $\dfrac{3}{y} = \dfrac{7}{63}$ (c) $\dfrac{4t}{12} = \dfrac{2}{3}$ (d) $\dfrac{2}{5k} = \dfrac{5}{50}$

42. Solve each of the following proportions:

(a) $\dfrac{k}{5} = \dfrac{9}{45}$ (b) $\dfrac{2}{7} = \dfrac{h}{28}$ (c) $\dfrac{72}{f} = \dfrac{9}{2}$ (d) $\dfrac{18}{6} = \dfrac{12}{y}$

43. Solve each equation for the missing value or values:

(a) $2:5:9 = x:20:y$ (b) $7:9 = h:54$ (c) $2:3:5:7 = 1:a:b:c$ (d) $2:5 = 5:d$

LEARNING GOAL 9

Solving Simple Density Problems

44. Given that $D = \dfrac{m}{V}$, solve this formula for (a) m and (b) V.

45. A substance has a mass of 25 g and a volume of 5 mL. What is its density?

LEARNING GOAL 10

Solving Problems Involving Percentages

46. A university has the following enrollments: (a) liberal arts, 200 students; (b) laboratory technology, 100 students; and (c) health science technology, 300 students. This accounts for all the students at the university. What percentage of the students are enrolled in each curriculum?

47. A survey of 2,000 people shows that 40% want to travel to Europe, 30% to Asia, 10% to Alaska, and 20% to Miami. How many people want to travel to each place?

LEARNING GOAL 11

Performing Mathematical Operations Using a Calculator

48. Perform the following operations using a calculator:

(a) $11 \times 6 = ?$

(b) $11 + 6 = ?$

(c) $30 \div 5 = ?$

(d) $30 - 5 = ?$

49. Perform the following operations using a calculator:
 (a) $(12 \times 5) + 5 = ?$
 (b) $(20 \div 4) - 2 = ?$
 (c) $(10 + 20) \div (5 \times 6) = ?$
 (d) $(100 \times 50) \div (10 \times 40) = ?$

50. Perform the following operations using a calculator:
 (a) $(50 \times 6) \div 12 = ?$
 (b) $(100 + 50) \div (15 \times 0.01) = ?$
 (c) $(0.05 \times 0.30) + 4 = ?$
 (d) $(10{,}000 \times 5) \div 4{,}000 = ?$

APPENDIX B

IMPORTANT CHEMICAL TABLES

Table B.1 Prefixes and abbreviations
Table B.2 The metric system
Table B.3 Conversion of units (English-Metric)
Table B.4 Solubilities

TABLE B.1 Prefixes and Abbreviations

nano- = 0.000000001	= 10^{-9}	nanometer = nm	centigram = cg
micro- = 0.000001	= 10^{-6}	micrometer = μm (Greek letter mu)	decimeter = dm
milli- = 0.001	= 10^{-3}	millimeter = mm	decigram = dg
centi- = 0.01	= 10^{-2}	milliliter = mL	kilometer = km
deci- = 0.1	= 10^{-1}	milligram = mg	kilogram = kg
deka- = 10	= 10^{1}	centimeter = cm	
kilo- = 1,000	= 10^{3}		

TABLE B.2 The Metric System

Length
1 millimeter = 0.001 meter = $\frac{1}{1000}$ meter   or   1 meter = 1,000 millimeters
1 centimeter = 0.01 meter = $\frac{1}{100}$ meter   or   1 meter = 100 centimeters
1 decimeter = 0.1 meter = $\frac{1}{10}$ meter   or   1 meter = 10 decimeters
1 kilometer = 1,000 meters   or   1 meter = 0.001 kilometer
Mass
1 microgram = 0.000001 gram   1 gram = 1,000,000 micrograms
1 milligram = 0.001 gram   1 gram = 1,000 milligrams
1 centigram = 0.01 gram   1 gram = 100 centigrams
1 decigram = 0.1 gram   1 gram = 10 decigrams
1 kilogram = 1,000 grams   1 gram = 0.001 kilogram
Volume
1 milliliter = 0.001 liter
1 milliliter = 1 cubic centimeter*

*Cubic centimeter is abbreviated cm^3 or cc.

TABLE B.3 Conversion of Units (English—Metric)

TO CONVERT	INTO	MULTIPLY BY
Length		
inches	centimeters	2.540 cm/in
centimeters	inches	0.3937 in/cm
feet	meters	0.30 m/ft
meters	feet	3.28 ft/m
Weight (mass)		
ounces	grams	28.35 g/oz
grams	ounces	0.035 oz/g
pounds	grams	454 g/lb
grams	pounds	0.0022 lb/g
Volume		
liters	quarts	1.057 qt/L
quarts	liters	0.9463 L/qt

Solubility table

Anions	\	Cations												
	H^+	NH_4^+	K^+	Na^+	Ag^+	Ba^{2+}	Ca^{2+}	Mg^{2+}	Zn^{2+}	Cu^{2+}	Pb^{2+}	Fe^{2+}	Fe^{3+}	Al^{3+}
OH^-	—	S	S	S	—	S	P	I	I	I	I	I	I	I
NO_3^-	S	S	S	S	S	S	S	S	S	S	S	S	S	S
Cl^-	S	S	S	S	I	S	S	S	S	S	P	S	S	S
S^{2-}	S	S	S	S	I	S	P	P	I	I	I	I	—	—
SO_3^{2-}	S	S	S	S	P	P	P	P	P	—	I	P	—	—
SO_4^{2-}	S	S	S	S	P	I	P	S	S	S	I	S	S	S
CO_3^{2-}	S	S	S	S	I	I	I	I	I	—	I	I	—	—
SiO_3^{2-}	I	—	S	S	—	I	I	I	I	I	I	I	—	—
PO_4^{2-}	S	S	S	S	I	I	I	I	I	I	I	I	I	I
CH_3COO^-	S	S	S	S	S	S	S	S	S	S	S	S	S	S

A Table for the Solubility of Salts, Acids, Basics in Water (t° 20-25° C)

Conventional signs

- S — Soluble in water
- I — Insoluble in water
- P — Partially Soluble in water
- — — compounds do not exist or decomposes in water

© Iaryna Turchyniak/Shutterstock.com

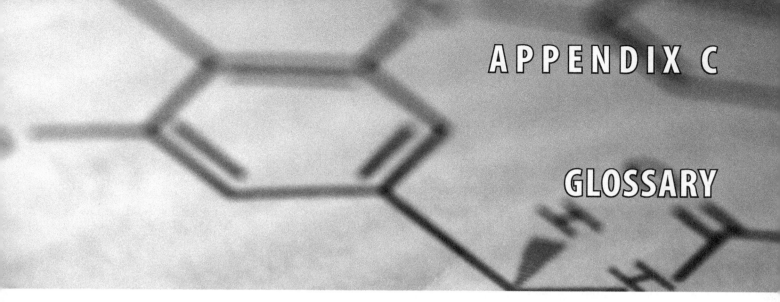

APPENDIX C

GLOSSARY

In each entry, the number in parentheses is the number and section of the chapter in which the term is first discussed.

accuracy (2.4) The closeness of a measurement to its true value.

acid (10.4) A substance that releases hydrogen ions in solution, counteracts bases, and donates protons.

activity series (10.5) The elements listed in decreasing order of their reactivity, or ability to react chemically.

alpha (α) particle (4.4) Particle with a mass of 4 atomic mass units and a charge of +2 emanating from an unstable nucleus. This particle is like a helium atom with its electrons removed.

amorphous solid (3.5) A solid that does not have a well-defined crystalline structure.

anion (6.7) A negatively charged ion.

anode (4.3) A positive electrode in an electrolytic cell.

area (2.6) A measure of the extent of a surface, equal to an object's length times its width.

atom (3.10) The smallest part of an element that can enter into chemical combinations.

atomic mass (3.15) The mass of an element in relation to the mass of an atom of carbon-12.

atomic mass unit (amu) (4.5) The unit used to compare the relative masses of atoms. One atomic mass unit is one-twelfth the mass of a carbon-12 atom.

atomic number (4.6) The number of electrons or protons in a neutral atom.

atomic radius (6.6) The distance from the center of the nucleus to the outermost electron.

Avogadro's number (9.1) The number of atoms whose mass is the gram-atomic mass of any element. It is equal to 6.02×10^{23}.

balanced equation (10.2) A chemical equation that has the same number of atoms of each element on the reactant side as on the product side.

base (10.4) A substance that releases hydroxide ions in solution, counteracts acids, and accepts protons.

base number (2.5) A number that is raised to a power, for example, 10 in 10^3.

binary compound (8.1) A compound composed of two elements.

boiling-point elevation (12.9) The temperature at which a solution boils, which is higher than the boiling point of the pure solvent alone.

287

boiling-point-elevation constant (12.10) The boiling point of a solution minus the boiling point of the pure solvent alone divided by the molality of the solution. This constant is unique for each solvent.

cathode (4.3) A negative electrode in an electrolytic cell.

cation (6.7) A positively charged ion.

Celsius (centigrade) temperature scale (2.11) A temperature scale on which the freezing point of water is 0 degrees and the normal boiling point of water is 100 degrees.

chemical bond (Chapter 7, Introduction) The binding force due to electron loss or gain that holds atoms or ions together in molecules or formula units.

chemical change (3.6) A change in the chemical composition of a substance.

chemical formula (3.14) The combination of the symbols of the particular elements that form a chemical compound, showing the number of atoms per element.

chemical nomenclature (Chapter 8, Introduction) A system for naming chemical compounds.

chemical property (3.6) Those properties that show how one substance reacts with another substance.

coefficient (10.3) A number placed before an element or a compound in a chemical equation to balance the equation.

colligative property (12.9) Any of the collective properties of solutions, such as boiling-point elevation and freezing-point depression, that depend mainly on the number of solute particles present in a given quantity of solvent.

combination reaction (10.4) A reaction in which two or more substances combine to form a more complex substance.

common name (Chapter 8, Introduction) A name for a chemical compound that is derived from common usage or has been handed down through chemical history.

compound (3.11) A chemical combination of two or more elements.

concentration (12.3) The amount of solute in a solution, which can be expressed in terms of percentage, molarity, normality, or molality.

continuous spectrum (5.1) A series of colors in which one color merges into the next, such as a rainbow.

coordinate covalent bond (7.3) A chemical bond in which one atom donates all the electrons used to form the bond.

covalent bond (7.2) A chemical bond formed by the sharing of electrons between two atoms.

crystalline solid (3.5) A solid that has a fixed, regularly repeating, symmetrical internal structure.

decomposition reaction (10.4) A reaction in which a complex substance is broken down into simpler substances.

density (2.10) The mass per unit volume of a substance.

diatomic element (7.2) An element that is found in molecules made up of two like atoms. For example, hydrogen, oxygen, and nitrogen exist in the diatomic forms H_2, O_2, and N_2, respectively.

diffusion (12.11) The process by which a solute moves from an area of high solute concentration to an area of low solute concentration.

dissociate (10.4) To separate an ionic substance into ions by the action of a solvent.

double covalent bond (7.2) A chemical bond in which two pairs of electrons are shared between two atoms.

double-replacement reaction (10.4) A reaction in which two compounds exchange their ions with each other.

electrolytic solution (12.7) A solution that conducts electric current.

electromagnetic spectrum (5.1) The full range of electromagnetic radiation, including radio waves and x rays.

electron (4.3) A particle with a relative negative charge of one unit and a mass of 0.0005486 atomic mass units.

electron affinity (6.8) The energy released when an additional electron is added to a neutral atom.

electron configuration (5.8) The positions of the electrons in the various energy levels of an atom.

electron-dot structure (Lewis electron-dot structure) (7.1) A notation that shows the symbol for an element and, by the use of dots, the number of outer electrons in an atom of the element.

electronegativity (7.6) The attraction that an atom has for the electrons it is sharing with another atom.

element (3.9) Any of the basic building blocks of matter that cannot be broken down physically or by chemical means into simpler substances.

empirical formula (9.2) A chemical formula showing the simplest whole-number ratio of the atoms that make up a molecule of a compound.

energy (3.2) The ability to do work. Energy appears in many forms—for example, as heat and as chemical, electrical, mechanical, and radiant (light) energy.

energy level (5.3) Any of the various regions outside the nucleus of an atom in which electrons move.

energy sublevel (subshell) (5.6) Any of the more specific regions within an energy level in which electrons move

equivalent (of an acid or a base) (12.5) The mass of the acid or base that contains Avogadro's number of hydrogen ions or hydroxide ions.

excited state (5.3) The state of an electron when it is at an energy level higher than its ground-state level. An electron is in an excited state after it has absorbed energy.

exponent (2.5) A number written to the right of and above another number (the base number) and indicating the operation of raising to a power, for example, 3 in 10^3.

Fahrenheit temperature scale (2.11) A temperature scale on which the freezing point of water is 32 degrees and the normal boiling point of water is 212 degrees.

family (6.2) A vertical column of elements in the periodic table.

formula mass (3.16) The sum of the atomic masses of all the atoms that make up a formula unit of a compound.

formula unit (3.13) For an ionic compound, the smallest part of the compound that retains the properties of that compound.

freezing-point depression (12.9) The temperature at which a solution freezes, which is lower than the freezing point of the pure solvent alone.

freezing-point depression constant (12.10) The freezing point of a solution minus the freezing point of the pure solvent alone, divided by the molality. This constant is unique for each solvent.

gas (3.5) The least compact of the three physical states of matter. Gases have no definite shape or volume; they are easily compressible and will spread to fill the container in which they are placed.

gram-atomic mass (9.1) The atomic mass of a mole of atoms of an element expressed in grams.

gram-formula mass (9.3) The formula mass of a substance expressed in grams.

ground state (5.3) The most stable state of an atom. In this state, the electrons are in their lowest possible energy levels.

group (of elements) (6.2) A vertical column of elements in the periodic table.

Haber process (Chapter 11, Introduction) A method used to synthesize ammonia from its elements, nitrogen and hydrogen, for commercial purposes.

heterogeneous matter (3.7) Matter made up of parts with different properties; nonuniform matter.

heterogeneous mixture (3.7) A mixture that consists of two or more substances that retain their own characteristic properties.

homogeneous matter (3.7) Matter made up of parts with similar properties; uniform matter.

homogeneous mixture (3.7) A mixture that consists of two or more substances but is uniform in composition—that is, every part of the mixture is exactly like every other part.

immiscible (12.1) Incapable of forming a solution when mixed.

ion (7.4) An atom or group of atoms that has gained or lost one or more electrons and therefore has a positive or negative charge.

ionic bond (7.4) A bond formed by the transfer of electrons from one atom to another. The atoms involved are always of different elements.

ionization (6.7) The process by which ions are formed from atoms or molecules by the transfer of electrons.

ionization potential (ionization energy) (6.7) The energy needed to remove an electron—generally an outer-shell electron—from an isolated atom.

isotope (4.7) One of two or more atoms that have the same number of electrons and protons but different numbers of neutrons.

Kelvin scale (2.11) An absolute temperature scale in which kelvins equal degrees Celsius + 273.

kinetic energy (3.4) Energy that an object possesses by virtue of its motion.

Law of Conservation of Energy (3.3) Energy can be neither created nor destroyed.

Law of Conservation of Mass (3.3) Matter can be neither created nor destroyed.

Law of Conservation of Mass and Energy (3.3) Matter and energy can be neither created nor destroyed, but they can change from one form to another, and the sum of all the matter and energy in the universe always remains the same.

Law of Definite Composition (3.11) A law that states every compound is composed of elements in a certain fixed proportion.

line spectrum (5.1) A series of colors that shows bright lines separated by dark bands.

liquid (3.5) The physical state of matter in which particles are held together but are free to move about. Liquids have a definite volume but take the shape of the container in which they are placed.

mass (2.9) A measure of the quantity of matter in an object.

mass number (4.7) The mass of a particular atom in atomic mass units. It is essentially the total number of protons and neutrons in the nucleus of the atom.

matter (3.2) Anything that occupies space and has mass.

metal (3.9) An element that conducts electricity and heat, has luster, and takes on a positive oxidation number when it bonds with another element.

metalloid (3.9) An element that has some of the properties of metals and some of the properties of non-metals.

metric system (2.8) A system of measurement based on multiples of 10.

mixture (3.7) A combination of two or more substances that can be separated by physical means.

molality (m) (12.8) A concentration unit for solutions: moles of solute per kilogram of solvent.

molarity (12.4) A concentration unit for solutions: moles of solute per liter of solution.

molar mass (9.3) A general term used to describe the gram-formula mass or gram-atomic mass of a substance.

mole (9.1) 6.02×10^{23} items

molecular formula (9.4) A chemical formula showing the number of atoms of each element in a molecule of a compound.

molecular mass (3.16) The sum of the gram-atomic masses of all of the elements that make up a molecule.

molecule (3.12) The smallest particle of a compound that can enter into chemical reactions and retain the properties of the compound.

monatomic ion (7.4) An ion consisting of a single atom that has taken on a positive or negative charge.

negative ion (anion) (6.7) An atom that has gained one or more electrons and thereby taken on a negative charge.

neutralization reaction (10.4) A reaction in which an acid and a base react to form a salt and water.

neutron (4.5) A particle with no electric charge and a mass of 1.0086650 atomic mass units.

nonelectrolytic solution (12.7) A solution that does not conduct electric current.

nonmetal (3.9) An element that is not a good conductor of heat or electricity and usually takes on a negative oxidation number when it bonds with a metal.

nonpolar covalent bond (7.8) A covalent bond in which the electrons are shared equally by the atoms forming the bond.

normality (N) (12.5) A concentration unit for solutions: equivalents of solute per liter of solution.

nucleus (4.4) The center of an atom, containing most of the mass and one or more units of positive charge.

octet rule (7.2) A general rule stating that atoms with eight valence electrons tend to be nonreactive.

orbital (5.4) A region of space, near the atomic nucleus, in which there is a 95% probability of finding an electron.

osmometry (12.11) A method for determining molecular masses that relates osmotic pressure to the molar concentration of a solute in a solution.

osmosis (12.11) The passage of a solvent through a semipermeable membrane.

osmotic pressure (12.11) The amount of pressure that must be applied to prevent the flow of a solvent through a semipermeable membrane.

outermost shell (5.9) The energy level farthest away from the nucleus of an atom.

oxidation (10.6) The loss of electrons by a substance undergoing a chemical reaction.

oxidation number (8.3) A number that expresses the combining capacity of an element or a polyatomic ion in a compound.

oxidation–reduction (redox) reaction (10.6) A reaction in which one chemical substance is oxidized and another chemical substance is reduced.

oxidizing agent (10.6) A substance that causes something else to be oxidized, or to lose electrons.

percentage composition (9.5) The percentage by mass of each element in a compound.

period (6.2) A horizontal row of elements in the periodic table.

physical property (3.6) A characteristic property such as color, odor, taste, boiling point, or melting point that can be measured by nonchemical means.

plasma (3.5) A form of matter composed of electrically charged atomic particles.

polar covalent bond (7.8) A covalent bond in which there is an unequal sharing of electrons by the atoms forming the bond.

polar molecule (dipole) (7.8) A molecule that is partially positive at one end and partially negative at another end.

polyatomic ion (8.4) A charged group of covalently bonded atoms.

positive ion (cation) (6.7) An atom that has lost one or more electrons and thereby taken on a positive charge.

potential energy (3.4) Energy that is stored in an object by virtue of its position or its inherent chemical energy.

precipitate (10.4) A solid substance that separates out of a solution in the course of a chemical reaction.

precision (2.4) The closeness of repeated measurements to each other.

product (10.1) A substance produced in a chemical reaction.

proton (4.4) A particle with a relative positive charge of one unit and a mass of 1.0072766 atomic mass units.

pure substance (3.7) Matter that has a definite and fixed composition. Elements and compounds are pure substances.

quantum (plural, *quanta*) (5.2) A specific bundle of energy emitted by an electron as it moves from one energy level to another.

quantum mechanics (5.4) A mathematical model of the atom based on the probability of finding electrons in a particular region of space surrounding the nucleus of an atom.

reactant (10.1) Any of the starting materials in a chemical reaction.

redox (10.6) An abbreviation for the term *oxidation-reduction*.

reducing agent (10.6) A substance that causes something else to be reduced, or to gain electrons, while it is oxidized.

reduction (10.6) The gain of electrons by a substance undergoing a chemical reaction.

representative element (6.2) An A-group element. For the A-group elements, the group number indicates how many outer electrons there are.

rounding off (2.4) A process in which one or more digits at the right end of a number are dropped in order to attain the correct number of significant figures.

salt (10.4) A compound composed of the positive ion of a base and the negative ion of an acid.

saturated solution (12.2) A solution in which no more solute can be dissolved.

saturation point (12.3) The level of concentration at which no more solute can dissolve in a given amount of solvent at a particular temperature.

scientific method (3.1) A series of logical steps used by scientists to approach a problem and solve it effectively.

scientific notation (2.5) A number expressed in exponential notation. The number 0.00625 can be 6.25×10^{-3} in scientific notation.

shells (5.3) See energy levels.

significant figures (2.4) Digits that express information that is reasonably reliable.

single covalent bond (7.2) A chemical bond in which a single pair of electrons is shared by two atoms.

single-replacement reaction (10.4) A reaction in which an uncombined element replaces another element that is in a compound is a single replacement reaction.

solid (3.5) The physical state of matter in which particles are held in a definite arrangement. Solids have a definite shape and definite volume.

solute (12.1) In a solution, the substance that is being dissolved.

solution (3.8) A homogeneous mixture.

solvent (12.1) In a solution, the substance that is doing the dissolving.

stoichiometry (11.1) The calculation of the quantities of substances involved in chemical reactions.

supersaturated (12.2) A solution that contains more solute than is needed for saturation for a given quantity of solvent at a particular temperature.

synthesize (Chapter 11, Introduction) To combine reactants to make a product.

systematic chemical name (Chapter 8, Introduction) Any of the names for chemical compounds derived from the naming system developed by the International Union of Pure and Applied Chemistry.

temperature (2.11) A measure of the intensity of heat.

ternary compound (8.1) A compound composed of three elements.

transition metal (6.2) A B-group element.

triple covalent bond (7.2) A chemical bond in which three pairs of electrons are shared by two atoms.

unsaturated solution (12.2) A solution that contains less solute than can be dissolved in it at a particular temperature.

vacuum (4.3) An enclosed space from which all matter has been removed.

volume (2.6) A measure of the capacity of a three-dimensional object.

volumetric flask (12.4) A type of flask used by chemists to prepare a solution with a particular concentration.

weight (2.9) The gravitational attraction of an object to the earth or any other body.

ANSWERS TO SELECTED EXERCISES

This part of the text contains answers to self-test exercises and extra exercises whose numbers are in underscored, boldfaced, italic type at the end of each chapter. In addition, answers to all practice exercises, cumulative review exercises, and exercises in appendix A are included.

Answers to Practice Exercises

2.1 27 **2.2** 53.44 **2.3** 6.3 cm² **2.4** 10 **2.5** 82 **2.6** (a) 2 (b) 4 (c) 6 (d) 4 (e) 2 (f) 5 **2.7** (a) 4.2×10^3 (b) 5.60×10^4 (c) 6.023×10^6 (d) 1.23×10^{-3} **2.8** 301 in³ **2.9** 3,500 cm **2.10** 0.350 g **2.11** 1,800 cm **2.12** 66 cm **2.13** (a) 5.5 cm (b) 0.055 m (c) 0.000055 km or 5.5×10^{-5} km **2.14** (a) 4,400 g (b) 44,000 dg (c) 440,000 cg (d) 4,400,000 mg or 4.4×10^6 mg **2.15** 13.8 ft by 4.10 ft **2.16** 140 L **2.17** 86,400 s **2.18** 2.7 g/cm³ **2.19** 8.0 g/cm³ **2.20** 5.50 g/cm³ liquid A, 2.00 g/cm³ liquid B. Liquid A is more dense. **2.21** $V = 1,300$ cm³ **2.22** 118°C **2.23** 302°F **3.1** (a) 2 carbon atoms, 7 hydrogen atoms, 1 nitrogen atom (b) 2 nitrogen atoms, 8 hydrogen atoms, 1 sulfur atom, 4 oxygen atoms **3.2** (a) 138.9 (b) 55.8 (c) 39.9 (d) 118.7 **3.3** (a) 106.0 (b) 129.9 (c) 87.0 (d) 92.0 **4.1** (a) 14 p, 14 e, 14 n (b) 15 p, 15 e, 16 n (c) 3 p, 3 e, 3 n **4.2** Average mass = 6.941 **4.3** $^{100}Y = 20.0\%$, $^{110}Y = 80.0\%$ **4.4** 909,200 **5.1** (a) 32 electrons (b) 98 electrons **5.2** (a) 32 electrons (b) 4 sublevels: s, p, d, and f (c) s can hold 2, p can hold 6, d can hold 10, and f can hold 14. (d) s sublevel has 1 orbital, p sublevel has 3 orbitals, d sublevel has 5 orbitals, and f sublevel has 7 orbitals. **5.3** (a) $1s^2\,2s^2\,2p^6\,3s^2$ (b) $1s^2\,2s^2\,2p^6\,3s^2\,3p^6\,4s^2\,3d^{10}\,4p^4$ **6.1** The order is S, B, N, and O from longest to shortest radius. **6.2** The order is O, N, B, and S from highest to lowest ionization potential.

7.1 (a) H:N̈:H (b) H:C̈l: (c) :Ö:S̈
                 H   :Ö:

7.2 H—O←S→O—H
              ‖
              O

7.3 (a) Covalent (b) Ionic (c) Covalent (d) Covalent **7.4** (a) 84% ionic, 16% covalent (b) 63% ionic, 37% covalent (c) 55% ionic, 45% covalent (d) 59% ionic, 41% covalent **7.5** CS_2 is most covalent, H_2S is next, then $AsCl_3$ and CO_2 tie for least covalent. **7.6** (a) Nonpolar (b) Polar (c) Polar **7.7** (a) BF_3 bonds are polar; molecule is nonpolar. (b) Cl_2O bonds are polar; molecule is polar. (c) CO_2 bonds are polar; molecule is nonpolar.

8.1 (a) SO_2 (b) N_2O_5 (c) N_2O_4 (d) PBr_5 **8.2** KF **8.3** MgO **8.4** SrS **8.5** K_2O **8.6** $AlBr_3$ **8.7** (a) HgO (b) $FeBr_3$ (c) CoI_2 (d) MnO_2 (e) Cu_2S (f) Fe_2S_3 **8.8** (a) AgCl (b) ZnO **8.9** (a) $Fe(NO_2)_3$

295

(b) $Ba_3(PO_4)_2$ (c) $CuSO_4$ (d) K_2CrO_4 **8.10** (a) Di-phosphorous pentoxide (b) Oxygen difluoride (c) Sulfur dioxide (d) Carbon monoxide **8.11** 2+ **8.12** (a) 4+ (b) 5+ **8.13** (a) Rubidium chloride (b) Gallium sulfide (c) Strontium oxide (d) Zinc iodide (e) Nickel(II) chloride (f) Iron(III) iodide (g) Mercury(II) sulfide (h) Copper(I) oxide **8.14** (a) Ammonium oxide (b) Magnesium cyanide (c) Aluminium nitrite (d) Zinc phosphate (e) Calcium oxalate (f) Nickel(II) borate **8.15** (a) Hydrochloric acid (b) Hydroselenic acid (c) Hydroiodic acid **8.16** (a) Phosphoric acid (b) Chloric acid **8.17** (a) Chlorous acid (b) Nitrous acid **8.18** (a) Perchloric acid (b) Chloric Acid (c) Chlorous Acid (d) Hypochlorous Acid **9.1** (a) 6.00ft (b) 120 in **9.2** (a) 10.0 moles (b) 32.0 g **9.3** (a) 0.500 mole (b) 0.100 mole (c) 20.0 moles (d) 1.00×10^{-4} mole **9.4** (a) 981 g (b) 16.0 g (c) $12\overline{0}$ g **9.5** (a) 9.03×10^{24} atoms (b) 1.20×10^{23} atoms (c) 1.81×10^{24} atoms **9.6** H_2O **9.7** Cr_2S_3 **9.8** ZnN_2O_4 **9.9** (a) 5.0 moles (b) 0.0010 mole **9.10** (a) 184 g (b) 5.0 g (c) 395 g (d) 7.9 g **9.11** (a) 2.41×10^{24} (b) 3.0×10^{22} (c) 1.51×10^{24} (d) 3.6×10^{22} **9.12** $C_9H_8O_4$ **9.13** 32.9% K 67.1% Br **9.14** (a) 85.7% C, 14.3% H (b) 39.3% Na, 60.7% Cl (c) 92.3% C, 7.7% H **9.15** 75.0 g **10.1** $2CuO \rightarrow 2Cu + O_2$ **10.2** $Zn + H_2SO_4 \rightarrow ZnSO_4 + H_2$ **10.3** (a) $2Al + 3Cl_2 \rightarrow 2AlCl_3$ (b) $2Na + 2H_2O \rightarrow 2NaOH + H_2$ (c) $2KNO_3 \rightarrow 2KNO_2 + O_2$ (d) $2HNO_3 + Ba(OH)_2 \rightarrow Ba(NO_3)_2 + 2H_2O$ **10.4** (a) $4K + O_2 \rightarrow 2K_2O$ (b) $2Ca + O_2 \rightarrow 2CaO$ (c) $H_2 + Br_2 \rightarrow 2HBr$ **10.5** (a) $Sr(OH)_2 \rightarrow SrO + H_2O$ (b) $SrCO_3 \rightarrow SrO + CO_2$ (c) $2KClO_3 \rightarrow 2KCl + 3O_2$ (d) $2KCl \rightarrow 2K + Cl_2$ **10.6** (a) $2Li + 2H_2O \rightarrow 2LiOH + H_2$ (b) $2Al + 3H_2SO_4 \rightarrow Al_2(SO_4)_3 + 3H_2$ (c) $3Mg + 2Al(NO_3)_3 \rightarrow 3Mg(NO_3)_2 + 2Al$ **10.7** (a) $2BiCl_3 + 3H_2S \rightarrow Bi_2S_3 + 6HCl$ (b) $2Fe(OH)_3 + 3H_2SO_4 \rightarrow Fe_2(SO_4)_3 + 6H_2O$ **10.8** (a) No reaction occurs. (b) Reaction does occur. **10.9** (a) Copper is oxidized: silver ion is reduced. (b) Sodium is oxidized; chlorine is reduced. (c) Bromide ion is oxidized; chlorine is reduced. (d) Zinc is oxidized; sulfur is reduced. **10.10** (a) Copper is the reducing agent; silver nitrate is the oxidizing agent. (b) Sodium is the reducing agent; chlorine is the oxidizing agent. (c) Sodium bromide is the reducing agent; chlorine is the oxidizing agent. (d) Zinc is the reducing agent; sulfur is the oxidizing agent.

11.1 2.43 g Mg, 1.60 g O **11.2** 0.900 g Al, 7.10 g $Al(NO_3)_3$, 3.27 g Zn **11.3** 49.0 g
12.1 $4\overline{0},000$ cal or $4\overline{0}$ kcal **12.2** 12,500 cal or 12.5 kcal **12.3** 5.380 cal or 5.38 kcal
12.4 2,680 cal or 2.68 kcal **12.5** $C_2H_6(g) + 3\frac{1}{2}O_2(g) \rightarrow 2CO_2(g) + 3H_2O(l) + 372$ kcal
12.6 $2C_2H_{14}(l) + 19O_2(g) \rightarrow 12CO_2(g) + 14H_2O(l) + 1,98\overline{0}$ kcal **12.7** 204.9 kcal heat are produced.
12.8 $PCl_5(g) \rightarrow PCl_3(g) + Cl_2(g) - 22.2$ kcal **12.9** $\Delta H = -75$ kcal **12.10** $\Delta H = -135.4$ kcal
12.11 $\Delta H = -216.6$ kcal **12.12** $\Delta H = -6$ kcal **12.13** $\Delta H = -74.4$ kcal **13.1** (a) 57 torr (b) 7,370 torr
13.2 (a) 0.010 atm (b) 12.0 atm **13.3** 250 mL **13.4** $50\overline{0}$ mL **13.5** 6,080 torr **12.2** (a) No (b) Yes (c) Yes
12.3 (a) Supersaturated (b) Unsaturated (c) Saturated **12.4** (a) 44.4% (b) 30.0% **12.5** (a) 20.0% (b) 55.6%
12.6 (a) 4.5% (b) 20.00% **12.7** (a) 0.200 M (b) 0.0164 M (c) 1.07 M **12.8** (a) 3.71×10^{-2} g **12.9** 49.9 mL
12.10 2.00 N and 2.00 M **12.11** 0.440 N and 0.440 M **15.12** 12.0 g **12.13** 2.00 L of the 6.00 N HCl solution are needed. **12.14** 16.7 mL of the 12.0 N H_3PO_4 solution are needed. **12.15** (a) 1.0 (b) 0.0400 (c) 0.160
12.16 (a) 100.83°C (b) 100.02° **12.17** (a) –2.98°C (b) –0.0744°C **12.18** Water will move out of the cell.

CHAPTER 2

1. (a) Five (b) One (c) Three (d) Four (e) Seven (f) Two (g) One (h) Four (i) Three **3.** (a) Seven (b) Two (c) One (d) Three (e) Seven (f) Three (g) Six (h) Four (i) Six (j) Four **5.** 17.7 cm; three significant figures **7.** 15.0 cm; three significant figures **9.** 20.60 cm **11.** 214 cm^2 **13.** (a) Two (b) Three (c) Four (d) Four (e) One (f) Two (g) three (h) Four (i) Four (j) Three **15.** (a) 6×10^2 (b) 6.00×10^2 (c) 6.000×10^2 (d) 3.2×10^{11} (e) 2×10^{-3} (f) 3.007×10^{-4} (g) 1.5×10^{-7} **17.** (a) 1.0581×10^4 (b) 2.05×10^{-3} (c) 1×10^6 (d) 8.02×10^2 **19.** (a) 8.5×10^8 (b) 6.07×10^{-6} (c) 6.308×10^6 (d) 6.005×10^{-2} (e) 5×10^2 (f) 5.0×10^2 (g) 5.00×10^2 (h) 5.000×10^2 (i) 2.300×10^7 (j) 9.30×10^{-8} **21.** 24.0 m^2, 96.0 m^3 **23.** 426 cm^2 or 66.1 in^2 (depending on how you measure the page) **25.** $6\overline{0}0$ m^2 **27.** $45\overline{0}$ cm^3 **29.** 42,400 cm^3 **31.** 2,560 ft^3, 7.27×10^7 cm^3 **33.** 7.27×10^4 L, 19,300 gal. **35.** 1,728 in^3 **37.** (a) 250 g (b) 2,500 dg (c) 250,000 mg (d) 250,000,000 μg **39.** (a) 31 dm (b) 310 cm (c) 3,100 mm **41.** (a) 14.9 cm (b) 0.149 cm (c) 0.000149 km **43.** (a) $7,85\overline{0}$ mm (b) 785.0 cm (c) 78.50 dm (d) 0.007850 km **45.** (a) 340 m (b) 3,400 dm (c) 34,000 cm (d) 340,000 mm **47.** (a) 218.5 cg (b) 21.85 dg (c) 2.185 g (d) 0.002185 kg **49.** (a) 3.500 L (b) 35.00 dL **51.** 1.00×10^6 cm^3 **53.** (a) 82.0 ft (b) 985 in **55.** (a) 0.32 gal (b) 1.3 qt **57.** (a) 0.0110 lb (b) 0.176 oz **59.** 35.40 ft^3/m^3 **63.** 10.8 ft^2/m^2 **69.** (a) 3.0 m (b) 3.0×10^2 cm (c) 3.0×10^3 mm **71.** (a) 1.83 m (b) 183 cm **73.** (a) 91.5 m (b) 9,150 cm **75.** (a) 18.9 L (b) 18,900 mL **79.** 4.80 g/cm^3 **81.** 203 cm^3 or $20\overline{0}$ cm^3 (for the proper number of significant figures) **83.** 53 g **85.** 0.019 g **87.** 6.83 g/cm^2 **89.** 1.25 g/cm^3 **91.** 110 g **93.** 162 g **95.** 9.32 g/cm^3 **97.** 28,400 g or $29,\overline{0}00$ (for the proper number of significant figures) **99.** 22.1 cm^3 **101.** 3.67 g/cm^3 **103.** 19,500 g **105.** 50.0 cm^3 **107.** 37.8° C, 310.8 K **109.** –40° C. At this temperature, degrees F equal degrees C. **111.** 5.4°F, 104°F **113.** (a) 10°C (b) –70°C (c) 215°C (d) –90.0°C **115.** (a) 203°F (b) –112°F (c) 176°F (d) $41\overline{0}$°F **118.** No; the person would have to be about 13.1 feet tall and weigh $44\overline{0}$ pounds. **119.** 48 cm^3 **120.** 218°C **123.** 820 **124.** –8.89°C **125.** (a) 61°F (b) 392°F **126.** (a) 149°C (b) –101°C **127.** 22.6 miles/hour

CHAPTER 3

5. (a) 2 (b) 5 (c) 3 (d) 4 (e) 1 **7.** (a) Physical (b) Chemical (c) Chemical (d) Physical (e) Chemical (f) Chemical **11.** (a) Metal (b) Metalloid (c) Non-metal (d) Metal (e) Metalloid (f) Metal (g) Metal **13.** Co is the element cobalt, and CO is the compound carbon monoxide. **15.** The present H and O are the same for each experiment. **17.** (a) 12 carbon atoms, 22 hydrogen atoms, 11 oxygen atoms, (b) 2 potassium atoms, 1 chromium atom, 4 oxygen atoms (c) 8 hydrogen atoms, 2 nitrogen atoms, 3 oxygen atoms, 2 sulfur atoms, (d) 1 zinc atom, 2 nitrogen atoms, 6 oxygen atoms **21.** The atomic mass of sulfur would be 2 amu. **23.** (a) 85.5 (b) 52.0 (c) 238.0 (d) 79.0 (e) 74.9 **25.** (a) 71.8 (b) 159.6 (c) 317.3 (d) 164.0 (e) 58.3 (f) 218.5 (g) 791.8 (h) 96.0 **27.** (a) 60.1 (b) 82.1 (c) 121.6 (d) 104.5 (e) 187.5 (f) 218.7 (g) 149.0 (h) 60.0 **35.** (a) Magnesium and chlorine (b) Nitrogen and oxygen (c) Nitrogen, hydrogen, sulfur, and oxygen (d) Hydrogen, phosphorus, and oxygen **37.** (a) Element (b) Mixture (c) Compound (d) Mixture (e) Compound **39.** (a) 254.2 (b) 63.0 (c) 89.8 (d) 601.9

CUMULATIVE REVIEW

CHAPTERS 1–3

1. True **2.** False **3.** False **4.** True **5.** False **6.** True **7.** False **8.** True **9.** True **10.** True **11.** 128 m^2 **12.** 4,548 cm^2 **13.** 16,540 cm^3 **14.** (a) 280 g (b) 2,800 dg (c) 280,000 mg **15.** (a) 68 dm (b) 680 cm (c) 6,800 mm **16.** (a) 12.5 cm (b) 0.125 m (c) 0.000125 km **17.** (a) 25.595 L (b) 255.95 dL **18.** (a) 164 ft (b) 1,970 in **19.** (a) 0.0562 lb (b) 0.899 oz **20.** 16.4 cm^3/in^3 **21.** 0.363 lb **22.** (a) 45.4 L (b) 1,540 fl oz **23.** 0.289 g/cm^3 **24.** 0.177 g/mL. The density of air is 1.18×10^{-3} g/mL, so this gas is heavier than air and will not float in air. **25.** 6.00 g/mL **26.** 4.00 cm **27.** (a) 31.8 (b) 9.29 (c) 79.3 (d) 98.3 **28.** (a) 5×10^3 (b) 5×10^{-4} (c) 6.023×10^8 (d) 3.5000×10^7 **29.** (a) 5 (b) 4 (c) 5 (d) 4 **30.** (a) 7.22°C (b) –23.3°C (c) 230°C (d) –73.33°C **31.** (a) $19\overline{0}$°F (b) 9.5°F (c) $15\overline{0}$°F (d) 752°F **32.** Amorphous solids have no definite internal structure or form. Crystalline solids have a fixed, regularly repeating, symmetrical inner structure. **33.** (a) Chemical

298 ESSENTIAL CONCEPTS OF CHEMISTRY

(b) Physical (c) Chemical (d) Physical **34.** Heterogeneous matter is made up of different parts with different properties. Homogeneous matter is made up of part with the same properties throughout **35.** (a) 1 Zinc atom, 4 carbon atoms, 6 hydrogen atoms, 4 oxygen atoms (b) 2 nitrogen atoms, 8 hydrogen atoms, 1 chromium atom, 4 oxygen atoms **36.** Molecular mass is used for compounds composed of Molecules. Formula mass is used for compounds composed of ions. **37.** Mercury would have a mass of 5 amu. **38.** (a) 173.0 (b) 210.0 (c) 31.0 (d) 107.9 **39.** (a) 232.8 (b) 132.1 (c) 180.0 (d) 251.1 **40.** The formula of $Al_2(SO_4)_3$ means that this formula unit contains 2 aluminum atoms, 3 sulfur atoms, and 12 oxygen atoms. **41.** (a) Mixture (b) Element (c) Compound (d) Compound (e) Element (f) Mixture (g) Mixture **42.** (a) Heterogeneous (b) Homogeneous (c) Heterogeneous (d) Homogeneous **43.** Actinium (Ac), aluminum (Al), americium (Am), argon (Ar), arsenic (As), astatine (At), gold (Au), and silver (Ag) **44.** Hydrogen (H), Helium (He), carbon (C), nitrogen (N), oxygen (O), fluorine (F), neon (Ne), phosphorus (P), sulfur (S), Chlorine (Cl), argon (Ar), selenium (Se), bromine (Br), Krypton (Kr), iodine (I), xenon (Xe), and radon (Rn) **45.** (a) Potassium and sulfur (b) Silver, chromium, and oxygen (c) Potassium manganese, and oxygen (d) Mercury, phosphorus and oxygen **46.** (a) N_2O (b) K_2CrO_4 (c) NH_3 (d) $SrSO_4$ **47.** (a) Chemical (b) Physical (c) Chemical (d) Physical **48.** No is the element nobelium and NO is the compound nitrogen monoxide. **49.** No! During the burning process, some of the atoms that compose paper react with oxygen in the air and form gaseous compounds. **50.** Add water to the mixture and shake. The salt dissolves and the sand settles to the bottom of the container. Filter the sand and collect the salt water (filtrate). To retrieve the pure salt, evaporate the water.

CHAPTER 4

9. 1,836 electrons **13.** (a) 94 p, 94 e, 150 n (b) 22 p, 22 e. 26 n (c) 103 p, 103 e, 159 n **15.** (a) 8 p, 8 e, and 8 n (b) 8 p, 8 e and 9 n (c) 8 p, 8 e, and 10 n (d) 10 p, 10 e, and 10 n (e) 10 p, 10 e, and 11n (f) 10 p, 10 e, and 12 n **17.** (b) is not an isotope of the others **19.** (a) 25 p, 25 e, and 32 n (b) 27 p, 27 e, and 33 n (c) 36 p, 36 e, and 44 n (d) 52 p, 52 e and 76 n **21.** 3 p, 3 e, and 8 n

23.

SYMBOL	PROTONS	ELECTRONS	NEUTRONS	MASS NUMBER	ATOMIC NUMBER
$^{174}_{70}Yb$	70	70	104	174	70
$^{141}_{59}Pr$	59	59	82	141	59
$^{104}_{44}Ru$	44	44	60	104	44
$^{45}_{21}Sc$	21	21	24	45	21
$^{50}_{22}Ti$	22	22	28	50	22
$^{25}_{12}Mg$	12	12	13	25	12

27. Protons, electrons, neutrons **29.** $^{200}_{80}Hg$ (b) $^{15}_{7}N$ (c) $^{27}_{13}Al$ (d) $^{262}_{107}Bh$ **31.** Atomic mass of Ga = 69.72. **33.** 35.46 amu **35.** 52.00 amu

37. The percentage abundances of ^{79}Br is 50.87%, and the percentages abundance of ^{81}Br is 49.13% **39.** ^{121}Sb = 57.696%, ^{123}Sb = 42.304% **41.** 8.82×10^7 atoms ^{22}Ne **43.** 2.0×10^7 oxygen atoms **45.** 7.42×10^6 atoms ^{6}Li

47. 5.59×10^5 atoms ^{26}Mg **49.** 222 atoms ^{13}C **51.** 3.7×10^{12} atoms ^{15}N **53.** 1.0×10^9 hydrogen atoms **56.** (a) C (b) D (c) 222 (d) D (e) Po **57.** They are the same.

59. the Lenard experiment using ultraviolet light **60.** (a) 6 p, 6 e, 7 n (b) 5 p, 5 e, 5 n (c) 7 p, 7 e, 7 n **61.** Isotopes exist. **62.** 2,570 Ne-21 atoms **63.** 1,836 electrons

CHAPTER 5

3. By examining the line spectrum of an extraterrestrial body with a spectroscope, scientists can determine which elements are present. **9.** Energy levels in the atoms **11.** Electrons **13.** Seven energy levels: K, L, M, N, O, P, Q or 1, 2, 3, 4, 5, 6, 7, respectively **23.** (a) 1 (b) 2 (c) 3 (d) 4 **25.** The number of electrons in an energy level is the sum of the electrons in the sublevels.
27. (a) 6 (b) 72 **31.** Total of four orbitals **35.** c and d **37.** (a) $1s^22s^22p^3$ (b) $1s^22s^22p^63s^23p^64s^23d^3$
(c) $1s^22s^22p^63s^23p^64s^23d^{10}4p^65s^24d^{10}5p^6$ (d) $1s^22s^22p^63s^23p^64s^23d^{10}4p^65s^24d^{10}5p^3$
39. (a) $_{56}$Ba (b) $_{20}$Ca (c) $_{38}$Sr (d) $_4$Be **41.** (a) 31 (b) 8 (c) 13 (d) 10 (e) 31 **43.** Elements 6 and 14, elements 3 and 19, elements 17 and 35 **45.** 98 electrons
47. (a) $1s^22s^22p^63s^23p^64s^23d^{10}4p^1$
(b) $1s^22s^22p^63s^23p^64s^23d^{10}4p^65s^24d^{10}5p^3$
(c) $1s^22s^22p^63s^23p^64s^23d^{10}4p^65s^24d^{10}5p^66s^24f^{14}5d^{10}6p^2$
(d) $1s^22s^22p^63s^23p^64s^23d^{10}4p^65s^24d^{10}5p^66s^24f^{14}5d^{10}6p^67s^2$
49. (a) $_{12}$Mg (b) $_{38}$Sr (c) $_{70}$Yb (d) $_{81}$Tl **51.** (a) 60 (b) 12 (c) 24 (d) 20 (e) 4 (f) 60 **55.** The Group VIIIA elements tend to be chemically unreactive. **65.** The atomic number would be 280.

CHAPTER 6

7. (a) Ga (b) S (c) Hg (d) Bh **8.** Period, group or family **9.** The same number of electrons in the outermost energy level **11.** The number of electrons in the outermost energy level **12.** (a) 8 (b) 8 (c) 18 (d) 18 (e) 32 (f) 23 **13.** A-group elements, B-group elements **15.** Ba, Sr, Te, I, Br, **17.** P, As, I, Sb, In
25. (a) decreases (b) increases (c) decreases **27.** Cl, S, Ca, Sr, Rb **29.** Sb, I, As, P **31.** (d) **33.** (c) **35.** (d)
39. (a) $_{11}$Na, $1s^22s^22p^63s^1$ (b) $_{13}$Al, $1s^22s^22p^63s^23p^1$ (c) $_8$O, $1s^22s^22p^4$ (d) $_{16}$S, $1s^22s^22p^63s^23p^4$ (e) $_5$B, $1s^22s^22p^1$ (f) $_3$Li, $1s^22s^1$. Therefore, (a) and (f) are similar, (b) and (e) are similar, and (c) and (d) are similar. **41.** Atomic number 118 **42.** (a) Group IA (b) Group VIIA **44.** (a) K (b) K (c) K **48.** (a) k (b) Se **49.** (a) K (b) Se **50.** Group IVA

CUMULATIVE REVIEW

CHAPTERS 4–6

1. Protons and neutrons in nucleus, electrons surrounding nucleus **2.** Cathode rays came from the cathode. **3.** A magnet would deflect particles, not waves. An electric field would easily deflect particles. **4.** UV light is directed onto a metal and makes it emit electrons. **5.** An atom is a sphere of positive electric charge in which electrons are embedded. The positive particles balance the negative electrons. **6.** Rutherford found that one alpha particle in 20,000 ricocheted. This led him to believe that an atom has a small but dense center of positive charge. **7.** New evidence disproved Thomson's theory. **8.** Neutrons **9.** Isotopes are atoms of the same element with different atomic masses. **10.** (a) 27 protons (b) 40 protons (c) 34 protons **11.** 20.17 **12.** 370 ^{15}N atoms **13.** (a) $^{35}_{17}$Cl (b) $^{28}_{14}$Si **14.** 75% ^{50}Z and 25% ^{52}Z **15.** 32 p, 32 e, 41 n **16.** 26 electrons **17.** ^{16}O is most abundant. **18.** 1900 ^{7}Li atoms **19.** 9.258×10^5 atoms **20.** 1.00×10^7 atoms **21.** Electrons **22.** (a) **23.** Six electrons **24.** (a) K has 1. (b) K has 2 and L has 1. (c) K has 2, L has 8, and M has 1. (d) K has 2, L has 8, M has 8, and N has 1. **25.** Two electrons **26.** sulfur **27.** (b) **28.** It has five electrons in its outermost energy level. **29.** (a) K has 1. (b) K has 2 and L has 1. (c) K has 2, L has 8, and M has 1. (d) K has 2, L has 8, M has 8, N has 1. **30.** $_{70}$Yb or any element with a higher atomic number. **31.** Raindrops act as prisms. **32.** Line spectra are used to analyze the sun and other extraterrestrial bodies. **33.** Line spectra of his hair indicated the presence of the element arsenic. **34.** Our major sources of radiant energy is the sun. **35.** Electromagnetic waves travels through the vacuum of space. **36.** (c) **37.** By producing electromagnetic waves that were

longer than visible light waves **38.** Luminous watch dials absorb light energy and reemit it when they glow. **39.** A sample of an element is heated and it begins to glow. **40.** Quanta **41.** The periodic law states that the chemical properties are periodic function of their atomic numbers. This means that elements with similar chemical properties recur at regular intervals and are placed accordingly on the periodic table. **42.** Moseley was able to determine the nuclear charge or the atomic numbers of the atoms of known elements.
43. $1s^2 2s^2 2p^6 3s^2 3p^6 3d^{10} 4s^2 4p^6 4d^{10}\ 4f^{14} 5s^2 5p^6 5d^{10} 5f^{14} 6s^2 6p^6 6d^5 7s^2$. Element 107 is a Group VIIB element.
44. $1s^2 2s^2 2p^6 3s^2 3p^6 3d^{10} 4s^2 4p^6 4d^{10} 4f^{14} 5s^2 5p^6 5d^{10} 5f^{14} 6s^2 6p^6 6d^5 7s^2$. Element 106 is a Group VIB element.
45. Elements 18–19, 27–28, 52–53, and 92–93. **46.** False **47.** True **48.** False **49.** Fe, Ru, Os **50.** $_{11}$Na **51.** True
52. False **53.** Fluorine **54.** $1s^2 2s^2 2p^6 3s^2 3p^2$ **55.** True **56.** (a) Group IIA (b) Group VIA **57.** (a) Na (b) Ca
58. Mg **59.** Cs **60.** Li

CHAPTER 7

1. (a) ·Äs· (b) ·C̈l: (c) Ra· (d) Ċs (e) :R̈n:
3. (a) ·Ö: (b) :Ö:²⁻ **5.** (a) Mġ· (b) Mg²⁺
9. (a) Mg²⁺ (b) Na¹⁺ (c) Al³⁺
11. :C̈l:¹⁻ or K¹⁺ are two of several examples.
15. Double

19. (a)
$$\text{H} - \underset{\overset{|}{\text{Br}}}{\overset{\overset{\text{Br}}{|}}{\text{C}}} - \text{Br}$$ (b) H—C≡C—H

(c) $\text{H} - \underset{\overset{|}{\text{H}}}{\text{P}} - \text{H}$ (d) H—C≡N̈

21. (a) $:\!\ddot{\text{F}} - \underset{\overset{|}{:\ddot{\text{F}}:}}{\overset{\overset{:\ddot{\text{F}}:}{|}}{\text{C}}} - \ddot{\text{F}}:$ (b) $\text{H} - \underset{\overset{|}{\text{H}}}{\text{C}} = \underset{\overset{|}{\text{H}}}{\text{C}} - \text{H}$

(c) $\text{H} - \underset{\overset{|}{\text{H}}}{\text{As}} - \text{H}$ (d) $\text{H} - \ddot{\text{S}}:$ with H below

23. Ten electrons surround the P in this compound. (PCl₅ structure shown)

25. (a) :Ö=N—Ö—H with :Ö: ↑ above N (b) H—Ö—S—Ö—H with :Ö: ↑ above S

29. (a) Ionic (b) Covalent (c) Covalent (d) Covalent **31.** (a) 4% ionic, 96% covalent (b) 0% ionic, 100% covalent (c) 51% ionic, 49% covalent (d) 63% ionic, 37% covalent **33.** HI, HBr, HCl, HF (Most covalent → Least covalent)

35. (a) 43% ionic, 57% covalent (b) 0.5% ionic, 99.5% covalent (c) 74% ionic, 26% covalent (d) 47% ionic, 53% covalent

37. H_2O, H_2S, H_2Se, H_2Te (Least covalent → Most covalent) **39.** CsCl is the most ionic. **41.** (a) Covalent (b) Covalent (c) Ionic (d) Covalent **45.** (a) Bonds are polar; molecule is polar. (b) Bonds are polar; molecule is polar. (c) Bonds are polar; molecule is polar. (d) Bond is nonpolar; molecule is nonpolar. **47.** (a) Bonds are polar; molecule is nonpolar. (b) Bonds are polar; molecule is polar. (c) Bonds are polar; molecule is nonpolar. (d) Bond is nonpolar; molecule is nonpolar.

50.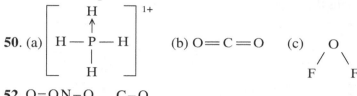

52. O=O N–O   C–O
  Nonpolar bond   Most polar bond of the three

53. C–S bond

54. (a) $[H-\overset{H}{\underset{H}{N}}-H]^{1+}$ (b) Cl–Br **55.** (a) MgO (b) Al_2O_3 (c) Cs_2S

59. There is no way that the duet rule can be satisfied for a molecule containing three hydrogen atoms.

CHAPTER 8

3. Hydrogen oxide or dihydrogen oxide **5.** (a) 7 (b) 2 (c) 3 (d) 8 (e) 1 **7.** (a) P_2S_5 (b) ClO_2 (c) N_2O_4 (d) Cl_2O_7 **9.** (a) Al_2S_3 (b) Li_2O (c) Na_3N (d) Sr_3P_2 (e) AgBr (f) ZnO **11.** (a) IX (b) VIII (c) VII (d) VI **13.** (a) Metal ion with the higher charge (b) Metal ion with the lower charge (c) The suffix for the atom or ion that has a negative charge in a chemical compound. **15.** (a) Cu_2S (b) $HgCl_2$ (c) FeO (d) SnI_2 (e) $CoBr_3$ (f) Hg_3N_2 **17.** (a) $(NH_4)^{1+}$ (b) $(OH)^{1-}$ (c) $(BO_3)^{3-}$ (d) $(HCO_3)^{1-}$ (e) $(C_2O_4)^{2-}$ (f) $(SO_3)^{2-}$ **19.** (a) Arsenate (b) Permanganate (c) Sulfite (d) Ammonium (e) Hydrogen sulfate (f) Acetate **21.** (a) $Hg_3(PO_4)_2$ (b) $Sn_3(AsO_4)_2$ (c) $Fe(C_2H_3O_2)_3$ (d) Li_3PO_4 (e) $Al_2(SO_3)_3$ (f) $Zn(NO_2)_2$ **23.** (a) $PbSO_4$ (b) $CO_3(PO_4)_2$ (c) $(NH_4)_2Cr_2O_7$ (d) CaC_2O_4 (e) $Sn(NO_3)_2$ (f) $Mg(HSO_4)_2$ **25.** (a) Diphosphorus pentasulfide (b) Carbon monoxide (c) Silicon dioxide (d) Chlorine dioxide **27.** (a) Aluminum oxide (b) Sodium iodide (c) Zinc chloride (d) Magnesium nitride (e) Silver Sulfide (f) Lithium iodide **29.** (a) 1– (b) 2.67+ (c) 4+ (d) 5+ (e) 1– **31.** (a) 5+ (b) 1+ (c) 1+ (d) 15+ (e) 3– (f) 1– **33.** (a) Osmium (VIII) oxide (b) Mercury(I) phosphide or mercurous phosphide (c) Iron(II) sulfide or ferrous sulfide (d) Cobalt(II) chloride or cobaltous chloride (e) Copper(I) nitride or cuprous nitride (f) Copper(I) oxide or cuprous oxide **35.** (a) Silver carbonate (b) Mercury(II) phosphate or mercuric phosphate (c) Iron(III) sulfate or ferric sulfate (d) Sodium nitrate (e) Copper(II) chromate or cupric chromate (f) Zinc hydroxide **37.** (a) Calcium sulfate (b) Potassium cyanide (c) Aluminum phosphate (d) Copper(I) oxalate or cuprous oxalate (e) Iron(III) chromate or ferric chromate (f) Copper(II) nitrite or cupric nitrite **39.** (a) HClO (b) HBr (c) HNO_3 (d) $HClO_2$ (e) $HBrO_4$ (f) H_2SO_4 **41.** (a) Hydrocyanic acid (b) Hydrosulfuric acid (c) Bromic acid (d) Sulfuric acid (e) hypochlorous acid (f) bromous acid **43.** (a) $Mg(OH)_2$ (b) H_2SO_4 (c) $NaNO_3$ (d) N_2O **45.** (a) Quicklime (b) Marble (c) Gypsum (d) Oil of vitriol **47.** (a) SrO (b) AlN (c) Rb_2S **49.** (a) Aluminum nitride (b) Vanadium(V) oxide (c) Iron(III) hydroxide or ferric hydroxide (d) Ammonium sulfide **53.** (a) GaF_3 (b) $Pd(NO_3)_2$ (c) AuP (d) $La(C_2H_3O_2)_3$ (e) PuO_2 (f) RuO_4

CUMULATIVE REVIEW

CHAPTERS 7-8

1. (a) Ċa· (b) [Ca]²⁺ 2. (a) ·C̈l̤: (b) [:C̈l̤:]¹⁻
3. An ionic bond is formed by the transfer of electrons from one atom to another. The atoms are always of different elements. A covalent bond is formed by the sharing of electrons between two atoms. A coordinate covalent bond is a bond in which one element donates the electrons to form the bond.

5. HCl = 0.9, CaCl₂ = 2.0, BaCl₂ = 2.1, CaF₂ = 3.0 6. CsF 7. (a) Cu₂SO₃ (b) OsO₄ (c) (NH₄)₂CO₃ (d) N₂O
8. (a) W⁵⁺ (b) Sn⁴⁺ (c) V⁵⁺ (d) In³⁺ 9. (a) Ammonium iodide (b) Calcium carbide (c) Copper(I) chromate or cuprous chromate 10. X₂Y₃ 11. False 12. True 13. True 14. False 15. True 16. MgO 17. C–H 18. Na₂SO₄
19. 2–
20. (a) [:Ö:N:Ö:]¹⁻ (b) [H:N:H]¹⁺
                                    H
21. (a) 6+ (b) 4+ (c) 7+ (d) 3+ 22. (a) Hydrogens have partial positive charges, and oxygen has a partial negative charge. (b) Hydrogen has a partial positive charge, and iodine has a partial negative charge. (c) Chlorines have partial positive charges, and oxygen has a partial negative charge. (d) Phosphorus has a partial positive charge, and chlorines have a partial negative charge.
23. (a) :F̈:F̈: (b) :C̈l:C::C:C̈l: (c) :Cl—P(Cl)(Cl)—Cl: (d) :C̈l:C::C:C̈l:
                                                                    :C̈l::C̈l:
24. (a) 1+ (b) 1– (c) 2+ (d) 3+ 25. (a) Fe₃(PO₄)₂ (b) Hg(CN)₂ (c) Zn(HCO₃)₂ (d) Fe₂(Cr₂O₇)₃ 26. (a) In₂S₃ (b) Mg₃(AsO₄)₂ (c) Ga₂O₃ (d) Rb₂Se (e) FeO (f) CoCl₂ (g) Cu₃N (h) Cu₂O 27. (a) HIO₄ (b) HClO₃ (c) HBr (d) HClO₂ 28. A coordinate covalent bond is a bond in which one atom donates both electrons to form the bond. 29. Yes. A molecule like carbon tetrachloride has four polar C–Cl bonds, but due to the shape of the molecule, the dipoles cancel out, because the center of positive charge in the molecule coincides with the center of negative charge. 30. (a) Group VIA (b) 6 electrons (c) 2 –

CHAPTER 9

3. 0.100 moles 5. 2.53 × 10⁸ dollars or 3 × 10⁸ dollars (for one significant figure) 7. (a) 0.600 mole (b) 5.00 moles (c) 0.00250 mole (d) 15.00 moles 9. 50.2 g 11. 44.2 g 13. (a) 27̄0 g (b) 0.479 g (c) 8,020 g (d) 1.28 g 15. 4.46 × 10²² atoms 17. 9 × 10⁻²⁰ dollars/atom 19. 1 × 10¹⁰ miles 21. NaCl, sodium chloride (table salt) 23. C₂₇H₄₆O 25. Empirical formula is CH₃; molecular formula is C₂H₆. 27. Empirical formula is C₂H₄O; molecular formula is C₄H₈O₂. 29. 0.7500 moles 31. 2.74 moles 33. 0.0500 mole 35. (a) 40.0 g CH₄ (b) 3.20 g SO₃ (c) 2.05 g Al₂(SO₄)₃ (d) 3.0 g (NH₄)₃PO₄ 37. 13 g of SO₂ 39. 2.270 g of propane or 5.00 pounds of propane 41. 2.71 × 10²⁴ molecules 43. 1.50 × 10²³ formula units 45. 3.25 × 10²⁴ formula units 47. C₆H₈O₆
49. C₆H₆ 51. C₃H₆N₆ 53. C₂H₄O₂, acetic acid; C₆H₁₂O₆, dextrose 55. K = 40.3%, Cr = 26.8%, O = 33.0%
57. (a) B = 15.9%, F = 84.1% (b) U = 67.61%, F = 32.39% (c) C = 39.1%, H = 8.7%, O = 52.2% 59. 34 g of sugar present 61. (a) S = 50.1%, O = 49.9% (b) C = 75.0%, H = 25.0% (c) C = 42.11%, H = 6.43%, O = 51.46% (d) Ca = 40.1%, C = 12.0%, O = 48.0% 63. 1,280 g of copper 65. 244.7 g of iron 67. 199 g of oxygen 69. 304 g of Na₂S and 196 g of Fe₂O₃ 71. 1.9 × 10¹⁶ years 72. CH₂Cl 73. C₂H₄Cl₂ 74. 362 g

75. 1.76×10^{22} atoms **76.** Hematite, 69.9% Fe, and magnetite, 72.3% Fe **77.** Cu_5FeS_4 and $CuFeS_2$
78. S = 50.1%, O = 49.9% **79.** 1.79×10^{-21} g **80.** 4.00 moles **81.** 124 g **82.** 1.66×10^{-24} g **83.** (a) 0.20 mole
(b) 1.2×10^{23} molecules (c) 0.40 mole C atoms (d) 1.2 moles H atoms (e) 0.20 moles O atoms (f) 2.4×10^{23}
C atoms (g) 7.2×10^{23} H atoms (h) 1.2×10^{23} O atoms **85.** (a) 0.400 mole H_2O (b) 3.00 moles MnO_2
(c) 2.5×10^{-3} mole N_2 (d) 14.99 moles $(NH_4)_2SO_4$ **87.** (a) 4.00 moles (b) 2.41×10^{24} molecules
(c) 8.00 moles C (d) 20.0 moles H (e) 8.00 moles O (f) 4.00 moles N (g) 96.0 g C (h) $2\overline{0}$ g H
(i) 128 g O (j) 56.0 g N (k) They should be the same.

CHAPTER 10

1. (a) $2K + 2H_2O \rightarrow 2KOH + H_2$ (b) $2HC_2H_3O_2 + Ca(OH)_2 \rightarrow Ca(C_2H_3O_2)_2 + 2H_2O$

(c) $Mg + Cu(NO_3)_2 \rightarrow Mg(NO_3)_2 + Cu$ (d) $Na_2O + H_2O \rightarrow 2NaOH$ (e) $2ZnS + 3O_2 \rightarrow 2ZnO + 2SO_2$

(f) $3KOH + Al(NO_3)_3 \rightarrow Al(OH)_3 + 3KNO_3$

3. (a) $4Fe + 3O_2 \rightarrow 2Fe_2O_3$ (b) $3H_2SO_4 + 2Al(OH)_3 \rightarrow Al_2(SO_4)_3 + 6H_2O$

(c) $2AgNO_3 + BaCl_2 \rightarrow 2AgCl + Ba(NO_3)_2$ (d) $Cu_2S + O_2 \rightarrow 2Cu + SO_2$

5. (a) Composition (b) Decomposition (c) Double-replacement (d) Single-replacement (e) Double-replacement **7.** (a) Decomposition (b) Decomposition (c) Decomposition (d) Combination (e) Combination
(f) Single-replacement (g) No reaction (h) Double-replacement **9.** (a) Combination (b) Double-replacement
(c) Double-replacement (d) Single-replacement

11. (a) $H_2 + Br_2 \rightarrow 2HBr$ (b) $BaO + H_2O \rightarrow Ba(OH)_2$

(c) $2Na + Cl_2 \rightarrow 2NaCl$

(d) $N_2 + 2O_2 \rightarrow 2NO_2$ (e) $CaCO_3 \rightarrow CaO + CO_2$ (f) $2KOH \rightarrow K_2O + H_2O$ (g) $Hg(ClO_3)_2 \rightarrow HgCl_2 + 3O_2$

(h) $PbCl_2 \rightarrow Pb + Cl_2$ (i) $2Li + 2H_2O \rightarrow 2LiOH + H_2$ (j) $Zn + H_2SO_4 \rightarrow ZnSO_4 + H_2$

(k) $Ni + Al(NO_3)_3 \rightarrow$ no reaction (Aluminum is above nickel in the activity series.)

(l) $2Al + 3Hg(C_2H_3O_2)_2 \rightarrow 2Al(C_2H_3O_2)_3 + 3Hg$

(m) $H_2SO_4 + 2NH_4OH \rightarrow (NH_4)_2SO_4 + 2H_2O$

(n) $2AgNO_3(aq) + BaCl_2(aq) \rightarrow Ba(NO_3)_2(aq) + 2AgCl(s)$ (o) $3H_2SO_3 + 2Al(OH)_3 \rightarrow Al_2(SO_3)_3 + 6H_2O$

(p) $NaNO_3(aq) + KCl(aq) \rightarrow$ no reaction

13. (a) $2N_2O \rightarrow 2N_2 + O_2$ (b) $H_2CO_3 \rightarrow H_2O + CO_2$ (c) $2NaNO_3 \rightarrow 2NaNO_2 + O_2$

(d) $H_2 + F_2 \rightarrow 2HF$ (e) $N_2 + 3H_2 \rightarrow 2NH_3$ (f) $Ca + 2HCl \rightarrow CaCl_2 + H_2$

(g) $Cu + NiCl_2 \rightarrow$ no reaction

(h) $3AgNO_3(aq) + K_3AsO_4(aq) \rightarrow 3KNO_3(aq) + Ag_3AsO_4(s)$

15. (a) $2K + Cl_2 \rightarrow 2KCl$ (b) $K_2O + H_2O \rightarrow 2KOH$ (c) $MgO + H_2O \rightarrow Mg(OH)_2$ (d) $CaO + CO_2 \rightarrow CaCO_3$

(e) $2NH_3 \rightarrow N_2 + 3H_2$ (f) $SrCO_3 \rightarrow SrO + CO_2$ (g) $2NaClO_3 \rightarrow 2NaCl + 3O_2$

(h) $Mg(OH)_2 \rightarrow MgO + H_2O$ **17.** (a) NaOH and HCl (b) KOH and H_2SO_4 (c) $Al(OH)_3$ and HNO_3

19. (a) $Zn + 2HNO_2 \rightarrow Zn(NO_2)_2 + H_2$ (b) $Ag + NiCl_2 \rightarrow$ no reaction (c) $Zn + 2AgNO_3 \rightarrow Zn(NO_3)_2 + 2Ag$
(d) $2Cs + 2H_2O \rightarrow 2CsOH + H_2$ (e) $3HCl + Al(OH)_3 \rightarrow AlCl_3 + 3H_2O$ (f) $KNO_3 + ZnCl_2 \rightarrow$ no reaction
(g) $Al(NO_3)_3 + 3NaOH \rightarrow Al(OH)_3(s) + 3NaNO_3$ (h) $K_2CrO_4 + Pb(NO_3)_2 \rightarrow PbCrO_4(s) + 2KNO_3$
23. (a) $2K + Cl_2 \rightarrow 2KCl$ (The K is oxidized; the Cl is reduced.) (b) $2NH_3 \rightarrow N_2 + 3H_2$ (The N is oxidized; the H is reduced.) (c) $CuO + H_2 \rightarrow Cu + H_2O$ (The H is oxidized; the Cu is reduced.) (d) $Sn + 2Cl_2 \rightarrow SnCl_4$ (The Sn is oxidized; the Cl is reduced.) **25.** (a) K is the reducing agent; Cl_2 is the oxidizing agent. (b) The NH_3 is both the oxidizing and the reducing agent. (c) H_2 is the reducing agent; CuO is the oxidizing agent. (d) Sn is the reducing agent; Cl_2 is the oxidizing agent. **27.** (a) Zn is the reducing agent: HNO_2 is the oxidizing agent. (b) No reaction took place. (c) Zn is the reducing agent; $AgNO_3$ is the oxidizing agent. (d) Cs is the reducing agent: H_2O is the oxidizing agent. **29.** (a) vanadium pentoxide (b) zinc (c) zinc (d) vanadium

30. $MgCO_3 \rightarrow MgO + CO_2$, $CaCO_3 \rightarrow CaO + CO_2$

31. (a) $N_2 + 3H_2 \rightarrow 2NH_3$ (b) $3Fe + 4H_2O \rightarrow Fe_3O_4 + 4H_2$ (c) balanced

33. (a) $2C + O_2 \rightarrow 2CO$ (b) $H_2 + Cl_2 \rightarrow 2HCl$ (c) $2Na + 2H_2O \rightarrow 2NaOH + H_2$ (d) no reaction

34. (a) no reaction (b) $Mg + Zn(NO_3)_2 \rightarrow Mg(NO_3)_2 + Zn$ (c) no reaction (d) $Zn + Cu(NO_3)_2 \rightarrow Zn(NO_3)_2 + Cu$

36. (a) $Ca(OH)_2 + 2HCl \rightarrow CaCl_2 + 2H_2O$ (double replacement) (b) $Zn + H_2SO_4 \rightarrow ZnSO_4 + H_2$ (single replacement) (c) $2Ba + O_2 \rightarrow 2BaO$ (combination) (d) $Cs_2O + H_2O \rightarrow 2CsOH$ (combination)

37. $4NH_3 + 7O_2 \rightarrow 2N_2O_4 + 6H_2O$ **38.** (a) Fe^{2+} (b) chromium in $Cr_2O_7^{2-}$ (c) $Cr_2O_7^{2-}$ (d) Fe^{2+}

39. $\underset{\text{Sodium}}{12Na} + \underset{\text{Phosphorus}}{P_4} \rightarrow \underset{\substack{\text{Sodium}\\\text{phosphide}}}{4Na_3P}$

CHAPTER 11

1. (a) $2Na_2S_2O_3 + AgBr \rightarrow Na_3Ag(S_2O_3)_2 + NaBr$ (b) 79.10 g sodium thiosulfate (c) 100.3 g $Na_3Ag(S_2O_3)_2$, 25.73 g NaBr **3.** (a) $NH_4NO_3 \xrightarrow{\text{Heat}} N_2O + 2H_2O$ (b) 4.0 g ammonium nitrate (c) 1.8 g water

5. (a) $C_7H_8 + 3HNO_3 \rightarrow C_7H_5N_3O_6 + 3H_2O$ (b) 405 g of toluene, 832 g nitric acid (c) 238 g water

7. (a) $2NH_4Cl + H_2SO_4 \rightarrow (NH_4)_2SO_4 + 2HCl$ (b) 14.6 g sulfuric acid (c) 19.7 g ammonium sulfate. 10.8 g hydrogen chloride **9.** Percent $KClO_3$ = 51.2% **11.** (a) $3BaO_2 + 2H_3PO_4 \rightarrow 3H_2O_2 + Ba_3(PO_4)_2$ (b) 68.0 g hydrogen peroxide (c) $13\overline{0}$ g phosphoric acid **13.** 11.9 g hydrogen **15.** 1,460 g air

17. (a) $2C_4H_{10} + 13O_2 \rightarrow 8CO_2 + 10H_2O$ (b) 83.2 g oxygen (c) 70.4 g CO_2 and 36.0 g H_2O **19.** 64.0 g ozone

21. (a) $2Al(OH)_3 + 3H_2SO_4 \rightarrow Al_2(SO_4)_3 + 6H_2O$ (b) 228 g $Al(OH)_3$ and 430 g H_2SO_4 (c) 158 g H_2O

23. (a) $Fe_2O_3 + 3CO \rightarrow 2Fe + 3CO_2$ (b) $65\overline{0}$ g Fe_2O_3 (c) 342 g CO and 536 g CO_2

25. (a) $4NH_3 + 5O_2 \rightarrow 4NO + 6H_2O$ (b) $1.\overline{000}$ g O_2 (c) $75\overline{0}$ g NO and 675 g H_2O

27. 502 g $Zn(NO_3)_2$ **29.** (a) $CaC_2 + 2H_2O \rightarrow Ca(OH)_2 + C_2H_2$ (b) 148 g $Ca(OH)_2$ and 52.0 g C_2H_2

31. (a) $Fe_2O_3 + 3CO \rightarrow 2Fe + 3CO_2$ (b) 223 g of iron, 264 g of carbon dioxide

33. (a) $2H_2 + O_2 \rightarrow 2H_2O$ (b) 72 g water 35. (a) $Ca_3P_2 + 6H_2O \rightarrow 2PH_3 + 3Ca(OH)_2$ (b) 136 g PH_3 (c) 445 g $Ca(OH)_2$ 37. $2\overline{0}$ moles $CaSiO_3$. 33 moles CO and 6.7 moles P_2

39. (a) $3Fe + 4H_2O \rightarrow Fe_3O_4 + 4H_2$ (b) 10.7 g H_2 (c) $31\overline{0}$ g Fe_3O_4

41. (a) $2K + Cl_2 \rightarrow 2KCl$, 0.128 mole of Cl. (b) 19.1 g

42. (a) $2Cu(NO_3)_2 \rightarrow 2CuO + 4NO_2 + O_2$ (b) 8.54 g (c) $11\overline{0}$ g 43. (a) oxygen (b) 206 g

44. (a) $2H_2 + O_2 \rightarrow 2H_2O$ (b) 0.8 g H_2 left and 72.0 g of H_2O produced

45. $39\overline{0}$ g 46. (a) $Zn + H_2SO_4 \rightarrow ZnSO_4 + H_2$ (b) 19,600 g

47. (a) The second experiment, $Pb + Cu(NO_3)_2 \rightarrow Pb(NO_3)_2 + Cu$ (b) 79.8 g $Pb(NO_3)_2$ and 15.3 g Cu

48. 6.7 g 49. 58.5 g NaCl produced and 40.0 g NaOH left 50. 112 g

CUMULATIVE REVIEW
CHAPTERS 9–11

1. 0.0100 mole 2. 0.40 mole 3. (a) 73.6 g H_2SO_4 (b) 90.0 g H_2O (c) 0.434 g H_2CO_3 (d) 3.000×10^3 moles $HC_2H_3O_2$ 4. 1.2×10^{23} atoms 5. SnO_2 6. $C_{40}H_{64}O_{12}$ 7. $C_6H_{12}O_6$ 8. (a) 0.200 mole (b) 1.20×10^{23} molecules (c) 1.80 moles C atoms (d) 1.60 moles H atoms (e) 0.800 moles O atoms 9. N_2O_4 10. (a) 310.3 (b) 16.0 (c) 46.0 (d) 255.3 11. (a) 5.9% H, 94.1% S (b) 39.3% Na. 60.7% Cl (c) 24.0% Fe. 30.9% C, 3.9% H. 41.2% O (d) 44.9% K. 18.4% S. 36.7% O 12. C_6H_{14} 13. 74.1% C. 8.6% H. 17.3% N 14. 2.41×10^{22} molecules

15. 2.50 g $CaCO_3$ 16. $CaBr_2 + H_2SO_4 \rightarrow 2HBr + CaSO_4$

17. $2Mg + O_2 \rightarrow 2MgO$ 18. $NH_3(g) + HCl(g) \rightarrow NH_4Cl(s)$

19. (a) Decomposition (b) Combination (c) Single replacement (d) Double replacement

20. (a) $H_2 + Cl_2 \rightarrow 2HCl$ (b) $2NaCl + Pb(NO_3)_2 \rightarrow 2NaNO_3 + PbCl_2$ (c) $2Al + 3H_2SO_4 \rightarrow Al_2(SO_4)_3 + 3H_2$ (d) $SrCO_3 \rightarrow SrO + CO_2$ 21. $CaCO_3 \rightarrow CaO + CO_2$

22. (a) $TiCl_4 + 2H_2O \rightarrow TiO_2 + 4HCl$ (b) $P_4O_{10} + 6H_2O \rightarrow 4H_3PO_4$

23. (a) $Fe(OH)_3 + H_3PO_4 \rightarrow FePO_4 + 3H_2O$ (b) $Pb(OH)_2 + 2HNO_3 \rightarrow Pb(NO_3)_2 + 2H_2O$

24. $SO_3 + H_2O \rightarrow H_2SO_4$ 25. $2C_8H_{18} + 25O_2 \rightarrow 16CO_2 + 18H_2O$ 26. $2C_3H_7OH + 9O_2 \rightarrow 6CO_2 + 8H_2O$

27. $CaSO_4 + Na_2CO_3 \rightarrow CaCO_3(s) + Na_2SO_4$ 28. $C_{12}H_{22}O_{11} \rightarrow 12C + 11H_2O$

29. True 30. False 31. True 32. True 33. True 34. True 35. $C + 2H_2SO_4 \rightarrow CO_2 + 2SO_2 + 2H_2O$ 36. 5 moles CuO 37. 0.20 mole NaOH 38. 88 grams 39. 728 grams HCl 40. 12.6 grams CO_2 41. 1,860 grams H_3PO_4 42. 77.6 grams $Ca_3(PO_4)_2$, 55.6 grams $Ca(OH)_2$ 43. 15.0 moles, 1,810 grams 44. 119 grams H_2O_2 45. 13.9 grams $PbCl_2$ 46. 96.1 grams LiOH 47. 72.9 grams Mg 48. 171.3 grams $Ba(OH)_2$, 2.0 grams H_2, 36.0 grams excess H_2O

CHAPTER 12

1. The matching pairs are 1. (c), 2. (a), 3. (d), 4. (b) 3. (a) HCl is the solute, water is the solvent. (b) CO_2 is the solute, water is the solvent. (c) Benzene is the solute, carbon tetrachloride is the solvent. 7. True. 9. Saturation 11. (a) Unsaturated (b) Supersaturated 15. (a) 6.25% by mass (b) 20.0% by mass (c) 8.00%

by mass **17.** 2̄0 g **19.** (a) 2̄0% by volume (b) 6% by volume **21.** 12.5 mL **23.** 2̄0 mL **25.** (a) 3.3% by mass–volume (b) 5.0% by mass–volume **27.** 15 g **29.** 0.05% by mass **31.** (a) 12.5% (b) 5.00% **33.** 3̄0 g **35.** (a) 5.0% (b) 0.833% **37.** 28 mL **39.** 2̄0̄0 mL **41.** (a) 3.13% (b) 5.00% **43.** 12.5% g **45.** (a) 0.30 M (b) 0.5 M **47.** 2.74 M **49.** 8.4 g **51.** 0.109 M **53.** 0.15 M **55.** (a) 0.167 M (b) 0.200 M **57.** 0.525 M **59.** 16.4 g **61.** (a) 0.10 equivalent (b) 0.20 equivalent (c) 0.0902 equivalent **65.** (a) 0.0299 equivalent (b) 0.750 equivalent **67.** (a) 1,120 g (b) 9.27 g **69.** (a) 2.97 N (b) 0.13 N (c) 0.99 M, 0.063 M **71.** 59 g **73.** 250 mL **75.** (a) 1.00 N (b) 0.500 N (c) 0.500 M Fe(OH)$_2$, 0.167 M H$_3$BO$_3$ **77.** 3.25 g **79.** 599 mL **81.** Take 125 mL of 16 N H$_2$SO$_4$; dilute to 5̄0̄0 mL with water **83.** Take 50 mL of 6 N NaOH; dilute to 3 L with water. **85.** Take 104 mL of 12.0 M HCl; dilute to 2̄50 mL with water. **87.** Take 42 mL of 3.00 M H$_2$SO$_4$; dilute to 5.00 L with water. **97.** 0.15 molal **99.** (a) 0.80 molal (b) 0.333 molal **105.** −33.1°C **107.** 109.26°C **109.** b.p. = 100.78°C, f.p. = −2.79°C **111.** (a) Osmosis (b) diffussion **116.** 1 × 10^{-2} M or 0.01 M **117.** 0.005 M and 0.01 N **118.** 5̄0̄0 g **119.** 0.510 M and 1.02 N **122.** Ethanol **123.** 41.3 mL. **124.** 1̄50 g/mole **125.** 2.00 *molal* solution **126.** 3.72 g/mole **127.** A 0.104°C increase in boiling point; therefore, if the boiling point of the pure water is 100.000°C, the boiling point of the solution is 100.104°C **128.** (a) 2.00 equivalents (b) 2.00 equivalents **129.** (a) 1.00 equivalent (b) 2.00 equivalents

INDEX

A

Accuracy, 12
Acetic acid ($HC_2H_3O_2$), 3, 149, 220, 296
Acid–base reaction, 192, 296
Acids
    equivalent of, 243–246, 296
    inorganic, 148–150, 296
    phosphoric, 118–119, 295
    reactions, 192, 296
Activity series, 192–193, 296
Alchemy, 3–4
Algebraic operations
    addition, 266
    equation, 67, 272–274
    subtraction, 266
Alpha particles, 62
Aluminum chloride ($AlCl_3$), 147, 186, 296
Aluminum chromate ($Al_2(CrO_4)_3$), 148
Aluminum nitrate ($Al(NO_3)_3$), 215, 296
Aluminum oxide (Al_2O_3), 49, 141, 212
Ammonia (NH_3), 116, 206, 208
Ammonium phosphate (($NH_4)_3PO_4$), 148
Amorphous solids, 42
Anaximenes, 3
Angstrom, 101
Anion, 103
Anode, 59–61
Antifreeze, 251
Aqueous, 191–192
Area, measurement, 18–20, 295
Atomic mass, 48–50, 64–67, 97, 295
Atomic mass unit (amu), 63–65, 67
Atomic radius, 100–104, 295
Atomic theory, 57–68, 73–90, 161

atomic number, 64
Bohr atom, 77–78
continuous spectrum, 74–75
electromagnetic spectrum, 75–76
electron configuration, 77, 83–89, 295
electron orbitals, 81–82, 295
electrons, 59–61
energy levels, 79–80, 295
energy subshells/sublevels, 80–81
isotopes, 64–68, 295
light energy, 76–77
line spectrum, 74–75
models, 58–59
neutrons, 63–64
protons, 61–63
quantum mechanical model, 78
spectroscope, 74–75
visible spectrum, 74
Atoms, 45–48, 59, 62, 77–78, 81, 105, 115, 296
Aufbau principle, 84–85
Avogadro's hypothesis, 162, 166, 296

B

Balanced chemical equation, 183–186, 206–207, 296
Base number, 16–17, 267–268, 295
Benzene (C_6H_6), 238–239, 296
Binary compounds
    metals, 138–144, 146–148, 296
    nonmetals, 137–148, 295–296
    polyatomic ions, 139–140, 144
    systematic names, 137
    ternary and higher compounds, 139–140, 144–145, 148, 295–296
Bohr atom, 77–78

Boiling-point–elevation constant, 251–253, 296
Boyle's influence, 3–6
Bromic acid ($HBrO_3$), 149, 150
Bromous acid ($HBrO_2$), 150

C

Carbon dioxide (CO_2), 42, 233
Cathode, 59–61
Cathode-ray tube, 59–60
Cation, 103
Celsius temperature scale, 21, 30–31, 295
Chemical bond, 113–129
    coordinate covalent bond, 118–119, 295
    covalent bond, 115–118, 120–122, 125–126, 295
    electron-dot structure, 114–115
    electronegativity, 120–122, 295
    ionic and covalent percentage, 122–124, 295
    ionic bonding, 119–120, 295
    octet rule, 115–118, 120, 295
    polar and nonpolar molecules, 126–128, 295
    three-dimensional characteristics of molecules, 124–125
Chemical compounds
    chemical formula (*see* Chemical formula)
    covalent bond, octet rule, 115–118, 295
    electron-dot structure, 114–115
    electronegativity, 120–122, 295
    formula mass, 49–50, 295
    formula unit, 46
    ionic compounds, 46
    Law of Definite Composition, 45–46
    molecular compounds, 46
    molecular mass, 49–50, 295
    nomenclature (*see* Chemical nomenclature)
    phosphoric acid, 118–119, 295
Chemical formula
    elements and compounds, 47–48, 140–142, 295
    empirical formula, 167–169, 296
    gram-atomic mass, 162–166, 296
    gram-formula mass, 169–172, 296
    mole, 163–166, 169–172, 296
    molecular formula, 172–173, 296
    percentage composition, 173–175, 296
    word equations, 182
Chemical industries, 6
Chemical nomenclature, 135–152
    binary compounds, 137–144, 146–148, 295–296
    common names, 151
    inorganic acids, 148–150, 296
    polyatomic ions, 139–140, 144
    ternary and higher compounds, 144–145, 148, 295–296

Chemical symbols, 45
Chloroform ($CHCl_3$), 29, 218–220, 295–296
Chromium(II) hydroxide ($Cr(OH)_2$), 139, 148
Cobalt(III) chloride ($CoCl_3$), 147
Combination reactions, 186–188, 296
Common names, 137, 151
Concentration
    molality, 249–250, 296
    normality, 243–246, 296
    percent-by-mass solutions, 237–238, 296
    percent-by-mass–volume solutions, 239–240, 296
    percent-by-volume solutions, 238–239, 296
    saturation point, 236–237
Continuous spectrum, 74–75
Conversion of units, 25–26, 284
Coordinate covalent bond, 118–119, 295
Covalent bond, 115–118, 120–122, 125–126, 295
Crookes tube, 59–60
Cupric ion (Cu^{2+}), 142
Cuprous ion (Cu^{1+}), 142

D

Dalton's atomic theory. *See* Atomic theory
Decimals, 12–17, 21, 48–50, 270–272, 295
Decomposition reactions, 188–189, 296
Density, 27–29, 275, 295
Dephlogisticated air, 205
Derived units, 21, 25, 295
Deuterium, 1–2, 64–65
Diatomic elements, 115–116, 124–127
Diffusion, 253–256, 296
Dinitrogen monoxide (N_2O), 138, 146, 295
Dipole molecule, 126
Double covalent bond, 117
Double-replacement/double-exchange reaction, 186, 191–192, 296

E

Electrodes, 59–60
Electrolytic solutions, 247–249
Electromagnetic spectrum, 75–76
Electronegativity, 120–124, 137–138, 295
Electronic calculators, 11, 277–278
Electrons
    atomic number, 64
    atomic radius, 101–103, 295
    Bohr atom, 77–78
    cathode-ray tube, 59–61
    configuration, 83–89, 100, 295
    coordinate covalent bond, 118–119
    electron affinity, 105

electron-dot structure, 114–115
electronegativity, 120–122, 295
elements, 100
energy levels/electron shell, 79–80, 295
energy subshells/sublevels, 80–81
isotopes, 64–68, 295
neutron, 63–64
orbitals, 81–82, 295
periodicity, 100
quantum mechanical model, 78
Thomson experiment, 61–63
Empirical formula, 167–169, 172–173, 296
Energy level, 77, 79–86, 100–102, 114–117, 119, 295
Equivalent mass, 243–244
Equivalent of a base, 243–246, 296
Equivalent of an acid, 243–246, 296
Excited state, 78
Exponents, 16–17, 267–268, 295

F

Factor-unit method, 22–23, 26, 164, 170, 235–236, 268–270, 295–296
Fahrenheit temperature scale, 30–31, 295
Ferric dichromate ($Fe_2(Cr_2O_7)_3$), 139, 147
Ferric ion (Fe^{3+}), 142
Ferric oxide (Fe_2O_3), 147
Ferrous oxide (FeO), 147
Fluid ounce, 20–21
Fluoride ion, 120
Food processing, 6
Formula equation, 182–183
Formula mass, 49–50, 169–172, 295–296
Formula unit, 46
Fractions, 266
Freezing-point–depression constant, 251–253, 296

G

Glass measurement, 12–16, 295
Gram-atomic mass, 162–166, 296
Gram-formula mass, 169–172, 296
Ground state, 77, 79, 83–84
Groups, 99–100, 140

H

Haber process, 206
Heterogeneous matter, 43–44
Heterogeneous mixture, 44
Homogeneous matter, 43–44
Homogeneous mixture, 43–44
Hydrogen, 45–47, 64–66, 83, 115, 119
Hypobromous acid (HBrO), 150

I

Immiscible, 233
Inches, 20–21
Inorganic acids, 148–150, 296
Inorganic chemistry, 6
International Union of Pure and Applied Chemistry (IUPAC), 99–100, 137, 151
Ion, 120
Ionic bond, 119–120, 295
Ionic compounds, 46–47
Ionization, 78
  energy, 103–104, 295
  potential, 103–104, 295
  solutions, 247–249
Iron(III) dichromate ($Fe_2(Cr_2O_7)_3$), 139, 147
Iron(II) oxide (FeO), 147
Iron(III) oxide (Fe_2O_3), 147
IUPAC. *See* International Union of Pure and Applied Chemistry (IUPAC)

K

Kelvin temperature scale, 30
Kinetic energy, 42

L

Law of Combining Volumes, 161
Law of Conservation of Energy, 42
Law of Conservation of Mass, 41
  balancing chemical equation, 183, 206–207
  decomposition reactions, 189
  formula equation, 182–183
Law of Definite Composition, 45–46
Law of Octaves, 97
Lewis structures, 114–115
Limiting-reactant, 215–220, 296
Line spectrum, 74–75
Liquid measurement, 11–12
Lithium sulfite (Li_2SO_3), 148

M

Magnesium nitride (Mg_3N_2), 147
Magnesium nitrite ($Mg(NO_2)_2$), 148
Magnetic bottle, 1
Manganese(IV) oxide (MnO_2), 147
Mass number, 64–68, 295
Matter and energy, 39–50
  atoms, 45
  chemical properties, 43
  compounds, 45–50, 295
  elements, 45, 47–49, 295

Matter and energy (*Continued*)
   gas, 42
   kinetic energy, 42
   Law of Conservation of Mass and Energy, 41–42
   liquid, 42
   mixtures and pure substances, 43–44
   molecular *vs.* ionic compounds, 46
   molecules, 46
   physical properties, 43
   plasma, 42–43
   potential energy, 42
   scientific method, 40–41
   solid, 42
   solutions, 44
Metalloids, 45, 100
Metals, 45, 138–144, 146–148, 295–296
Metastasis, 95
Metric system, 21–27, 283, 295
Mixtures, 43–44
Molar mass, 169
Mole, 163–166, 169–172, 296
Molecular compounds, 46
Molecular formula, 172–173, 296
Molecular mass, 49–50, 172–173, 243, 255, 295
Monatomic ion, 120

N

Negative ion, 46, 103, 139–140, 191
Neutral beam heating, 1
Neutralization reaction, 192
Nitric acid (HNO_3), 149
Nitrogen monoxide (NO), 145
Nitrous acid (HNO_2), 150, 295
Nitrous oxide (N_2O), 151
Noble gases, 97, 115
Nonelectrolytic solutions, 247–249
Nonmetals, 45, 137–144, 146–148, 295–296
Nonpolar covalent bond, 125, 295
Nonpolar molecules, 127–128, 295
Nonredox reactions, 194
Nuclear fission, 1
Nuclear fusion, 1
Nuclear model, 61–63
Nucleus, 62, 64, 77–79, 81, 84, 86, 103–105

O

Octet rule, 115–118, 120, 295
Orbital, 78, 81–82, 84
Organic chemistry, 6
Osmometry, 255
Osmosis, 253–256, 296

Osmotic pressure, 254–256, 296
Outermost shell, 86–89, 99–100
Oxidation number, 138–143, 146–148, 194–195, 295–296
Oxidation–reduction (redox) reaction, 194–196, 296
Oxidizing agent, 195–196, 296
Oxygen, 45–47, 64–66, 83, 115, 119

P

Perbromic acid ($HBrO_4$), 150
Percentage composition, 173–175, 276–277, 296
Periodic law, 98
Periodic table, 95–106
   atomic radius, 101–103, 295
   electron affinity, 105
   electron configuration, 100
   groups, 99–100
   history, 96–98
   ionization potential, 103–104, 295
   IUPAC, 99–100
   periodicity, 100
   periods, 99–100
   properties, 100
   representative elements, 99
   transition metals, 99
Pharmaceutical industries, 6
Phase, 44
Phlogiston theory, 205
Phosphorous acid (H_3PO_3), 150
Photoelectric effect, 60
Plum-pudding theory, 61–63
Polar covalent bond, 125–127, 295
Polar molecules, 126–128, 295
Polya's method, 10–11
Polyatomic ions, 139–140, 144, 148, 191, 296
Polyvinyl chloride (PVC), 214–215, 296
Positive ion, 103, 137, 139, 191
Potassium acetate, 220, 296
Potassium phosphide (K_3P), 147
Potential energy, 42
Precipitate, 191
Precision, 12
Princeton Plasma Physics Laboratory (PPPL), 1–2
Propane gas (C_3H_8), 209–211
Proportion, 3, 44, 274
Protium, 1, 64–65
Pure substances, 43–44

Q

Quanta, 77–78
Quantum mechanics, 78

R

Radio frequency heating, 1
Ratios, 236, 274
Reducing agent, 195–196, 296
Reduction reactions, 194
Rounding off, 13–16, 163, 295
Rutherford model, 61–63

S

Scientific method, 40–41
Shells, 77–79
Significant figures, 11–16, 295
Single covalent bond, 115–116
Single-replacement reaction, 190–191, 296
Sodium ion (Na^{1+}), 120, 139, 234, 248
Sodium oxide (Na_2O), 147
Solubility, 235, 285
Solute, 232–233, 235, 249, 251, 255, 296
Solution, 231–256
    boiling-point–elevation constant, 251–253, 296
    colligative properties, 251
    diffusion and osmosis, 253–256, 296
    dilutions, 246–247, 296
    freezing-point–depression constant, 251–253, 296
    immiscible, 233–234, 296
    ionization, 247–249
    molality, 249–250, 296
    molarity, 240–243, 296
    normality, 243–246, 296
    percent-by-mass solutions, 236–238, 296
    percent-by-mass–volume solution, 236, 239–240, 296
    percent-by-volume solution, 236, 238–239, 296
    saturated solutions, 234–236, 296
    saturation point, 236
    solute, 232–233, 296
    solvent, 232–233, 296
    supersaturated solutions, 234–236, 296
    unsaturated solutions, 234–236, 296
Spectroscope, 73–75
Stock system, 142
Stoichiometry
    balanced chemical equation, 206–207
    dephlogisticated air, 205
    Haber process, 206
    limiting-reactant problem, 215–220, 296
    phlogiston theory, 205
    reactants/products, 207–215, 296
    synthesize, 206
Sulfate ion, 144–145, 295
Sulfate ion (($SO_4)^{2-}$), 144–145, 149
Sulfur dioxide (SO_2), 138, 167, 295
Sulfur trioxide (SO_3), 145, 182–183, 233, 296
Sulfuric acid (H_2SO_4), 149, 182
Sulfurous acid (H_2SO_3), 119, 150
Systematic chemical name, 137
Systems of measurement, 9–31, 44
    area and volume, 18–20, 295
    density, 27–29, 295
    English system of units, 20–21
    mass and weight, 27
    metric system, 21–27, 295
    problem solving, 10–11
    scientific notation, 16–17, 295
    significant figures, 11–16, 295
    temperature, 30–31, 295

T

Temperature measurement, 30–31, 295
Ternary compounds, 137, 144–145, 148, 295–296
Thermite reaction, 212–213, 296
Thomson model, 61–63
Tokamak Fusion Test Reactor (TFTR), 1–2
Transition metals, 99–101, 104
Trial-and-error method, 173
Triple covalent bond, 117
Tritium, 1–2, 64, 65

V

Vacuum, 59
Volume, measurement, 18–20, 295

W

Word equations, 182, 275–276

Table of Atomic Numbers and Atomic Masses

Name	Symbol	Atomic number	Atomic mass
Actinium	Ac	89	(227)
Aluminum	Al	13	26.98
Americium	Am	95	(243)
Antimony	Sb	51	121.8
Argon	Ar	18	39.95
Arsenic	As	33	74.92
Astatine	At	85	(210)
Barium	Ba	56	137.3
Berkelium	Bk	97	(247)
Beryllium	Be	4	9.012
Bismuth	Bi	83	209.0
Bohrium	Bh	107	(264)
Boron	B	5	10.81
Bromine	Br	35	79.90
Cadmium	Cd	48	112.4
Calcium	Ca	20	40.08
Californium	Cf	98	(251)
Carbon	C	6	12.01
Cerium	Ce	58	140.1
Cesium	Cs	55	132.9
Chlorine	Cl	17	35.45
Chromium	Cr	24	52.00
Cobalt	Co	27	58.93
Copernicium	Cn	112	(285)
Copper	Cu	29	63.55
Curium	Cm	96	(247)
Darmstadtium	Ds	110	(261)
Dubnium	Db	105	(262)
Dysprosium	Dy	66	162.5
Einsteinium	Es	99	(252)
Erbium	Er	68	167.3
Europium	Eu	63	152.0
Fermium	Fm	100	(257)
Flerovium	Fl	114	(289)
Fluorine	F	9	19.00
Francium	Fr	87	(223)
Gadolinium	Gd	64	157.3
Gallium	Ga	31	69.72
Germanium	Ge	32	72.59
Gold	Au	79	197.0
Hafnium	Hf	72	178.5
Hassium	Hs	108	(265)
Helium	He	2	4.003
Holmium	Ho	67	164.9
Hydrogen	H	1	1.008
Indium	In	49	114.8
Iodine	I	53	126.9
Iridium	Ir	77	192.2
Iron	Fe	26	55.85
Krypton	Kr	36	83.80
Lanthanum	La	57	138.9
Lawrencium	Lr	103	(260)
Lead	Pb	82	207.2
Lithium	Li	3	6.941
Livermorium	Lv	116	(293)
Lutetium	Lu	71	175.0
Magnesium	Mg	12	24.31
Manganese	Mn	25	54.94
Meitnerium	Mt	109	(266)
Mendelevium	Md	101	(258)